ADVANCES IN MOLECULAR SIMILARITY

Volume 1 • 1996

ADVANCES IN MOLECULAR SIMILARITY

Editors: RAMON CARBÓ-DORCA
Institute of Computational Chemistry
University of Girona
Girona, Spain

PAUL G. MEZEY
Department of Chemistry and
Department of Mathematics and Statistics
University of Saskatchewan
Saskatoon, Canada

VOLUME 1 • 1996

JAI PRESS INC.

Greenwich, Connecticut *London, England*

55 Old Post Road, No. 2
Greenwich, Connecticut 06836

JAI PRESS LTD.
38 Tavistock Street
Covent Garden
London WC2E 7PB
England

ISBN: 0-7623-0131-7

Transferred to digital printing 2006

CONTENTS

LIST OF CONTRIBUTORS

Neil L. Allan	School of Chemistry University of Bristol Bristol, England
Lluís Amat	Institute of Computational Chemistry University of Girona Girona, Spain
E. Besalú	Institute of Computational Chemistry University of Girona Girona, Spain
Ramon Carbó-Dorca	Institute of Computational Chemistry University of Girona Girona, Spain
Jerzy Cioslowski	Department of Chemistry Florida State University Tallahassee, Florida
Pere Constans	Institute of Computational Chemistry University of Girona Girona, Spain
David L. Cooper	Department of Chemistry University of Liverpool Liverpool, England
Miquel Duran	Institute of Computational Chemistry University of Girona Girona, Spain
Xavier Fradera	Institute of Computational Chemistry University of Girona Girona, Spain

J. Galvez — Departamento Química Física
Universidad de Valencia
Valencia, Spain

R. García-Domenech — Departamento Química Física
Universidad de Valencia
Valencia, Spain

C. de Gregorio Alapont — Departamento Química Física
Universidad de Valencia
Valencia, Spain

J.V. de Julián-Ortiz — Departamento Química Física
Universidad de Valencia
Valencia, Spain

Peter T. Measures — School of Chemistry
University of Bristol
Bristol, England

Jordi Mestres — Institute for Computational Chemistry
University of Girona
Girona, Spain

Paul G. Mezey — Department of Chemistry and
Department of Mathematics and Statistics
University of Saskatchewan
Saskatoon, Canada

Josep M. Oliva — Institute for Computational Chemistry
University of Girona
Girona, Spain

Robert Ponec — Institute of Chemical Process Fundamentals
Academy of Sciences of the Czech Republic
Prague, Czech Republic

M.T. Salabert-Salvador — Departamento Química Física
Universidad de Valencia
Valencia, Spain

Guido Sello — Department of Organic and Industrial Chemistry
University of Milano
Milano, Italy

Miquel Solà — Institute of Computational Chemistry
University of Girona
Girona, Spain

R. Soler-Roca — Departamento Química Física
Universidad de Valencia
Valencia, Spain

Boris B. Stefanov — Department of Chemistry
Florida State University
Tallahassee, Florida

Manuela Termini — Department of Organic and Industrial Chemistry
University of Milano
Milano, Italy

Maricel Torrent — Institute of Computational Chemistry
University of Girona
Girona, Spain

INTRODUCTION TO THE SERIES: AN EDITOR'S FOREWORD

The JAI series in chemistry has come of age over the past several years. Each of the volumes already published contains timely chapters by leading exponents in the field who have placed their own contributions in a perspective that provides insight to their long-term research goals. Each contribution focuses on the individual author's own work as well as the studies of others that address related problems. The series is intended to provide the reader with in-depth accounts of important principles as well as insight into the nuances and subtleties of a given area of chemistry. The wide coverage of material should be of interest to graduate students, postdoctoral fellows, industrial chemists and those teaching specialized topics to graduate students. We hope that we will continue to provide you with a sense of stimulation and enjoyment of the various sub-disciplines of chemistry.

Department of Chemistry
Emory University
Atlanta, Georgia

Albert Padwa
Consulting Editor

PREFACE

Molecular similarity is a fundamental concept of chemistry. From the very origins of the evolution of chemical knowledge, similarity has played an important role. During the early history of chemistry, our knowledge was mostly phenomenological; models and theories for actually explaining chemical properties and reactions were either nonexistent or rather simplistic. However, even in this early stage, it was already possible to invoke the concept of similarity in a meaningful and predictive way, since chemicals exhibiting similar properties within one context often showed similar properties within some different context. Using similarity, predictions could be made even without much understanding of molecular behavior, or even without yet knowing anything about the existence of molecules. Relying exclusively on similarity, a great deal of the accumulated chemical knowledge could be organized in a systematic manner; that had a major role in the eventual recognition of trends, relations, rules, and many of the fundamental laws of chemistry. These relations, rules, and laws had been combined into chemical theories of increasingly more sophistication and reliability, where, again, similarity among relations found in seemingly very different fields of natural sciences provided the motivation and the basis for the further development of chemical theories.

The most fundamental aspect of similarity in chemistry is molecular similarity; all other aspects of chemical similarity, involving chemical reactions and various other interactions ultimately involve molecular similarity. Both in the context of physical properties and with respect to chemical reactions, molecular similarity provides a basis for the classification, characterization, and detailed scientific description of molecules.

The recognition and analysis of molecular similarities are fundamental for an understanding of molecular structures and properties. Detection and interpretation of similarities among molecules represent the first steps in the process of explaining chemical behavior and in the construction of theoretical models of chemistry. Molecular similarity provides the very first layer of the foundation of all predictive models in chemistry.

The goal of this new book series, *Advances in Molecular Similarity*, is to provide our readers with timely reviews and monographs on topics involving molecular similarity, ranging from the fundamental physical properties underlying molecular behavior to applications in industrially important fields such as pharmaceutical drug design and molecular engineering. The recent advances in the development of a better understanding of the fundamental electronic nature of molecules, the discoveries of powerful new synthetic methodologies, innovative experimental techniques for the determination of molecular properties, and the spectacular advances in computational methodologies provide strong motivation for new studies in molecular similarity. As a consequence of all these developments, molecular similarity can now be studied on a much deeper level, providing both qualitative and quantitative information useful in a wide range of practical applications. It is the hope of the Editors that *Advances in Molecular Similarity* will serve chemists well, motivate new ideas and approaches, help to systematize the rapidly accumulating new chemical information, and, ultimately, make chemistry better understood and better applied in its ever widening role in modern society.

Ramon Carbó-Dorca
Paul G. Mezey
Series Editors

QUANTUM MOLECULAR SIMILARITY MEASURES: CONCEPTS, DEFINITIONS, AND APPLICATIONS TO QUANTITATIVE STRUCTURE–PROPERTY RELATIONSHIPS

Ramon Carbó-Dorca, E. Besalú,
Lluís Amat, and Xavier Fradera

Advances in Molecular Similarity
Volume 1, pages 1–42

ISBN: 0-7623-0131-7

ABSTRACT

"Quantum molecular similarity measures" (QMSM) and the possibility of constructing a discrete n-dimensional representation of arbitrary electronic structures is discussed and the consequent applications presented. The dual nature of the QMSM molecular description is emphasized in the present paper. This duality consistently produces the following representation couple: (a) A *∞–dimensional* representation, usable when associated with quantum theory of molecular structure, and (b) a *n-dimensional* representation, appearing when QMSM are computed over a given molecular set. The approximate forms of QMSM are described. The "atomic shell approximation" (ASA) is used to produce QMSM surfaces, beside the direct computation of fast QMSM integrals. "Quantum molecular similarity indices" (QMSI) are also presented, studying them from a new perspective. They are shown to constitute, besides the original measures, a possible transformation of the initial QMSM, intended to be useful in a great variety of applications, mainly related to "quantitative structure–property relationships" (QSPR). A rational classification, directly based on QMSI definitions, is given. A comparison of QMSI, obtained by means of the quantum mechanical, ∞-dimensional electronic density distributions, with these derived from the QMSM discrete, n-dimensional, representation of molecules, leads to a handful of useful results. The new relationships obtained in this way allow a mathematical connection between the initial description of the Carbó and the Hodgkin–Richards QMSI. From the discussion of this kind of comparative reasoning, a description of new index forms can be deduced. Within another application branch of the QMSM discrete molecular representation, we present here the interesting fact that QSPR procedures may provide an algorithm to obtain the discrete approximate representation vector elements of some unknown operator. The operator expectation values can be associated with a chosen observed experimental property measure and the connected linear equation constitutes the theoretical fundament of QSPR. Several assorted application examples are presented.

I. INTRODUCTION

In our laboratory, and during the past 15 years, a rigorous definition of "quantum similarity" (QS) has been developed and some applications have been described.[1–15] Also, other research groups[17–41] have been active in the field, producing a great deal of interesting results. Independently of the QS formalism, and following an older tradition, other authors have focused their work on studying structure–activity relationships between molecules, as indicated by a recent example.[42] Among many useful chemical applications of QS published in the literature, our laboratory has been mainly involved with the manipulation and representation of *theoretical* results to find some order and rules for "quantum object sets" (QOS), whose elements are molecular structures.

The present study describes the possible construction of periodic tables, extended to molecular sets, using a point of view based on "quantum similarity measures" (QSM). When QOS are chosen as molecular structures, then "quantum molecular similarity measures" (QMSM) lead to the definition of formal *point-molecules* as n-dimensional vectors. A point-molecule assembly defines a *molecular point-cloud.* A molecular point-cloud may be seen as a collection of vertices forming some kind of n-dimensional geometrical body: a *quantum similarity polyhedron.* From this geometrical point of view, a quantum similarity polyhedron can be translated, rotated, and projected in such a manner as to obtain a visual picture of the molecular point cloud inside a subspace with reduced dimensions.

Another aspect of the question, which has been studied since the appearance of the initial papers dealing with the subject, is related to the description of "quantum molecular similarity indices" (QMSI). In the opinion stated many times by the authors of the present paper (see for example refs. 3,9,13), the fundamental ideas of "molecular similarity" (MS) studies should be based on QMS. QMSI are simple manipulations of the QMSM, and being so defined they depend essentially on the similarity measures formalism. As a consequence, QMSI are related to the derived quantities obtained from the QMSM, as calculated over molecular sets, leading to an n-dimensional representation of molecular structures. QMSI can thus be related to the discretization of the quantum molecular description. The presence of this characteristic in the QMSM framework also has consequences in the relationships between the QMSI. This problem will be covered in this work. Following the description of QMSM and the derived n-dimensional molecular description, QMSI are classified and connected through the dual molecular description, simultaneously based in the quantum mechanical ∞-dimensional picture and in the related QMSM

n-dimensional discretization mentioned before. This point of view permits one to obtain new QMSI definitions, as well as new tools to classify and analyze the various forms of these similarity indices.

The actual QMSM formalism permits one to enunciate the *Mendeleev conjecture*, which in turn leans upon the so-called *Mendeleev postulates*.[12–14] This conceptual framework, fundamental to the building up of an order over QOS and permitting the visualization of molecular sets, represents a collection of points embedded in n-dimensional vector spaces. Also, the Mendeleev postulates allow us to predict molecular properties from theoretical parameters coming from QMSM or QMSI and, in this manner, open the way to connect all the QMSM ideas with the "quantitative structure-property relationships" (QSPR) techniques. In fact, QMSM may be seen as a set of computational rules allowing the construction of n-dimensional representations of QOS once a set of attached electronic density functions is known. This main characteristic of QSM permits us not only to visualize molecular sets from the molecular point-cloud point of view and, in this manner, find some order within their relative geometrical positions, but to also use the components of point-molecules as theoretical parameters in a QSPR computational structure.

Taking into account all this preliminary information, the present paper will be structured as follows. An introduction to the QMSM theoretical background leading to a n-dimensional representation of QOS will be given first. Then, a classification and an analysis of the forms and meaning of QMSI, performed within a dual ∞-dimensional versus n-dimensional framework, will be developed. Description of new QMSI will be given as well as a relationship between two of the most usual QMSI definitions. A theoretical discussion on the QMSM, through the induced n-dimensional discrete molecular description, considering its role as a natural foundation of the "quantitative structure-activity relationships" (QSAR) technique will follow. For this purpose, a brief overview of the mathematical form of the, more general than QSAR, QSPR techniques will precede the definition of the *discrete expectation value of an operator* concept. The theoretical foundation of QSPR will be defined at the end of this discussion. Finally, some assorted application examples will be given in order to illustrate the usefulness of the theoretical background.

II. DESCRIPTION OF QUANTUM OBJECTS

According to the usual quantum mechanical principles,[43] "quantum objects" (QO)—systems formed by a numerable assembly of microscopic particles—are described by means of particularly attached state wavefunctions. In general, the Schrödinger description of a QO by means of a N-particle wavefunction may be written as,

$$\Psi(\mathbf{r},\mathbf{p}) = \Psi(r_1,r_2,\ldots,r_N,\mathbf{p}) \tag{1}$$

where the vector **r** collects the particle coordinates, while the symbol **p** describes the wavefunction dependence upon a parameter set. Usually the vector **p** can be chosen, in the case of molecular systems, as the system's nuclear positions within the usual Born–Oppenheimer[44–46] approximation. In this particular situation, **p** is composed of a set of constant nuclear coordinates. Knowing the details of a N-particle system wavefunction, the associated n-th order "density matrix elements" (DME) can be easily derived. This connection can be done using the theoretical development described years ago by McWeeny and Löwdin.[47–52] DME can be defined by means of the following integral,

$$\rho^{(n)}(\mathbf{r},\mathbf{u},\mathbf{p}) = \binom{N}{n} \int \Psi^*(r_1,r_2,\ldots,r_n,r_{n+1},\ldots,r_N,\mathbf{p})$$

$$\Psi(u_1,u_2,\ldots,u_n,r_{n+1},r_{n+2},\ldots,r_N,\mathbf{p})\, dr_{n+1}\, dr_{n+2}\ldots dr_N \tag{2}$$

where **r** and **u** are n-dimensional vectors.

According to the canonical interpretation, the set of DME contains *any kind of information* attached to a given QO. Mathematical transformations of such DME (2) can be used to obtain n-th order "density integral transforms" (DIT). These new mathematical objects can be defined by means of the following integral,

$$P^{(n)}(\mathbf{r},\mathbf{s},\mathbf{p}) = \int \Omega(\mathbf{r},\mathbf{s},\mathbf{u},\mathbf{p})\, \rho^{(n)}(\mathbf{r},\mathbf{u},\mathbf{p})\, d\mathbf{u} \tag{3}$$

where the operator $\Omega(\mathbf{r},\mathbf{s},\mathbf{u},\mathbf{p})$ is the "transformation kernel" (TK).[53,54]

Some particular cases, which have an immediate physical meaning from the quantum mechanical point of view, are worth mentioning. They arise from the analysis of the definition (Eq. 3), when the TK takes precise particular forms; for example the following ones illustrate the DIT definition:

1. *DME and Density Functions*. When the TK is defined as $\Omega(\mathbf{r},\mathbf{s},\mathbf{u},\mathbf{p}) = \Omega(\mathbf{s},\mathbf{u}) = \delta(\mathbf{u} - \mathbf{s})$, then the DIT becomes $P^{(n)}(\mathbf{r},\mathbf{s},\mathbf{p}) = \rho^{(n)}(\mathbf{r},\mathbf{s},\mathbf{p})$, and the transformation leaves invariant the DME. If the TK is defined as $\Omega(\mathbf{r},\mathbf{s},\mathbf{u},\mathbf{p}) = \delta(\mathbf{u} - \mathbf{r})$, then $P^{(n)}(\mathbf{r},\mathbf{s},\mathbf{p}) = \rho^{(n)}(\mathbf{r},\mathbf{p})$ and the integral transform produces a diagonal element of the density matrix: the n-th order density function.[47–52]
2. *Generalized Electrostatic Potentials*. When within the DIT definition, the TK is built up by means of the Coulomb-like operator: $\Omega(\mathbf{r},\mathbf{s},\mathbf{u},\mathbf{p}) = |\mathbf{u} - \mathbf{s}|^{-1}$, then it is obtained using the transform: $P^{(n)}(\mathbf{r},\mathbf{s},\mathbf{p}) = V^{(n)}(\mathbf{r},\mathbf{s},\mathbf{p})$. That is, a generalized form of the electrostatic potential is obtained. One can call this general formulation a: n-th order electrostatic potential.
3. *System Energy*. When the TK is made by the Hamiltonian operator, that is: $\Omega(\mathbf{r},\mathbf{s},\mathbf{u},\mathbf{p}) = \delta(\mathbf{u} - \mathbf{r})H(\mathbf{u},\mathbf{s},\mathbf{p})$, then $P^{(n)}(\mathbf{r},\mathbf{s},\mathbf{p}) = E(\mathbf{r},\mathbf{s},\mathbf{p})$ and the integral transform, after integration, produces the *system energy*, provided that the transform order n is chosen as to be the same as the particle number. It is

assumed that the differential operators present in $H(\mathbf{u},\mathbf{s},\mathbf{p})$ only act over the **s** coordinate vector.

III. QUANTUM SIMILARITY MEASURES (QSM)

The DIT obtained in the way described in Section II can be compared by means of the so-called QSM.[11–14] A QSM constitutes a simple but fundamental way to obtain well-defined QO relationships. An n-th order QSM between two QO with respect to an operator Ω, usually definite positive, can be constructed as the following integral,[13]

$$Z_{AB}^{(n)}(\Omega,\mathrm{p}) = \int\int\int\int \Omega(\mathbf{r}_1,\mathbf{r}_2,\mathbf{s}_1,\mathbf{s}_2,\mathbf{p})$$

$$P_A^{(n)}(\mathbf{r}_1,\mathbf{s}_1,\mathbf{p})\, P_B^{(n)}(\mathbf{r}_2,\mathbf{s}_2,\mathbf{p})\, d\mathbf{r}_1\, d\mathbf{r}_2\, d\mathbf{s}_1\, d\mathbf{s}_2 \tag{4}$$

where $P_A^{(n)}(\mathbf{r}_1,\mathbf{s}_1,\mathbf{p})$ and $P_B^{(n)}(\mathbf{r}_2,\mathbf{s}_2,\mathbf{p})$ are the DIT related to the systems A and B respectively. When the involved QO are molecules, this measure is named a quantum molecular similarity measure (QMSM).

Some particular but interesting cases of QMSM will be studied in the following:

- An overlap-like QSM arises when the operator Ω is defined as the following Dirac δ functions product,

$$\Omega(\mathbf{r}_1,\mathbf{r}_2,\mathbf{s}_1,\mathbf{s}_2,\mathbf{p}) = \delta(\mathbf{r}_1 - \mathbf{r}_2)\, \delta(\mathbf{s}_1 - \mathbf{s}_2) \tag{5}$$

  then, this kind of QSM takes the form:

$$Z_{AB}^{(n)} = Z_{AB}^{(n)}(\delta(\mathbf{r}_1 - \mathbf{r}_2)\delta(\mathbf{s}_1 - \mathbf{s}_2),\mathbf{p})$$

$$= \int\int P_A^{(n)}(\mathbf{r},\mathbf{s},\mathbf{p})\, P_B^{(n)}(\mathbf{r},\mathbf{s},\mathbf{p})\, d\mathbf{r}\, d\mathbf{s} \tag{6}$$

- An electrostatic potential QSM is obtained if Ω is a product of two Coulomb operators,

$$\Omega(\mathbf{r}_1,\mathbf{r}_2,\mathbf{s}_1,\mathbf{s}_2,\mathbf{p}) = \left[\, |\mathbf{r}_1 - \mathbf{r}_2|\ |\mathbf{s}_1 - \mathbf{s}_2| \,\right]^{-1} \tag{7}$$

  then, Eq. 4 acquires the following form:

$$Z_{AB}^{(n)}(r_{12}^{-1}\, s_{12}^{-1},\mathbf{p}) = \int\int\int\int P_A^{(n)}(\mathbf{r}_1,\mathbf{s}_1,\mathbf{p})\, P_B^{(n)}(\mathbf{r}_2,\mathbf{s}_2,\mathbf{p})$$

$$\left[\, |\mathbf{r}_1 - \mathbf{r}_2|\ |\mathbf{s}_1 - \mathbf{s}_2| \,\right]^{-1} d\mathbf{r}_1\, d\mathbf{r}_2\, d\mathbf{s}_1\, d\mathbf{s}_2 \tag{8}$$

- One can consider any other system's appropriate DIT as the positive definite operator Ω in Eq. 4. A triple density transform similarity measure[11,13] is then built up in terms of the n-th order DIT's; for example:

$$T^{(n)}_{AB,C} = \int\int\int P^{(n)}_{A}(\mathbf{r}_1,\mathbf{r}_2,\mathbf{p})$$

$$P^{(n)}_{C}(\mathbf{r}_2,\mathbf{r}_3,\mathbf{p})\, P^{(n)}_{B}(\mathbf{r}_3,\mathbf{r}_1,\mathbf{p})\, d\mathbf{r}_1\, d\mathbf{r}_2\, d\mathbf{r}_3 \qquad (9)$$

  Nothing opposes to construct multiple density transform similarity measures, except the growing complexity of the involved integrals.
- When the involved DIT are density functions, the counterparts of Eqs. 6 and 8 can be written[8,9] as,

$$Z^{(n)}_{AB}(\delta(\mathbf{r}_1 - \mathbf{r}_2),\mathbf{p}) = \int \rho^{(n)}_{A}(\mathbf{r},\mathbf{p})\, \rho^{(n)}_{b}(\mathbf{r},\mathbf{p})\, d\mathbf{r} \qquad (10)$$

  and,

$$Z^{(n)}_{AB}(\mathbf{r}^{-1}_{12},\, \mathbf{p}) =$$

$$\int\int P^{(n)}_{\mathrm{A}}(\mathbf{r}_1,\, \mathbf{p})\, P^{(n)}_{B}(\mathbf{r}_2,\, \mathbf{p})\, |\mathbf{r}_1 - \mathbf{r}_2|^{-1}\, d\mathbf{r}_1\, d\mathbf{r}_2. \qquad (11)$$

  respectively. The measure (Eq. 11) is named a Coulomb-like QSM.
- When the DME are transformed into n-th order electrostatic potentials, using the operator $\Omega(\mathbf{r}_1,\mathbf{r}_2,\mathbf{r}_3)$ definition as follows,

$$\Omega(\mathbf{r}_1,\mathbf{r}_2,\mathbf{r}_3) = 1/4\ \delta(\mathbf{r}_3 - (\mathbf{r}_1 + \mathbf{r}_2)/2) \qquad (12)$$

  then a similarity measure is obtained involving a *gravitational* operator structure:

$$Z^{(n)}_{AB}(\delta(\mathbf{r}_3 - (\mathbf{r}_1 + \mathbf{r}_2)/2)/4,\, \mathbf{p}) = \int\int \rho^{(n)}_{A}(\mathbf{r}_1,\, \mathbf{p})\, \rho^{(n)}_{B}(\mathbf{r}_2,\, \mathbf{p})\, |\mathbf{r}_1 - \mathbf{r}_2|^{-2}\, d\mathbf{r}_1\, d\mathbf{r}_2 \quad (13)$$

This measure may be referred as a gravitational-like QSM.[13,14]

Other examples can be defined, but the presented ones give a sufficient idea of the multiple possibilities associated with the QSM. There will follow a discussion on the open possibilities which can be deduced from the QSM framework. One of them will be the possibility of the n-dimensional description of molecular sets.

IV. DISCRETE *n*-DIMENSIONAL MATRIX REPRESENTATION OF QUANTUM OBJECTS

The basic knowledge attached to the QMSM framework has a very simple form. Suppose there is a known set, $M = \{ m_I \}$, composed by n molecules. Suppose there is also known a set of density matrices or the chosen DIT, $P = \{\rho_I\}$,[47–52] somehow associated in a one-to-one correspondence with the set M; that is:

$$\forall m_I \in M \rightarrow \exists \rho_I \in P \Rightarrow m_I \Leftrightarrow \rho_I;\ \forall I \tag{14}$$

Adopting this situation and looking from the quantum mechanical point of view at every molecule in M, one can consider that such molecular element is represented by a density matrix element included in the density set P.

Thus, within this context, a molecule is represented by a vector belonging to an ∞-dimensional space. The definition of QMSM offers no difficulty whatsoever. Once a positive definite operator Ω is chosen, a QMSM between a given pair of molecules $\{m_I, m_J\} \in M$ is obtained by choosing the corresponding density couple, $\{\rho_I, \rho_J\} \in P$, and computing the integral based on the definition presented in Eq. 4:

$$z_{JI}[\Omega] = \langle \rho_J | \Omega | \rho_I \rangle \tag{15}$$

For simplicity, the operator following the symbol of the QMSM is removed from the right side of Eq. 15, unless one wants to stress the nature or the role of the positive definite operator Ω. Then, from the definition (Eq. 15) one can construct a $(n\times n)$ *similarity matrix*: $\mathbf{Z} = \{z_{JI}\}$:

$$\mathbf{Z} = \{z_{JI}^{(n)}(\Omega, \mathbf{p})\} = \{z_{JI}(\Omega)\} = \{z_{JI}\} \tag{16}$$

The matrix $\mathbf{Z}$ contains information about the relationships of the elements of the set. The column partition set of matrix $\mathbf{Z}$,

$$\mathbf{Z} = (\mathbf{z}_1, \mathbf{z}_2, \ldots, \mathbf{z}_I, \ldots, \mathbf{z}_m) \tag{17}$$

can be interpreted as the matrix representation of every element in $\mathbf{P}$.

Thus, the similarity matrix can be partitioned, as in Eq. 17, taking into account the fact one can consider the matrix $\mathbf{Z}$ as a row hypervector whose components are the matrix columns, defined easily as the elements of a well-defined column vector set, $\{\mathrm{z}_I = [z_{JI}], \forall J,I\} \in \mathbf{Z}$. In this way, there can be established a new correspondence between the density set $\mathbf{P}$ and the column vector QMSM set $\mathbf{Z}$:

$$\forall \rho_I \in P \rightarrow \exists z_I \in \mathrm{Z} \Rightarrow \rho_I \Leftrightarrow \mathrm{z}_I;\ \forall I \tag{18}$$

In this manner one can construct a n-dimensional representation of the molecules belonging to the initial set M. The discrete n-dimensional description of a given molecule has been called a *point-molecule*,[1,9,12–14] and the name *molecular point-cloud* is used as a synonym for the column vector set $\mathbf{Z}$, collecting the point-molecules. Consequently, QMSM, computed over a given molecular set, constitutes a natural vehicle leading towards the discrete n-dimensional description of molecular structures.

V. PRACTICAL IMPLEMENTATION OF QSM: LCAO MO EXPRESSION OF QSM AND QUANTUM MOLECULAR SIMILARITY MEASURES (QMSM)

A. Quantum Molecular Similarity Measures

Let us resume the trivial basic idea underlying the concept of QMSM. Given two molecules, $\{m_A, m_B\}$, we assume that the Schrödinger equation is solved at an arbitrary level for both molecules. The respective wavefunctions, $\{\Psi_A, \Psi_B\}$, for a given state of both electronic structures are also supposed to be known. A density matrix function or DIT couple, $\{\rho_A, \rho_B\}$, connected with the respective wavefunction pair can be computed in the usual way.[47–52]

Using a positive definite operator Ω as a weight, a QMSM involving the molecules $\{m_A, m_B\}$ is defined as the integral,

$$z_{AB}[\Omega] = \int\int \rho_A(\mathbf{r}_1)\, \Omega\,(\mathbf{r}_1,\mathbf{r}_2)\, \rho_B(\mathbf{r}_2)\, d\mathbf{r}_1 d\mathbf{r}_2 \tag{19}$$

where $\{\mathbf{r}_1,\mathbf{r}_2\}$ are sets of electron coordinates associated with the corresponding density functions. Within this precise definition, the QMSM $z_{AB}[\Omega]$ are non-negative real numbers. Originally,[1] the weighting operator was chosen as a Dirac δ function, $\Omega = \delta(\mathbf{r}_1 - \mathbf{r}_2)$, and the involved densities $\{\rho_A, \rho_B\}$ as the first-order density functions. When using this kind of integrands, the QMSM as defined in Eq. 19 becomes the so called *overlap-like* measure:

$$z_{AB} = \int \rho_A(\mathbf{r})\, \rho_B(\mathbf{r})\, d\mathbf{r} \tag{20}$$

Many other QMSM can be defined following Section III rules, even within a more general conceptual context (see for example Refs. 13 and 14 for more details). From all the various possibilities the most conspicuously used has been the so-called *Coulomb-like* measure, defined as,

$$z_{AB}[\mathbf{r}_{12}^{-1}] = \int\int \rho_A(\mathbf{r}_1)\, |\mathbf{r}_1 - \mathbf{r}_2|^{-1}\, \rho_B(\mathbf{r}_2)\, d\mathbf{r}_1\, d\mathbf{r}_2 \tag{21}$$

which transforms into the Coulomb molecular energy when considering the self-similarity measure, $z_{AA}[\mathbf{r}_{12}^{-1}]$. Also, triple[11] or multiple[13] QMSM can be constructed in a similar manner as in Section II, without problems other than those of increasing the computational difficulty and time. Triple QMSM are easily constructed, when considering a third molecule $\{m_C\}$, besides the initial QMSM pair $\{m_A, m_B\}$. Then, using the third molecule's density $\{\rho_C\}$ instead of the operator Ω in Eq. 19, the following measure is obtained,

$$z_{AB;C} = z_{AB}[\rho_C] = \int \rho_A(\mathbf{r})\, \rho_C(\mathbf{r})\, \rho_B(\mathbf{r})\, d\mathbf{r} \tag{22}$$

giving one of the five possible definitions of QMSM involving three density functions. As another example of the possible alternative forms of triple density measures, consider the operator Ω substituted by the off-diagonal element of the density matrix, $\{\rho_C(\mathbf{r}_1,\mathbf{r}_2)\}$. Another QSM form, different from the previous QMSM Eq. 22, is obtained as the following integral is computed:

$$z'_{AB;C} = \int\int \rho_A(\mathbf{r}_1)\, \rho_C(\mathbf{r}_1,\mathbf{r}_2)\, \rho_B(\mathbf{r}_2)\, d\mathbf{r}_1\, d\mathbf{r}_2 \tag{23}$$

In any case, the previous discussion shows that a wide collection of QMSM can be defined in a unique way, when the molecular density matrices are known.

B. LCAO MO Expression of the Density Function

In a LCAO MO framework,[13] when the DIT resulting from Eq. 3 are simply first-order density functions, one can write,[47–49]

$$\rho^{(1)}(\mathbf{r}) = \sum_{\mu} \sum_{\nu} D_{\mu\nu}\, \chi_\mu(\mathbf{r})\, \chi_\nu(\mathbf{r}) \tag{24}$$

where the parameter vector $\mathbf{p}$ has been omitted for simplicity. $\{D_{\mu\nu}\}$ is the *charge-bond order matrix* and $\{\chi_\mu\}$ are the AO basis set functions.

Using this approach, the LCAO form of the previous measures can easily be written. But, this is not the practical way if QSM are computed. In fact, they are evaluated using an approach which reproduces the exact values found within the LCAO framework in a faster way. Efficient computational methods and algorithms are compulsive because the pair of QO's which are compared need to be oriented in such a way that QSM reaches maximal values. For example, when dealing with molecules, the QMSM definition has embedded the idea of optimizing the relative position of both QO in order to attain the maximal available value of the measure.[1–15] This optimization process becomes the bottleneck of all QSM computations and some efforts have been devoted to circumvent this problem.[16] LCAO density functions as those outlined in Eq. 24 are expanded as the linear combination,

$$\rho(\mathbf{r}) \approx D(\mathbf{r}) = \sum_i c_i\, g_i(\mathbf{r}) \tag{25}$$

where $\{c_i\}$ are positive coefficients and $\{g_i(\mathbf{r})\}$ simple spherically symmetric functions. Within this approach, the general expressions of the integrals involving the QMSM can be optimized within a reasonable computational cost.

Equation 25 is a generalization of the CNDO approach,[55] which has been used in some of our previous work. In this CNDO-like framework it is also easy to deduce that the first-order density function may be expressed as,

$$\rho(\mathbf{r}) \sim D(\mathbf{r}) = \sum_{I \in M} Q_I \, |S_I(\mathbf{r})|^2 \tag{26}$$

where $\{Q_I\}$ are Mulliken gross atomic overlap populations[56–59] and $\{S_I(\mathbf{r})\}$ *ns*-type functions, usually STO or GTO, centered at the molecular nuclei (for more details see Refs. 4,6,9,11,13).

C. Atomic Shell Approximation (ASA)

Sometimes a very simple expression like Eq. 26 is unable, in very special situations, to get the appropriate maximal value of the QMSM. Recently, it has been found that an expression like Eq. 26, where at each atom as many *ns*-type functions as atomic shells are needed, produces excellent results. In this case the $\{Q_I\}$ coefficients are taken as a convenient atomic charge partitioning into the charge fraction belonging to every atomic shell.

The previous form was named "atomic shell approximation" (ASA). A more sophisticated ASA technique has proved to give very accurate results too. This time, a set of s-type GTO were used to be fitted to an ab initio density function like the one in Eq. 24. A restricted least-squares technique was used[16] to optimize the exponents and the coefficients of the fitting function. The $\{Q_I\}$ fitting coefficients were confined to a definite positive form without the appearance of precision problems. The final fitted function behaves like a real probability density distribution and the definite positive set $\{Q_I\}$ must be taken as a charge partition of the total electronic cloud.

The following tables and graphical relationships show how the ab initio precision ASA may be, in many instances, substituted with a simpler form obtained by using for each atom as many STO *ns*-type function shells as classical atomic shells of type $\{K,L,M,\ldots\}$. Within this approach, a C atom, for example, will be represented in the ASA density expression with two shell *ns* STO functions, $\{1s,2s\}$. Exponents were taken from Clementi–Roetti tables,[81] and the atomic number is partitioned into the shells, according to the Aufbau principle to produce the coefficients $\{Q_I\}$ if they cannot be computed in other ways. Using again the C atom as an example, the ASA coefficients will be in this case, $\{Q_{1s} = 2, Q_{2s} = 4\}$.

D. QMSM Maps

QMSM maps based on the simplified ASA approach can be easily obtained in various ways. The most immediate one may consist of the following algorithm, obtained when observing the QMSM computed using a given molecular density function $\{\rho_I(\mathbf{r})\}$ constructed using ASA approach and the same density form of an atom $\{\rho_A(\mathbf{r}-\mathbf{R})\}$,

$$z_{IA}[\mathbf{R}] = \langle \rho_I(\mathbf{r}) \,|\, \Omega(\mathbf{r}) \,|\, \rho_A(\mathbf{r}-\mathbf{R}) \rangle \tag{27}$$

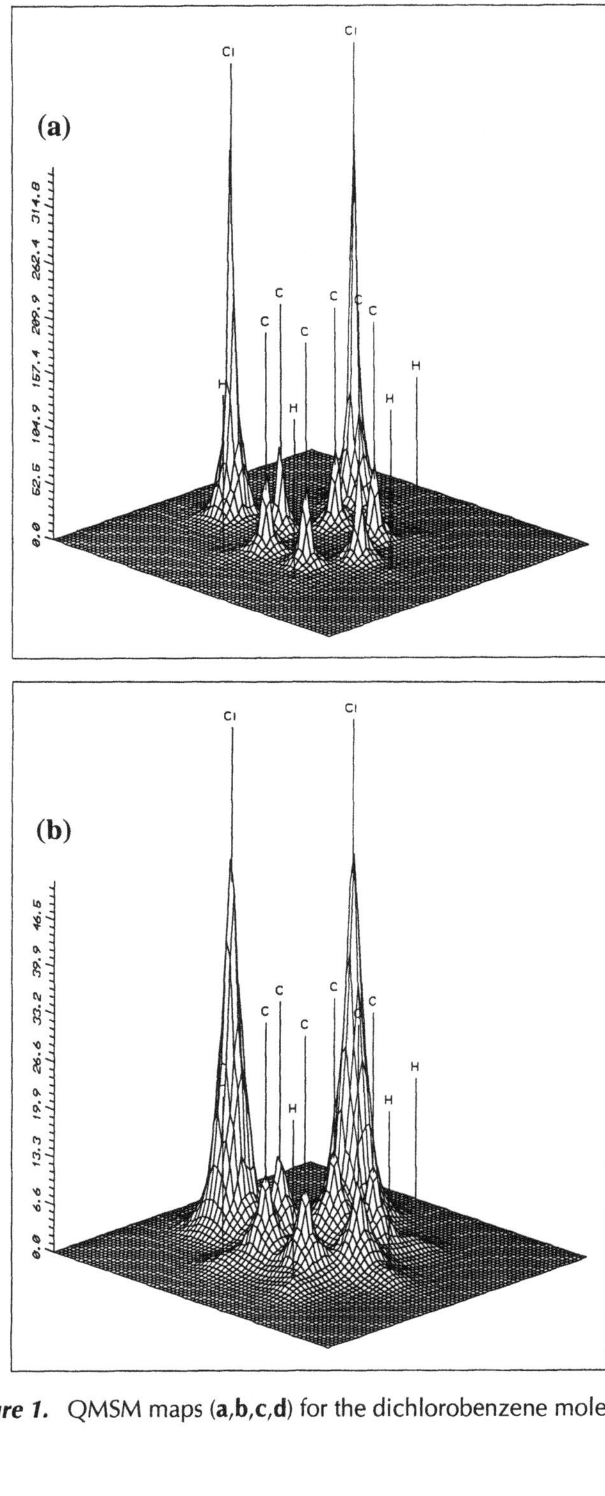

Figure 1. QMSM maps (**a**,**b**,**c**,**d**) for the dichlorobenzene molecule.

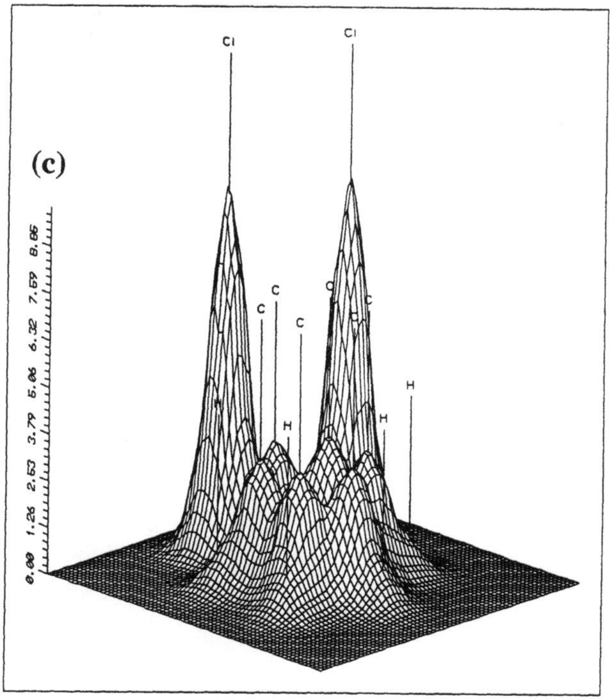

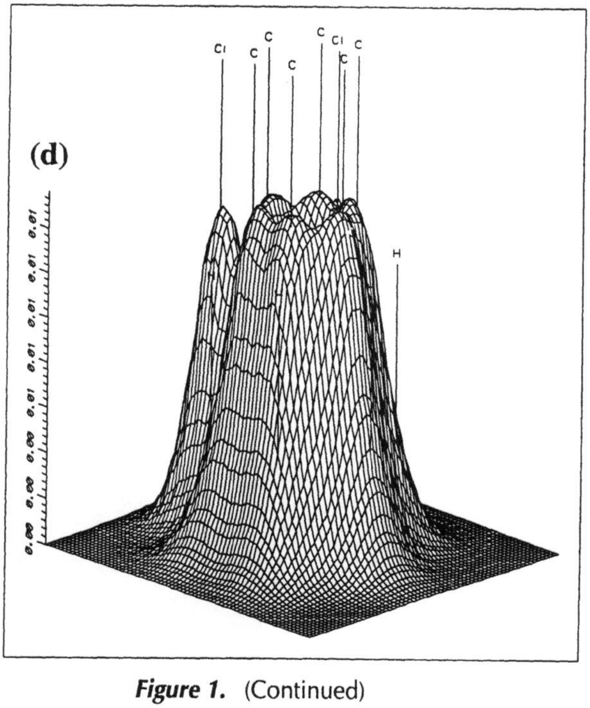

Figure 1. (Continued)

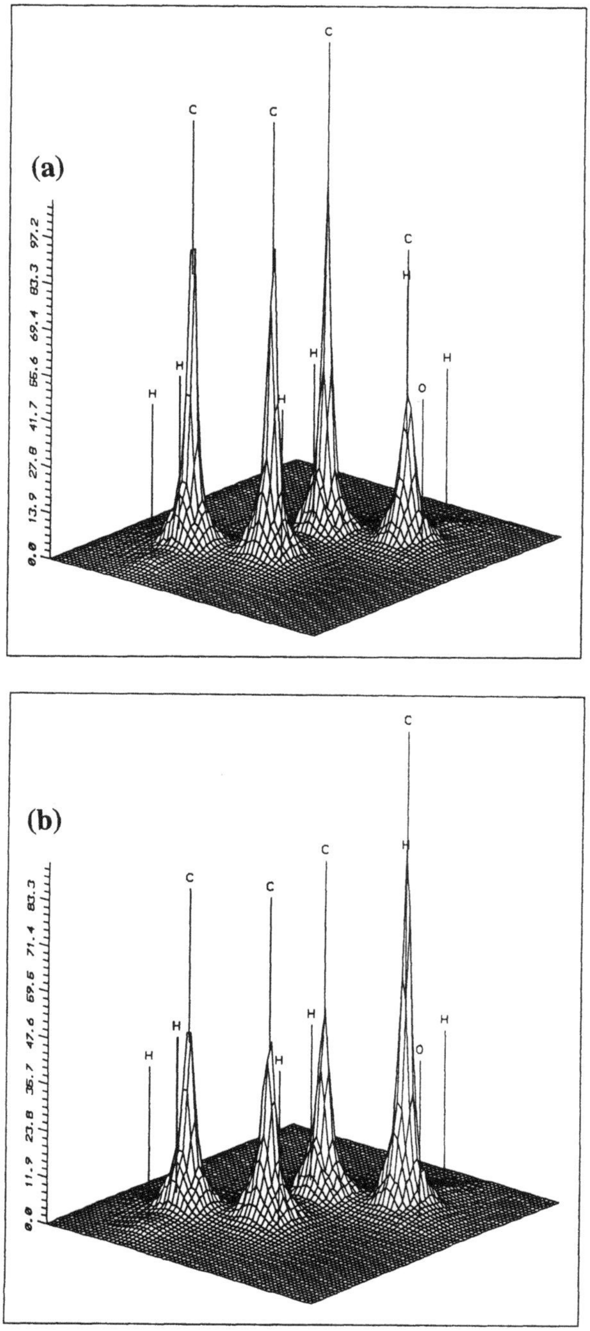

Figure 2. QMSM maps (**a**,**b**,**c**,**d**) for the butanol molecule.

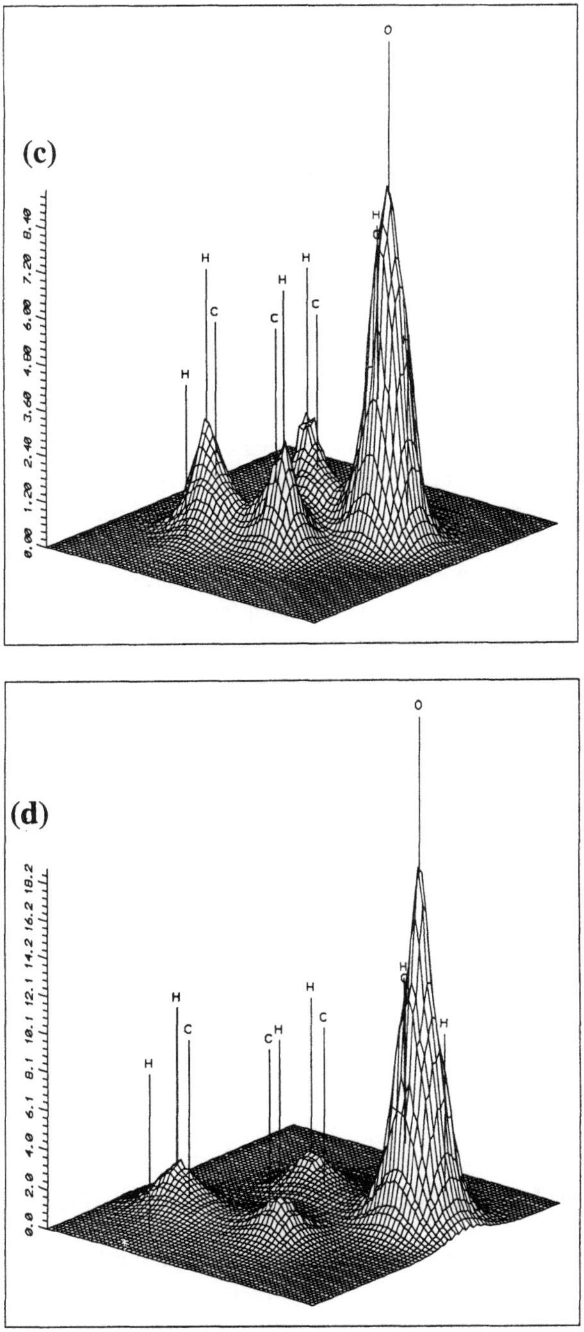

Figure 2. (Continued)

where the integration has been performed over the electron position **r**. In this way, the QMSM z_{IA} will depend on the atom position **R**.

Two examples of QMSM maps are presented here. Fully optimized geometries for each studied molecule have been obtained using the AMPAC program with the AM1 methodology. QMSM grids of overlap-like measures have been calculated within the ASA approach, taking a *ns*-STO function per period in each atom. Points in the grids have been calculated at distances of 0.4 au. Calculations at every 0.1 au have been added in the regions near the heavy atoms.

In the first example, a chlorine atom has been used to map the 1,2-dichloro-benzene molecule. Figure 1 shows four maps made in this way. Figure 1a shows the similarity surface in the plane defined by the benzene ring. Figures 1b, 1c, and 1d represent surfaces parallel to the benzene ring but at distances of 0.5, 1.0 and 4.0 au from it. As can be expected, these maps display strong and sharp peaks where the heavy atoms are located, while peaks due to the hydrogen atoms appear as very low and rounded peaks. As the atomic number augments, the peak becomes higher. By looking successively at the maps a, b, c, and d in Figure 1, it can be seen that as the surface is moved upwards, the peaks are lowered and rounded. At a distance of 4 au from the benzene plane, the carbon atom peaks are nearly fused forming a volcano-like shape, and the chlorine peaks have been lowered nearly to the carbon level. This is consistent with the well-known fact that similarities between atoms quickly decay with the interatomic distance.

The second example shows four maps of the butanol molecule made in a similar way, using a chlorine atom as the moving structure (Figure 2). The map in Figure 2a was made in the plane defined by the three carbon atoms in the $CH_3CH_2CH_2$– fragment. These atoms produce the three strong peaks which can be seen at the left, while the $-CH_2OH$ carbon atom, being out of the plane gives a lower peak, at the right of Figure 2. In Figure 2b, which maps a parallel plane 0.2 au over and parallel to the one in Figure 2a, the situation is inverted: the $-CH_2OH$ carbon is located on the plane and gives the strongest peaks, while the other atoms produce lower peaks. By increasing the vertical distance to the original plane and setting it to 1.71 au as in Figure 2c, the oxygen atom peak becomes the strongest one, and low peaks appear due to the presence of hydrogen near the map plane. It can be seen how these hydrogen atom peaks are broadened by carbon contributions. At a distance of 2.0 au (Figure 2d), the oxygen is the only atom in the molecule to give a remarkable peak, while the peaks due to the hydrogen and carbon atoms have been transformed into slight protuberances.

VI. QUANTUM MOLECULAR SIMILARITY INDICES (QMSI)

Once a QOS and the DIT set are chosen and the operator related to the QSM in the integral described in Eq. 4 defined, the QSM related to the QOS elements is unique. However, the elements of the similarity matrix **Z** can be transformed or combined in order to obtain new kind of matrix elements, which can be named "quantum similarity indices" (QSI). A vast number of possible QSM manipulations exist leading to diverse QSI definitions.

Simultaneously with the definition of QMSM, the "quantum molecular similarity indices" have evolved. In the seminal paper[1,2] on the subject two indices were defined. They can be named: *correlation or cosine-like* and as *Euclidian distance-like*, constituting a pair of similarity and dissimilarity indices respectively. Following this early perspective, Hodgkin and Richards[31] have described a new index, claiming a better performance within molecular comparison purposes for this new index form than the behavior of the original correlation-like one.

A thorough discussion on the meaning and usefulness of QMSM and QMSI has been carried out by Carbó and Domingo,[3] later on by Carbó and Calabuig[10] and in recent reviews by various authors.[13–16] The study in these last references was performed in such a way that points out new possibilities when the nature of QMSM and QMSI is described. Despite this, the connections between the diverse QMSI forms has not been discussed in the literature. Thus, here we attempt to find a way to relate QMSIs one to another. A methodology will be described to obtain the possible relationships between various index definitions, as well as to use the newest QMSM theoretical developments in order to construct new QMSI. For this purpose the discrete n-dimensional representation of molecular electronic systems, presented early in this chapter, will be used as a background. A comparison and study on the n-dimensional representation of molecular systems as point-molecules will lead us to an interesting relationship between the aforementioned indices.

A. QMSM and QMSI

A theoretical remark must be recalled before introducing the present subject of discussion. Once the systems to study and an appropriate computational framework as well as a weighting operator have been chosen, QMSM *are uniquely defined.* By contrast, QMSI, after the QMSM computational step, can be chosen using a great number of various mathematical manipulations, and can be considered as the result of some arbitrary transformation from the known QMSM as a starting point.[12–14]

Once the molecular point-cloud for the molecular set M has been obtained, as presented in Section IV, QMSI can be obtained through mathematical manipulations performed over the elements of the similarity matrix $\mathbf{Z}$. In the first paper discussing the nature of QSM,[1] two index classes were described, as mentioned at the beginning of this section. Here a tentative dual classification will be given as follows:

1. *C-class*: A similarity index, commonly referred to as the *Carbó index*, which is nothing more than a member of the *correlation-like index* class. In fact, the mathematical interpretation of such an index is the generalized form—most suitable in ∞-dimensional functional spaces—of the cosine of the angle subtended by two density distribution functions, weighted by the chosen positive definite operator

Ω. The concrete form of this similarity index (C) is usually written, using the pertinent similarity matrix elements, as:

$$C_{JI} = z_{JI}(z_{JJ}\, z_{II})^{-1/2} \tag{28}$$

The Carbó similarity index has values in the interval, [0,1]. The interval extreme values represent complete dissimilarity or total similarity, respectively. These two extremal situations correspond to a couple of orthogonal or colinear density distribution functions. A fuzzy set point of view[3] can be invoked at this moment because the correlation-like similarity index may be interpreted as a fuzzy membership function defined over the density function set **P** cartesian product, $\mathbf{P} \otimes \mathbf{P}$.

2. *D-class*: A dissimilarity index, taking the form of an euclidian distance belonging to a *distance-like index (D)* class. The mathematical interpretation of this alternative manipulation of the QMSM matrix elements is such that it represents a distance, defined in ∞-dimensional space, between two density distributions. The dissimilarity index may be defined as:

$$D_{JI} = (z_{JJ} + z_{II} - 2z_{JI})^{1/2} \tag{29}$$

The interval where the dissimilarity index values can be found is now, $[0,+\infty]$. The lower value now corresponds to complete similarity, while the higher the index numerical value is found, then less similarity can be attached between both densities.

In the following discussion all the descriptions of possible QMSI will belong, without exception, to one of the above described two classes: C-class or D-class, being complementary to each other. Inverse relationships between the two index classes will be defined later.

B. Generalized QMSI

There are many alternatives for the description of generalized definitions of QMSI. Here are given some possible choices within the two described classes. They will be given in reverse order because of the fact that a D-class generalized index may serve to define a C-class one.

D-Class Generalized Indices

A generalized Euclidian distance-like index can have the following form,

$$^{(g)}D_{JI}(K,X) = (K[z_{JJ} + z_{II}] - X\, z_{JI})^{1/2};\ X \in [0,2K] \tag{30}$$

which transforms into the Euclidian distance dissimilarity index as defined in Eq. 28 when using $K = 1$ and $X = 2$.

Another D-class index can be defined with the simple form,

$$^{(\infty)}D_{JI} = \max(z_{JJ}, z_{II}) \quad (31)$$

constituting a distance of infinite order.

C-Class Generalized Indices

The following QMSI form has the structure of a C-class family of indices. It has been proposed[12] in order to generalize the Hodgkin–Richards[31] and Tanimoto[60] indices. The general function can be cast in the next formula, which may be called the *Girona index*,

$$^{(g)}C_{JI}(K,X) = (2K - X)z_{JI}(D_{JI}(K,X))^{-2};\ K \in [0,1] \quad (32)$$

where the generalized distance index described in Eq. 30 has been used too. When the parameters in the Eq. 32 take the values $K = 1$ and $X = 0$ the Hodgkin–Richards index is obtained, whereas the Tanimoto index appears naturally when $K = X = 1$.

As a function of the D-class index of infinite order described in Eq. 31, the Petke index[61] can also be defined as having the form:

$$^{(\infty)}C_{JI} = z_{JI}(^{(\infty)}D_{JI})^{-1} \quad (33)$$

C. QMSI in the Molecular Point-Cloud *n*-Dimensional Representation

The polyhedron nature of the molecular point-cloud has not been used so far. Here, the columns of the similarity matrix $\mathbf{Z}$ can be taken directly to obtain new index forms. In fact, within this n-dimensional discrete representation of the molecular electronic structures, one can even consider the possibility of constructing point-molecules of larger dimensionality. Besides the sets used up to now as shown in Section IV, augmented sets may be gathered to obtain a great deal of information for the original molecular set M.

A New C-Class QMSI

One can augment the initial dimension of the molecular point-cloud $\mathbf{Z}$ (see Eqs. 16,17) by using the following procedure:

1. Choose a new molecular set $A = \{a_I\}$ composed of m molecules and compute the associated density functions set $\mathbf{D} = \{d_I\}$, such that:

$$\forall a_J \in A \rightarrow \exists d_J \in D \Rightarrow a_J \Leftrightarrow d_J \quad (34)$$

2. From here a set of column vectors $V = \{\mathbf{v}_I\}$ can be obtained by computing the appropriate QMSM:

$$v_{JI} = \langle d_J | \Omega | \rho_I \rangle;\ \forall d_J \in D \wedge \rho_I \in P \quad (35)$$

3. Then, a new augmented molecular point-cloud U may be constructed simply by building the direct sum of the original molecular point-cloud $\mathbf{Z}$ and the new discrete vector set V, that is:

$$U = Z \oplus V = \{ \mathbf{u}_I = z_I \oplus \mathbf{v}_I \} \tag{36}$$

4. Also, a new rectangular similarity matrix $\mathbf{U}$ of dimension $(d \times n)$ where $d = n + m$, whose columns are the augmented point-molecules $\{\mathbf{u}_I\}$ can be constructed, and a Gram matrix computed in the usual way:

$$\mathbf{S} = \mathbf{U}^{\mathrm{T}}\mathbf{U} \tag{37}$$

5. Knowing the Gram matrix (Eq. 37), a new C-class index may be computed using the auxiliary quotient, which bears a D-class structure,

$$^{(d)}\theta_{JI} = K\,(S_{JI})^{-1} \tag{38}$$

where K is a scale factor. Equation 38 can be cast into the C-class index,

$$^{(d)}C_{JI} = (1 + (^{(d)}\theta_{JI})^{r})^{-1/r} \tag{39}$$

where r is a positive integer.

Origin of the New C-Class QMSI

The origin of such a C-class index as the one defined finally in Eq. 39 may be easily seen when a $[2 \times 2]$ similarity submatrix is studied as a source of discrete molecular information. Once two molecules $\{A,B\}$ are chosen, such a matrix can be defined as,

$$^{(2)}\mathbf{Z}_{(A,B)} = (^{(2)}\mathbf{z}_A, {}^{(2)}\mathbf{z}_B) \tag{40}$$

where the two column vectors appearing on the right side of Eq. 40, and describing a couple of two-dimensional point-molecules, are written as,

$$^{(2)}\mathbf{z}_A = \begin{pmatrix} z_{AA} \\ z_{BA} \end{pmatrix} \wedge {}^{(2)}\mathbf{z}_B = \begin{pmatrix} z_{AB} \\ z_{BB} \end{pmatrix} \tag{41}$$

where the QMSM similarity matrix nondiagonal elements are equal: $z_{AB} = z_{BA}$. Then a C-class similarity index may be found for these two vectors as the correlation index,

$$^{(2)}c_{AB} = (^{(2)}\mathbf{z}_A \cdot {}^{(2)}\mathbf{z}_B)\,(\left|{}^{(2)}\mathbf{z}_A\right|\,\left|{}^{(2)}\mathbf{z}_B\right|)^{-1} \tag{42}$$

and it is very easy to see, after a simple manipulation, that it can be written as in Eq. 39 above by means of the appropriate two-dimensionally defined D-class index,

$$^{(2)}\theta_{AB} = Det\left|{}^{(2)}\mathbf{Z}_{(A,B)}\right|\,(^{(2)}\mathbf{z}_A \cdot {}^{(2)}\mathbf{z}_B)^{-1} \tag{43}$$

where Det $\left| {}^{(2)}\mathbf{Z}_{(A,B)} \right|$ is, in this case, the value taken by the scale factor K of Eq. 38 above.

Thus, this simple case shows the importance of the dual representation, linked to the use of QMSM, and involving the ∞-dimensional and n-dimensional point-molecules description.

D. Relationships between *C*- and *D*-Class QMSI

After the previous discussion on the many possible QMSI forms, one can present various connections between the indices, describing the relationships between the members of *C*- and *D*-classes and how they can be transformed from one class to another.

Knowing a set of *D*-class indices, $\{D_{JI}\}$, then it is easy to obtain a new set of *C*-class indices, $\{C_{JI}\}$, and vice versa using any of the following rules:

1. Transforming *D*-class to *C*-class indices:

$$\begin{aligned} &(a)\quad {}^{(m)}C_{JI} = 1 - (D_{JI}[Max[\forall\{J,I\}](D_{JI})^{-1}) \\ &(b)\quad {}^{(t)}C_{JI} = 1 - \tanh(D_{JI}) \\ &(c)\quad {}^{(p)}C_{JI} = (1 + (D_{JI})^{p})^{-1/p};\, p > 0 \end{aligned} \tag{44}$$

2. Transforming *C*-class to *D*-class indices. Defining the factor K as a scale factor, one also can describe the transformations:

$$\begin{aligned} &(a)\quad {}^{(a)}D_{JI} = K\,\pi^{-1}\,arcos(C_{JI}) \\ &(b)\quad {}^{(l)}D_{JI} = K\,(1 - C_{JI}) \\ &(c)\quad {}^{(c)}D_{JI} = K\,(C_{JI})^{-1}\,(1 - C_{JI}) \end{aligned} \tag{45}$$

   Another interesting possibility related with *C*- to *D*-class transformation may be obtained, when connecting the usual entropy definition with the QMSI. An entropy-like index is defined as:

$$S_{IJ} = -K\frac{\log(C_{IJ})}{C_{IJ}} \tag{46}$$

One can see in this way that, using the previous rules, a set of one class of indices can be easily transformed into the complementary class without problems. This allows a great freedom in the use of QMSI sets to obtain information, coming from the molecular point-cloud sets Z or U, which can be correlated with the characteristic properties of the molecular electronic structure set M.

Some Relationships Related to C-Class QMSI

In the previous Section VI.C a very helpful but simple situation has been analyzed. This preparatory discussion may be used to find the connection between the Hodgkin–Richards index and the initial C-class index defined by Carbó. Despite the apparent diversity of these indices, it can be proved that they are connected by the dual structure of the QMSM. Precisely, the presence in the theory of the duality between the ∞-dimensional and n-dimensional representation of molecular electronic structures is the clue allowing one to find the connection between both QMSI.

Consider the two-dimensional case discussed above in Eq. 40. The C-class index appearing in Eq. 42 can be also interpreted as the cosine of the angle subtended by the two-dimensional representations presented in Eq. 41. One can take this correspondence allowed for any C-class index, even if computed in a discrete two-dimensional scenario, as, ${}^{(2)}c_{AB} = \cos(\gamma_{AB})$. One can take into account the cosine of the angle subtended by the two ∞-dimensional density function distributions, associated at the same time to the two molecular electronic structures involved. The ∞-dimensional cosine can be computed by means of the C-class Carbó index c_{AB} as:

$$c_{AB}^2 = \cos^2(\alpha_{AB}) = (z_{AB})^2 \, (z_{AA} \, z_{BB})^{-1} \tag{47}$$

The expression of the two-dimensional D-class index ${}^{(2)}\theta_{AB}$ in Eq. 42 may be also rewritten defining the parameter,

$$\omega_{AB} = z_{AB}(z_{AA} + z_{BB})^{-1} \tag{48}$$

which is nothing but half of the Hodgkin–Richards C-class index. Then the ${}^{(2)}\theta_{AB}$ D-class index, defined before, can be written in terms of the parameter defined in Eq. 48 above as:

$${}^{(2)}\theta_{AB} = \omega_{AB}(1 - \cos^{-2}(\alpha_{AB})) \tag{49}$$

The cosine involving the two-dimensional representation: $\cos(\gamma_{AB})$, may be also written as in Eq. 42 using the parameter ${}^{(2)}\theta_{AB}$. After a trivial manipulation, the final relationship between Hodgkin–Richards and Carbó indices is found to be:

Table 1. Ordering Numbers of the Methane and Their Four Chloro Derivatives and QMSM (Ω = I) Values Normalized Using the Number of Electrons

	CH_4	CH_3Cl	CH_2Cl_2	$CHCl_3$	CCl_4
CH_4	0.160445				
CH_3Cl	0.136320E-01	0.148110E-01			
CH_2Cl_2	0.841200E-02	0.954900E-02	0.104050E-01		
$CHCl_3$	0.475400E-02	0.711100E-02	0.746000E-02	0.791400E-02	
CCl_4	0.475500E-02	0.571900E-02	0.608200E-02	0.625700E-02	0.635700E-02

Table 2. Numerical Values of MQSI for Every Molecular Pair of Table 1

Pair	DST^{a}	D_{∞}^{b}	${}^{(2)}\theta_{pq}^{c}$	S^{d}	CAR^{e}	HR^{f}	TAN^{g}	PET^{h}	${}^{(2)}C_{pq}^{i}$
1-1	0.000000	1.604453	0.000000	0.000000	1.000000	1.000000	1.000000	1.000000	1.0000000
2-1	2.127863	10.011905	0.085052	0.139028	0.884313	0.610222	0.439079	0.354006	0.996403
2-2	0.000000	10.011905	0.000000	0.000000	1.000000	1.000000	1.000000	1.000000	1.000000
3-1	3.590625	18.354544	0.240580	0.659099	0.651079	0.354046	0.215101	0.192498	0.972259
3-2	2.740807	18.354544	0.253690	0.341144	0.769198	0.735179	0.581252	0.568100	0.969295
3-3	0.000000	18.354544	0.000000	0.000000	1.000000	1.000000	1.000000	1.000000	1.000000
4-1	4.765732	26.622654	0.451099	2.045383	0.421909	0.195376	0.108264	0.103575	0.911546
4-2	3.897292	26.622654	0.385830	0.640074	0.656789	0.585395	0.413822	0.402771	0.932965
4-3	2.937639	26.622654	0.193738	0.238210	0.822142	0.808131	0.678037	0.682642	0.981745
4-4	0.000000	26.622654	0.000000	0.000000	1.000000	1.000000	1.000000	1.000000	1.000000
5-1	5.420353	34.813347	0.339256	1.599915	0.470822	0.193246	0.106957	0.101076	0.946987
5-2	4.776742	34.813347	0.461139	0.896876	0.589412	0.490973	0.325357	0.316085	0.908097
5-3	3.919696	34.813347	0.280305	0.388728	0.747759	0.711028	0.551625	0.542951	0.962888
5-4	2.779788	34.813347	0.124659	0.142220	0.882098	0.874223	0.776551	0.771382	0.992319
5-5	0.000000	34.813347	0.000000	0.000000	1.000000	1.000000	1.000000	1.000000	1.000000

Notes: [a]DST: Euclidean distance index (Eq. 29).

[b]D_{inf}: Distance index of infinite order (Eq. 31).

[c]${}^{(2)}\theta_{pq}$: D-class index (Eq. 38) ($k = 1$).

[d]S:Entropy-like index, (Eq. 38) ($k = 1$).

[e]CAR: Carbó index (Eq. 28).

[f]HR: Hodgkin-Richards index (Eq. 32) with $K = 1$ and $X = 0$.

[g]TAN: Tanimoto index (Eq. 32) with $K = X = 1$.

[h]PET: Petke index (Eq. 33).

[i]${}^{(2)}C_{pq}$: C-class index (Eq. 39) ($r = 2$) obtained from ${}^{(2)}\theta_{pq}$.

$$\omega_{AB} = tg(\gamma_{AB}) c_{AB}^2 (1 - c_{AB}^2)^{-1} \tag{50}$$

This means that there exists a function directly connecting the two QMSIs. This relationship involves the two subtended angles of the dual molecular representation. In fact, Eq. 50 above may also be written like a ratio between the tangents of the angles of both representations:

$$\omega_{AB} = tg(\gamma_{AB}) tg^{-2}(\alpha_{AB}) \tag{51}$$

Finally, an inverse relationship will give the result:

$$c_{AB}^2 = \omega_{AB} (tg(\gamma_{AB}) + \omega_{AB})^{-1} \tag{52}$$

It can be seen, within this dual representation context, how QMSI, appearing very different at first glance, can be related in simple ways.

A Numerical Example

Table 1 contains ordering information for five molecules: the isomers of methane and their four chloro derivatives. Molecular geometries for these compounds have been obtained by means of the AMPAC program.[65] Full geometry optimization has been carried out using the AM1 methodology.[63] After the mentioned optimization, the QMSM between all the molecular pairs has been computed. Each of the QMSM has been normalized by dividing it by the number of electrons of the involved molecules. The resulting values are listed in Table 1. Also, for every molecular pair, the most representative QMSI are reported in Table 2.

It can be seen from Table 2 that the distance index of infinite order, column D_{inf}, has a quite different behavior with respect to the other distance indices. The main difference is found in the diagonal elements of the matrix: there no null elements appear. It can be seen how the Carbó and Hodgkin–Richards indices give related values for every molecular pair. The Petke index attached to every molecular couple is lower bound to the Carbó index, as it can be easily deduced from the respective definitions. The Tanimoto index gives the lowest *C*-class index values while the $^{(2)}C_{pq}$ index bears the highest ones. One can consider the application of each index for specific purposes. In any case, the Carbó index appears to be a robust choice.

VII. QUANTITATIVE STRUCTURE–ACTIVITY RELATIONSHIPS (QSAR) AND QMSM

In the last 15 years the theoretical and practical formalism of the QMSM has been developed[1–15] in our laboratory and by other authors;[17–43] however, there appears a much older idea dating from the end of nineteenth century of obtaining empirical relationships between parameters and molecular properties.[66] Recent procedures seem to be very successful as tools for predicting new molecular structures with tailor-made properties.[67] Since recently QMSM have been used as parameters in "quantitative structure–activity relationships" (QSAR),[68] it seems worthwhile to search for the possible practical formalism allowing QMSM to be used in "quantitative structure–property relationships" (QSPR) or QSAR environments.

Beyond this initial landscape, the success of QSAR in the realm of molecular design appears certain. The fact is also certain that no comprehensive justification, other than the empirical evidence and pragmatism, has so far been given for this prediction. The continued successful use of QSPR techniques cannot be a product based on statistical factors only: it seems to preclude the evidence of an existing solid theoretical reason not yet described. A new idea, associated both to the QMSM theoretical framework and the quantum mechanical operator expectation value concept, will provide a solid basis in the following pages.

A. Mendeleev's Postulates, Molecular Set Order, and Visualization

The molecular point-cloud, $U = \{\mathbf{u}_I\}$, as defined in Eq. 36 may be manipulated afterwards, in order to extract information from its elements or to obtain new values which, in turn, can be used by other algorithms as in the computation of the Gram matrix (Eq. 37). Visualization of the molecular point-cloud U may be very helpful as a tool for gathering information on the relationships between members of the molecular set M.[7,10] This possibility has been used in various ways, as well as the related option to employ QMSM or derived QMSI, obtained from the manipulation of QMSM matrix elements, to look for the existence of some ordering among the elements of the set M.

B. Mendeleev's Postulates and Conjecture

From these previous considerations, a résumé can be structured in terms of four statements. The principles governing the QMSM application possibilities have been called *Mendeleev's postulates*,[13,14] in hommage to the first chemist who has sought order between QOS. The four postulates can be summarized as follows:

1. Every QO in a given state can be described by their DIT.
2. QO can be compared by means of a QSM or a QSI.
3. Projection of a QOS into some n-dimensional space is always feasible.
4. A QOS ordering exists.

Mendeleev's postulates (see Ref. 13 for more details) describe the fact that it is always possible to extract information, in the way previously described, from the studied QOS. The postulates can be connected to the following points of the theory: Postulate 1 is a usual quantum mechanical assumption. Postulate 2 describes the starting point of the use of QMSM allowing the definitions of Section III. Postulate 3 describes the reasoning carried out in the previous sections. Postulate 4 is nothing more than the application of *Zermelo's theorem*[54] into the developed QMSM theoretical context.

Moreover, postulates 3 and 4 permit a pictorial visualization of the molecular set M, using the representation form of every molecule within the set M which is contained in the molecular point-cloud U, as described previously. Reference 9 has founded the basic concepts of these procedures. Sketching the whole formulation again, one can say that a DME may be transformed according to Eq. 1 into a DIT before computation of the QSM. QSM can be transformed a posteriori into QSI. Postulate 3 also means that projection of a QOS into some finite-dimensional space is always feasible. QO ordering can be achieved by manipulating in the appropriate way a particular QSM or QSI computed over the QOS elements.

The above steps justify the use of QSM as a tool to order QOS elements by means of their discrete n-dimensional quantum representations: the DME or DIT. QSM values can be ordered and this order may be transferred into the compared QO using the *Mendeleev algorithm*.[5,6,13,14] This procedure justifies how once a molecular ordering has been established, this order may also be transferred to the QO properties. This possibility can be stated by means of the *Mendeleev conjecture*: "Object ordering induces order over the implicit relationships between QOS elements and QO properties." Unknown properties of ordered QOS may be evaluated in this way: by inspecting the relative position of the QO in the ordered molecular point-cloud sequence, a numeric interval where the associated property will take a value that can be easily obtained. This result precludes the use of QMSM as a basis of QSPR.

C. ND-CLOUD and MENDELEEV Programs

In order to apply all the previous statements, in our laboratory, we have constructed two computer programs which use the Mendeleev conjecture. They are based on the basic formalism present in the Mendeleev postulates. These codes can use any molecular point-cloud and predict any molecular property interval. The program input of both programs is the QMSM matrix related to a given set of molecules. A set of known molecular properties for some of the elements of the molecular set M is also given. The output is a diagram in the form of a tree or a graph in the ND-CLOUD program case. The estimation of the corresponding molecular properties attached to the remaining molecules is presented in the MENDELEEV program.

The ND-CLOUD program has been described elsewhere (see for example Refs. 9,10). The algorithm which is implemented by the MENDELEEV program is based on the following assumptions.

It is always possible to construct, for a set M of n molecules and, if necessary, the molecular pattern extensions, a $(d \times n)$ similarity matrix $\mathbf{U}$, which contains the QMSM or QMSI concerning all the involved molecules. Then, at this stage, it is supposed that every one of the n molecules in the set is represented by means of a d-dimensional vector. It is assumed that a $(m \times p)$ property matrix $\mathbf{P}$ for $m < n$ molecules of the molecular set M is known. Assuming that for each molecule a known number of properties are tabulated, the goal is to estimate the property values for the remaining $n - m$ molecules. The estimation is made by means of a similarity matrix **U** transformation. Usually, in order to obtain a non-negative definite matrix, a new Gram matrix **S** is constructed, as in Eq. 37. Performing the diagonalizaton of the Gram matrix and using a principal component expansion, defined as the matrix equation,

$$\mathbf{S} = \mathbf{C}\,\mathbf{D}\,\mathbf{C}^T \tag{53}$$

where **C** is a unitary matrix and **D** is a diagonal one, it is found that,

$$\mathbf{X} = (\mathbf{U}^T\mathbf{U})^{1/2} = \mathbf{C}\,\mathbf{D}^{1/2}\,\mathbf{C}^T \tag{54}$$

and any function of **X** matrix, $\mathbf{F} = f(\mathbf{X})$, can be computed as,

$$\mathbf{F} = f(\mathbf{X}) = \mathbf{C}\,f(\mathbf{D}^{1/2})\,\mathbf{C}^T \tag{55}$$

assuming the function $f(\mathbf{X})$ has a Taylor series expansion. The function **F** acts as a bridge between the space of the QMSM or QMSI and the space of the molecular properties: a linear transformation **T** relates **F** with the property values. With respect to the known properties, collected in the matrix **P**, it is presumed that,

$$\mathbf{P} = \mathbf{T}\mathbf{F}_k \tag{56}$$

or:

$$\mathbf{T} = \mathbf{P}\mathbf{F}_k^{-1} \tag{57}$$

Supposing that $\mathbf{F}_k$ is nonsingular and contains the information present in the matrix **F**, connected in turn with the m molecules with known property values. Once the matrix **T** is known, Eq. 56 can be considered as a general rule for reproducing molecular properties from theoretical parameters; that is the transformation of the QMSM matrix will produce a molecular parameter set.

In this case, with respect to the remaining molecules with unknown property values, it is possible to extract the related theoretical parameters from **F**, collect them into the matrix $\mathbf{F}_u$, and assume that the estimation of the property values can be obtained in the same way as it was done in Eq. 56:

$$\mathbf{P}_u = \mathbf{T}\mathbf{F}_u \tag{58}$$

Another possible method for calculating QSPR may be based on the discrete n-dimensional representation of the molecular point-cloud, as discussed above. The following pages will deal with the way to obtain information from the discretization procedure inherent to the QMSM calculation procedures.

D. QSPR

Having described the discrete representation of molecular structures and their possible use, one can realize that it also connects the previous QMSM formalism with parent theoretical procedures used to obtain information on QSPR or particularly on QSAR.

A typical QSPR procedure consists of assigning to every element $m_I \in M$ of the molecular set M, a vector $\mathbf{q}_I \in Q$, whose elements are chosen in an empirical way from various considerations. Some are chosen as molecular atomic charges or quantum chemical related parameters, but others come from empirical sources like octanol–water partition coefficients or may even constitute a purely binary information variable; others, finally, bear empirical structural intuitive bonding schemes like the connectivity related indices.[69]

However, the fact is that, although in a quite different way, QMSM and QSPR techniques both assign a vector to every element of the molecular set M. In the QMSM case one calls this vector a point-molecule. The next step in QSPR framework consists in connecting a given molecular property value π with the point-molecule representation $\mathbf{q}$ throughout a linear equation, such as,

$$\mathbf{x}^T \mathbf{q} = \pi \tag{59}$$

which can be also observed as a linear functional transformation of the discrete point-molecule $\mathbf{q}$ by means of a dual space vector $\mathbf{x}^T$, a vector whose set of coefficient elements can be easily obtained using a standard least-squares calculation. In QSPR, unless one chooses, in a very restricted way, the elements of the point-molecule $\mathbf{q}$, as discussed some years ago[2,70] no direct meaning can be whatsoever attached to the elements of the vector $\mathbf{x}$.

E. Discrete Expectation Values

The form of Eq. 59 in a QMSM environment may be written in a parallel manner as,

$$\mathbf{w}^T \mathbf{u} = \pi \tag{60}$$

where the constant π role, as a molecular property, is preserved here too. However, contrary to Eq. 59, to the elements of coefficient vector $\mathbf{w}$ which may be obtained by a least-squares technique, as in the QSPR context, one can *always* attach a coherent theoretical meaning related to the whole QMSM theory so far developed.

To prove this, let us consider again the point-molecules $\mathbf{u}_I \in U$ which, as defined above, are nothing but a discrete representation of the associated density functions or DIT, $\rho_I \in P$. The representation of the molecular point-cloud vectors $\{\mathbf{u}_I\}$ is obtained in the space where the density function basis set $P \oplus D$ is active. At the same time, since it has been employed when defining triple QMSM in Eqs. 9 or 22 and 23, the density ρ_I also has the structure of a positive definite operator, which in the QMSM context can be attached to the matrix representation of the point-molecule $\mathbf{u}_I$.

From the quantum mechanical point of view, given any observable O, and the associated hermitian operator Ω, the expectation value $<\Omega>_I$ of the system described by the density function ρ_I may be formally obtained as:

$$<\Omega>_I = <\Omega \mid \rho_I> = \int \Omega\, \rho_I\, d\mathbf{r} \tag{61}$$

Then, to the operator Ω one can assign the discrete vector representation $\mathbf{w}$, using the same basis set contained in $P \oplus D$, in such a way that both vectors $\mathbf{u}_I$ and $\mathbf{w}$ belong to the same discrete n-dimensional space representation. Using these results, the following scalar product,

$$<\Omega>_I \sim \mathbf{w}^T \mathbf{u}_I \tag{62}$$

can be associated to the approximate expectation value computed within the discrete space where the molecular point-cloud belongs.

The contents of this section, and the related ideas coming from the previous discussion, are a consequence of the usual computational practice in quantum chemistry and related quantum mechanical applications. Although they may appear unfamiliar to a reader used to square matrix representations of operators, it must be kept in mind that square matrix vector spaces may be made isomorphic to column matrix vector spaces of the appropriate dimension. A very good exposition of these ideas, in a somewhat different context mainly attributable to different applications, can be found in the monograph by Bohm and Gadella.[71]

F. Theoretical Foundation of QSPR

As has been stated before, every molecular property can be seen as some expectation value of an operator W whose matrix representation elements $\mathbf{w}$ may be evaluated by means of Eq. 60 using a least-squares technique. A more general form of Eq. 60 may be considered here. Let us define a new vector of QMSM origin obtained by some, even nonlinear, transformation of the original point-molecules vector space,

$$\mathbf{g} = R(\mathbf{u}) \tag{63}$$

where $R(\mathbf{u})$ represents any possible mathematical manipulation of the point-molecule $\mathbf{u}$ elements; then the equation,

$$\mathbf{w}^T \mathbf{g} = \pi \tag{64}$$

constitutes a QSPR-like equation, deduced from purely QMSM theoretical considerations. There is, however, a capital difference between Eqs. 59 and 64. Equation 64 has been deduced from quantum mechanical considerations, while the equations like 59 are produced in a pure empirical context. The interesting thing is that Eq. 64 somehow justifies Eq. 59, while considering that QSAR-like parameters are nothing but rough approximations to QMSM or some appropriate transform.

The nature of Eq. 63 can be observed from many points of view. Two of them, among many possibilities, will be briefly described.

As a first example, let us suppose that the property or biological activity π, appearing in Eq. 64, has a macroscopic character. Then, if this is so, within the quantum framework, where Eqs. 61 and 62 have been deduced, they are not so correct as in a microscopic environment. In this case the point-molecule $\mathbf{u}_I$ elements can be transformed in some statistical mechanics fashion into $\mathbf{g}_I$ elements. Using as transformation, for instance, a Boltzmann-like rule,

$$g_{JI} = \theta \exp[(u_{JI} - u_{II})/kT];\ \forall J,I \tag{65}$$

where θ is some normalization constant.

The second example may serve to present a generalization of the molecular connectivity and related parameters. The main idea may be based upon the description and calculation algorithms of a new quantum-related molecular topological descriptor parameter set.[72] Using this idea, it is possible to define the counterparts of many classical topological indices within the framework of the QMSM theory. For example, the elements of the topological matrix can be replaced by atomic ns shell orbital overlap integrals or more sophisticated measures like the ones described in Section III. Also, tridimensional distances can be used instead of the topological ones. Effective charge parameters may provide the definition of new indices, and so on. Essentially MO QMSM, as discussed in Ref. 3 or the related molecular self-similarities, may be considered good candidates to new QSPR parameters, replacing other doubtful concepts of empirical origin. These new quantum-related topological indices may contain three-dimensional information of the molecular structures and chemical structure information as well. In this context, these kind of indices may be able to distinguish rotamers, conformers, etc., contrary to the classical ones which, definitely, cannot.

As a final choice, one may use QMSI to manipulate the original information on QMSM as was done in Eq. 63.

VIII. SOME APPLICATION EXAMPLES

This section presents an assorted collection of QMSM calculations involving several molecular sets. It will be shown here how the application of the theory allows one to order the molecular set in such a way that molecular properties can be predicted. References 14, 15, 82, and 83 give more information about the applications of the QMSM: molecular ordering, prediction of the activity for a series of metal-substituted enzyme models, or the use of QMSM as an interpretative tool in chemical reactivity, among other problems. References 7, 10, 12, and 82 present a large amount of ordering examples.

The working scheme for the QSPR analysis has been the same for all the families studied: once the similarity matrix is constructed, each one of its column vectors or point-molecules is mean-centered and standardized to unit variance. From the resulting matrix, a factor-score matrix can be computed,

$$\mathbf{F} = \mathbf{Z}\mathbf{T}_S \tag{66}$$

such that each of the orthogonal factor-score vectors $\{\mathbf{fi}\}$, the columns of matrix **F**, is a linear combination of the original point-molecules, ordered according to their importance in explaining the variance of the original variables. The classical way to obtain these factors is through a "principal components analysis" (PCA),[74] but slightly better results are obtained using the "partial least-squares method" (PLS).[75,76] This method, which has recently become a widely used technique in other types of QSAR models, takes into account the property to be modeled when computing the factors, while the PCA analysis does not. Regardless of the method used to obtain the factors, they can be used to perform a multilinear regression analysis[73] and regression coefficients can be computed using a least-squares algorithm. A decision has to be made concerning the number of factors in the regression model: it would be desirable to include as few factors as possible, while keeping the maximum of the original information of the similarity matrix. The criterion used has been to always take the model associated with the lowest regression coefficient for prediction (Q). This ensures the maximal predictive capacity for the model and avoids the formation of overfitted models (for more technical details, see Refs. 73 and 76).

Fully optimized geometries of the studied molecular sets have been obtained using the Gaussian 92 program[65] under a STO-3G basis set for the first two examples (heptane isomers and pheromones). For the rest of the molecular sets, the AMPAC program[62] using the AM1 methodology[63] has been employed. When no information about an active conformation has been available, we have attempted to compute a minimum energy conformation, and this has been included in QMSM calculations.

Once the appropriate molecular geometry is obtained, a unique s function can be associated with each atom and the molecular density is reproduced in an approximate form using the ASA model. This procedure speeds up the whole similarity study, while preserving the quality of reliable results. STO functions have been found to fit better to the exact density than GTO ones, though the later ones are computationally cheaper. The overlap-like similarity measures have been systematically used. In every case, a PLS and multilinear regression analysis has been performed over the obtained similarity matrix, and the best predictive model has been chosen. Normally, two or three PLS factors yield the best model.

Table 3. Approximate Overlap-Like MQSM Values for the Heptane Isomers[a]

	1	*2*	*3*	*4*	*5*	*6*	*7*	*8*	*9*	*10*	*11*
1	10.89										
2	9.58	10.88									
3	9.38	9.64	10.88								
4	9.37	9.46	8.98	10.88							
5	8.18	8.25	9.63	9.63	10.88						
6	8.27	9.57	9.54	9.54	8.35	10.88					
7	8.11	9.50	9.50	9.56	9.53	9.28	10.87				
8	8.25	9.38	9.42	9.38	9.45	9.33	9.08	10.87			
9	8.25	9.55	9.55	8.23	8.39	9.61	9.25	9.42	10.88		
10	8.33	8.21	9.47	9.47	9.52	9.63	9.23	9.39	8.39	10.87	
11	6.98	8.23	8.37	8.25	8.40	9.58	9.51	9.44	8.39	9.59	10.86

Note: [a]See Table 4 for order number–molecular structure association.

A. Prediction of Boiling Points for the Heptane Isomers

Here, in the first place, a study of the boiling points for the heptane isomers is presented. Table 3 contains the approximate ASA STO overlap-like QMSM values obtained for all the possible pairs of molecules. Table 4 lists the experimental boiling points and the fitted ones using two PLS factors in the regression equation. Both the fitted and the predictive regression coefficients (R and Q) are excellent. Note that, although they have the same experimental value, different enantiomers

Table 4. Computed and Experimental Values for the Boiling Points of the Heptane Isomers

Ordering Numbers	*Molecule*	*Comp. Values*	*Exp. Values*[a]
1	7	97.996	98.427
2	2M6	90.943	90.052
3	3M6 (S)	91.501	91.850
4	3M6 (R)	92.267	91.850
5	3E5	94.762	93.475
6	22MM5	78.574	79.197
7	23MM5 (S)	88.745	89.784
8	23MM5 (R)	89.290	89.784
9	24MM5	81.570	80.500
10	33M5	86.786	86.064
11	223MM4	80.431	80.882

Regression Model Statistical Parameters:
$R = 0.989$ $s = 1.145$ $Q = 0.955$

Note: [a]See Ref. 77.

Table 5. Approximate Overlap-Like MQSM Values for the Pheromones[a]

	1	*2*	*3*	*4*	*5*	*6*	*7*	*8*	*9*	*10*	*11*	*12*	*13*
1	18.41												
2	14.91	18.41											
3	20.14	17.46	35.59										
4	17.46	20.14	32.89	35.59									
5	15.77	15.76	17.07	17.07	16.78								
6	15.74	13.62	17.33	15.21	14.37	15.50							
7	13.62	15.74	15.21	17.33	14.37	13.26	15.50						
8	14.55	14.56	15.86	15.87	15.46	14.21	14.21						
9	13.33	13.33	14.62	14.62	14.14	13.16	13.16	14.14					
10	16.78	16.77	18.04	18.03	17.60	15.73	15.73	16.61	15.45				
11	15.13	15.13	16.39	16.38	15.91	14.87	14.87	15.81	14.68	19.84			
12	12.33	12.61	13.83	14.11	12.81	12.20	12.51	12.81	12.84	14.44	13.67		
13	11.93	11.93	13.33	13.33	12.53	11.79	11.79	12.53	12.67	13.67	13.21	12.44	

Note: [a]See Table 6 for order number–molecular structure association.

give slightly different predictions. This is due to the fact that QMSM have the ability to distinguish between different enantiomers.

B. Prediction of the Activity for Several Pheromones

In second place, the alarm activity produced in a certain insect species (Iridormyrmex pruinosus) by a group of pheromones has been studied. This example has been chosen because of the fact it was the first biological example studied with a QMSM technique.[1] Table 5 contains the approximate QMSM values. These measures are overlap-like QMSM obtained within the ASA model using STO. The similarity matrix obtained in this way has been used as input to the ND-CLOUD program and for developing a QSPR model.

Visualization Example

Figure 3 shows an example of the results which can be obtained with the ND-CLOUD program. A descending nearest-neighbor graph was generated using the distances between the point-molecules. Note how molecules are clustered in groups of similar alarm activity. A successful ND-CLOUD graph shows that a good correlation between similarity and property exists.

QSPR Model

The same similarity matrix shown in Table 5 was used to develop a regression model. Table 6 shows the computed and the experimental values.[78] The fitted values were obtained from a regression equation with two PLS factors. A good agreement

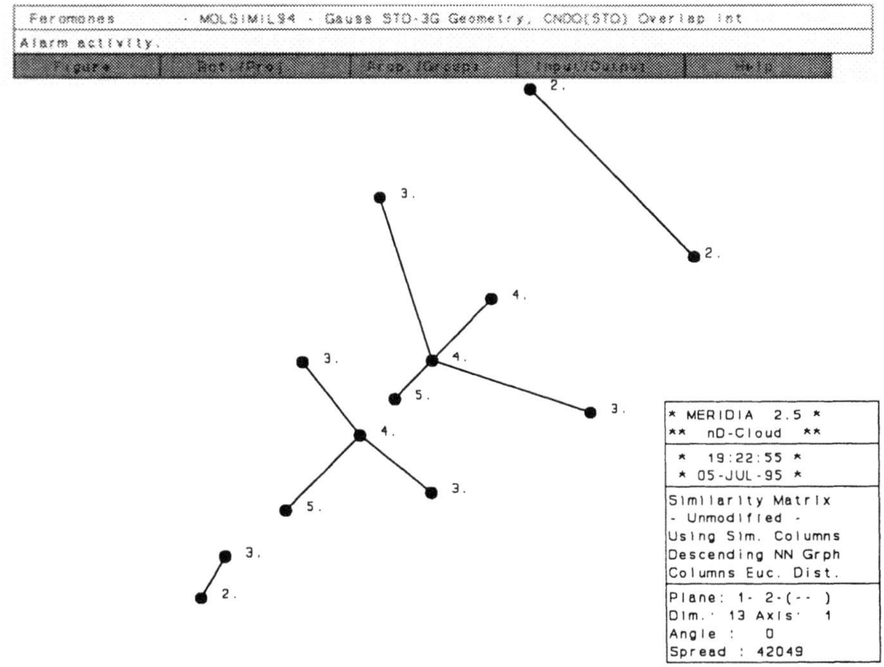

Figure 3. Descending nearest neighbor graph for the pheromones family.

Table 6. Computed and Experimental Values for the Alarm Activity of the Studied Set of Pheromones

Ordering Numbers	*Pheromones*	*Comp. Activity*	*Exp. Activity*[a]
1	Methyl-*sec-n*-octyl ether (R)	2.8	3.0
2	Methyl-*sec-n*-octyl ether (S)	2.8	3.0
3	2-Bromooctane (R)	1.9	2.0
4	2-Bromooctane (S)	1.9	2.0
5	2-Octanone	3.7	4.0
6	2-Heptanol (R)	3.1	3.0
7	2-Heptanol (S)	3.1	3.0
8	2-Heptanone	3.8	4.0
9	3-Heptanone	4.2	5.0
10	2-Ethoxyethyl acetate	4.8	4.0
11	*n*-Butyl acetate	4.5	5.0
12	5-Hexen-2-one	3.1	3.0
13	2-Pentanone	3.1	2.0
Regression Model Statistical Parameters:			
$R = 0.867$	$s = 0.563$	$Q = 0.608$	

Note: [a]See Ref. 78.

Table 7. Approximate Overlap-Like MQSM Values for the Indole Derivatives (**a,b,c**)[a]

(a)	*1*	*2*	*3*	*4*	*5*	*6*	*7*	*8*	*9*	*10*	*11*	*12*	*13*
1	65.14												
2	57.95	60.73											
3	52.29	52.73	52.38										
4	58.47	51.27	45.60	58.60									
5	59.00	51.83	46.11	52.97	58.60								
6	51.35	54.06	45.96	51.45	45.77	54.20							
7	45.46	46.09	45.71	45.78	39.93	46.08	45.85						
8	51.90	54.60	46.60	45.86	51.46	48.57	40.56	54.20					
9	46.16	46.62	46.26	40.27	45.79	40.61	40.23	46.20	45.85				
10	62.44	55.30	49.60	57.76	57.74	50.62	44.80	50.56	44.83	62.29			
11	57.62	50.46	44.78	57.30	52.97	49.91	44.30	45.90	40.30	57.62	57.17		
12	57.67	50.48	44.83	52.93	57.26	45.78	40.14	50.10	44.40	57.61	52.96	57.17	
13	52.92	45.67	40.08	52.47	52.52	45.19	39.66	45.39	39.74	52.93	52.49	52.53	52.07

(b)	*1*	*2*	*3*	*4*	*5*	*6*	*7*	*8*	*9*	*10*	*11*	*12*	*13*
14	55.28	58.04	50.05	50.41	50.57	53.36	45.31	53.34	45.30	55.14	50.45	50.21	45.54
15	49.65	50.06	49.69	44.98	44.96	45.30	45.01	45.35	45.00	49.52	44.82	44.76	40.14
16	50.36	53.22	45.26	50.01	45.70	52.89	44.84	48.56	40.60	50.39	50.00	45.73	45.13
17	44.91	45.27	44.89	44.54	40.15	44.86	44.55	40.52	40.23	44.87	44.35	40.16	39.61
18	50.56	53.27	45.20	45.92	50.18	48.54	40.55	52.88	44.78	50.54	45.88	50.03	45.35
19	45.04	45.36	44.94	40.35	44.35	40.62	40.19	44.92	44.49	44.82	40.29	44.32	39.80
20	45.89	48.55	40.54	45.41	45.33	48.11	40.08	48.12	40.14	45.89	45.41	45.30	44.83
21	37.15	37.48	37.11	36.95	37.14	37.28	36.83	37.38	37.15	37.06	36.84	37.04	36.82
22	45.25	45.57	45.08	40.30	44.84	40.53	40.06	45.15	44.70	45.03	40.32	44.49	39.80
23	40.26	40.61	40.17	39.80	39.67	40.16	39.74	40.14	39.77	40.27	39.85	39.69	39.19

(c)	*14*	*15*	*16*	*17*	*18*	*19*	*20*	*21*	*22*	*23*
14	57.89									
15	49.83	49.54								
16	53.22	45.22	52.77							
17	45.23	44.88	44.73	44.42						
18	53.19	45.20	48.54	40.51	52.77					
19	45.26	44.87	40.62	40.20	44.75	44.42				
20	48.57	40.57	48.11	40.11	48.12	40.16	47.66			
21	37.49	37.10	37.21	36.83	37.41	37.05	37.18	47.66		
22	45.20	44.74	40.53	40.12	44.72	44.27	40.08	36.87	47.67	
23	40.64	40.22	40.17	39.76	40.12	39.78	39.68	36.77	39.67	39.31

Note: [a]See Table 8 for order number–molecular structure association.

between fitted and experimental activity values is found, except in molecules 9 and 10. However, it must be noted that experimental values are quite arbitrarily defined in this case.

C. Prediction of Biological Activity for a Group of Indole Derivatives

The molecular set studied next consists of a group of 23 indole derivatives. The activity studied in this case is the displacement of flunitrazepam from binding to bovine brain membranes.[80] As usual, ASA overlap-like QMSM with STO functions were computed for the whole set, and the similarity matrix is presented in Table 7. Table 8 lists the experimental and the fitted activity values, from a model made using the first two PLS factors.

Table 8. Computed and Experimental Activity Values for the Studied Set of Indole Derivatives

Mol. Numbers	*Subs.* (R_1,R_2,R_3)	*Exp. Activities*	*Calc. Activities*[a]
1	H,H,H,H	6.93	6.26
2	Cl,H,H,H	6.21	6.48
3	NO_2,H,H,H	6.93	7.12
4	H,OCH_3,H,H	6.78	6.77
5	Cl,OCH_3,H,H	6.68	6.99
6	NO_2,OCH_3,H,H	7.27	7.63
7	H,H,OCH_3,H	6.54	6.31
8	Cl,H,OCH_3,H	6.79	6.53
9	NO_2,H,OCH_3,H	7.42	7.17
10	H,OCH_3,OCH_3,H	7.03	6.84
11	Cl,OCH_3,OCH_3,H	7.52	7.05
12	NO_2,OCH_3,OCH_3,H	7.96	7.69
13	H,Cl,H,H	7.17	6.80
14	H,H,H,Cl	5.59	5.79
15	H,OH,H,H	6.37	6.71
16	Cl,OH,H,H	6.82	6.92
17	NO_2,OH,H,H	7.92	7.56
18	H,H,OH,H	6.09	6.31
19	Cl,H,OH,H	6.24	6.52
20	NO_2,H,OH,H	7.19	7.17
21	H,OH,OH,H	6.46	6.76
22	Cl,OH,OH,H	6.75	6.97
23	NO_2,OH,OH,H	7.32	7.62

Regression Model Statistical Parameters:

$R = 0.833$ $s = 0.333$ $Q = 0.754$

Note: [a]See Ref. 79.

Table 9. Approximate Overlap-Like MQSM Values for the Baker Triazines (**a,b,c**)[a]

(a)	*1*	*2*	*3*	*4*	*5*	*6*	*7*	*8*	*9*	*10*	*11*	*12*	*13*
1	45.76												
2	32.89	40.32											
3	32.90	29.63	38.84										
4	32.83	31.96	29.44	38.84									
5	32.87	31.24	32.41	29.45	40.32								
6	33.75	32.33	29.41	34.41	29.41	57.21							
7	37.81	32.49	29.67	32.46	29.67	33.34	37.44						
8	42.83	32.93	36.20	32.91	32.43	33.79	37.87	52.21					
9	36.26	34.53	29.67	32.62	29.67	41.57	35.89	36.31	55.94				
10	37.79	32.58	31.50	32.21	31.50	34.55	33.25	38.02	35.34	42.44			
11	35.21	32.42	29.49	31.81	29.49	42.49	34.81	35.25	41.96	33.60	41.85		
12	32.87	29.49	30.83	29.42	30.69	29.39	29.66	31.98	29.66	31.44	29.48	34.34	
13	29.89	29.27	29.27	29.26	29.27	29.23	29.50	29.95	29.49	29.70	29.31	29.25	29.09

(b)	*1*	*2*	*3*	*4*	*5*	*6*	*7*	*8*	*9*	*10*	*11*	*12*	*13*
14	33.20	32.21	29.45	31.80	29.45	34.00	32.81	33.27	35.04	37.76	33.19	29.42	29.27
15	33.35	31.82	29.43	31.01	29.43	37.68	33.10	33.43	38.90	33.13	37.88	29.42	29.25
16	37.20	32.39	29.50	32.37	29.50	33.91	36.91	37.22	35.53	32.92	34.45	29.53	29.35
17	33.42	31.11	29.40	31.29	29.41	37.79	33.16	33.47	36.26	32.33	36.30	29.41	29.24
18	34.99	31.65	29.71	31.63	29.72	33.20	34.57	35.05	34.91	32.72	34.69	29.71	29.55
19	33.73	29.52	30.82	29.42	30.84	29.39	29.66	32.17	29.66	31.59	29.48	30.81	29.26
20	30.32	29.74	29.72	29.70	29.72	29.66	29.91	30.34	29.89	30.06	29.71	29.74	29.55
21	33.22	32.16	30.98	29.49	32.22	29.41	29.67	33.10	29.76	31.73	29.51	31.20	29.27
22	30.02	28.93	28.88	28.87	28.88	28.82	29.07	29.46	29.05	29.15	28.86	28.93	28.72
23	29.77	29.21	29.17	29.19	29.18	29.12	29.36	29.78	29.33	29.51	29.14	29.19	29.01
24	30.28	29.69	29.64	29.68	29.63	29.80	29.86	30.30	30.10	30.09	29.85	29.65	29.46
25	33.61	29.45	29.39	29.41	29.39	29.36	29.63	30.07	31.72	29.70	29.39	29.42	29.23

(c)	*14*	*15*	*16*	*17*	*18*	*19*	*20*	*21*	*22*	*23*	*24*	*25*
14	37.35											
15	32.75	36.97										
16	32.53	33.31	45.75									
17	31.76	35.21	33.12	45.30								
18	32.29	32.78	34.22	32.63	35.63							
19	29.42	29.41	29.49	29.40	29.71	32.84						
20	29.66	29.73	35.04	29.76	30.05	29.72	37.88					
21	29.59	29.47	29.52	29.42	29.72	31.32	29.73	37.35				
22	28.78	28.98	42.02	29.07	29.31	28.89	35.76	28.90	52.71			
23	29.09	29.23	37.80	29.28	29.54	29.17	34.61	29.19	41.67	37.43		
24	29.67	30.02	34.58	30.05	30.31	29.64	35.54	29.65	36.42	34.19	35.60	
25	29.33	29.47	38.19	29.54	29.81	29.39	34.76	29.41	42.36	37.75	34.37	45.77

Note: [a]See Table 10 for order number–molecular structure association.

D. Prediction of DHFR Inhibition Activity for a Group of Baker Triazines

Finally, a group of 25 Baker triazines acting as inhibitors of the dihydrofolate reductase enzyme is studied.[79] The bioactive conformation proposed by Hopfinger was used.[79] Table 9 shows the ASA overlap-like QMSM obtained using STO functions. As an example of the predictive power of our method, a model was constructed using 21 triazines as a training set and the remaining as a predictive set. Then, a regression model was made with the training set, using two PLS factors. Table 10 shows the fitted and experimental activities for this set. Property predictions were made for the molecules considered to have *unknown* property values. These are in good agreement, at least qualitatively, with the experimental ones, as can be seen in Table 11.

Table 10. Computed and Experimental Values for the Activity of the Baker Triazines Training Set

Mol. Numbers	*Exp. Activities*[a]	*Calc. Activities*
1	8.54	7.76
2	8.19	7.65
3	8.05	6.36
4	8.00	7.27
5	7.89	6.66
6	7.76	7.40
8	7.52	7.89
9	7.27	7.69
10	7.14	7.91
11	7.07	7.27
12	6.92	6.38
13	6.92	6.10
15	6.79	6.96
16	6.52	5.53
17	6.21	6.75
19	5.14	6.43
21	4.70	6.88
22	4.25	3.44
23	4.15	4.45
24	3.68	5.04
25	3.43	4.34
Regression Model Statistical Parameters:		
$R = 0.849$	$s = 0.890$	$Q = 0.787$

Note: [a]See Ref. 80.

Table 11. Calculated and Experimental Values for the Activity of the Baker Triazines in the Predictive Set

Mol. Numbers	*Exp. Activities*[a]	*Calc. Activities*
7	7.76	7.26
14	6.85	7.40
18	6.17	7.03
20	4.74	4.92

Note: [a]See Ref. 80.

IX. CONCLUSIONS

QMSM has been described as a tool for comparing molecular structures. The dualistic point of view $\{\infty\text{-D},n\text{-D}\}$ associated with the QMSM representation of molecular sets has interesting applications and flexibility, implying freedom to describe new QMSI. This freedom permits one to find conversion relationships between *C*-class and *D*-class indices, and hidden connections between Hodgkin–Richards and Carbó *C*-class index definitions.

Quantum molecular similarity measures form a nonempirical theoretical basis where QSPR or QSAR can be justified as scientific procedures. Although QSPR had been a very useful tool since early times in chemistry, a proof of the appropriate theoretical foundations has not yet been described. The present work provides this foundation, using a robust structure based on quantum chemical considerations.

The discrete representation of both an electronic density distribution and a convenient operator, connected with a quantum mechanical definition of the expectation value concept and, subsequently, with the evaluation of molecular properties, has been described.

Successful examples illustrate these points.

ACKNOWLEDGMENTS

This work has been partially financed by the CICYT-CIRIT, Fine Chemicals Programme of the "Generalitat de Catalunya" through a grant: #QFN91-4606. One of us (Ll.A.) benefits from a grant from the "Ministerio de Educación i Ciéncia". The authors have benefited from lively discussions with Mr. P. Constans, Mr. J. Mestres, and Dr. M. Solà.

REFERENCES

1. Carbó, R.; Arnau, M.; Leyda, L. *Int. J. Quantum Chem.* **1980**, *17*, 1185.
2. Carbó, R.; Arnau, C. *Medicinal Chemistry Advances*; de las Heras, F.G.; Vega, S., Eds.; Pergamon Press: Oxford, 1981.
3. Carbó, R.; Domingo, Ll. *Int. J. Quantum. Chem.* **1987**, *23*, 517.
4. Carbó, R.; Calabuig, B. *Comp. Phys. Commun.* **1989**, *55*, 117.

5. Carbó, R.; Calabuig, B. *Concepts and Applications of Molecular Similarity*; Johnson, M.A.; Maggiora, G., Eds.; John Wiley & Sons: New York, 1990, Ch. 6.
6. Carbó, R.; Calabuig, B. Proceedings del XIX Congresso Internazionale dei Chimici Teorici dei Paesi di Espressione Latina, Roma, Italy, September 10–14, 1990. *J. Mol. Struct. (Teochem)* **1992**, *254*, 517.
7. Carbó, R.; Calabuig, B. *J. Chem. Inf. Comput. Sci.* **1992**, *32*, 600.
8. Carbó, R.; Calabuig, B. In *Structure, Interactions and Reactivity*; Fraga, S., Ed.; Elsevier Pub.: Amsterdam, 1992.
9. Carbó, R.; Calabuig, B. *Int. J. Quantum Chem.* **1992**, *42*, 1681.
10. Carbó, R.; Calabuig, B. *Int. J. Quantum Chem.* **1992**, *42*, 1695.
11. Carbó, R.; Calabuig, B.; Besalú, E.; Martínez, A. *Molecular Engineering* **1992**, *2*, 43.
12. Carbó, R.; Besalú, E.; Calabuig, B.; Vera, L. *Adv. Quant. Chem.* **1994**, *25*, 253.
13. Carbó, R.; Besalú, E. *Molecular Similarity and Reactivity: From Quantum Chemical to Phenomenological Approaches*; Carbó, R., Ed.; Kluwer Acad., Amsterdam, 1995.
14. Besalú, E.; Carbó, R.; Mestres, J.; Solà, M. *Topics in Current Chemistry*; Sen, K., Ed.; Springer-Verlag: Berlin, 1995, Vol 173, pp. 31–62.
15. Mestres, J.; Solà, M.; Duran, M.; Carbó, R. *J. Comp. Chem.* **1994**, *15*, 1113.
16. Constans, P.; Carbó, R. *J. Chem. Inf. Comput. Sci.* **1995** (in press).
17. Cooper, D.L.; Allan, N.L. *J. Chem. Soc., Faraday Trans.* **1987**, *83*, 449.
18. Cooper, D.L.; Allan, N.L. *J. Computer-Aided Mol. Design* **1989**, *3*, 253.
19. Cooper, D.L.; Allan, N.L. *J. Am. Chem. Soc.* **1992**, *114*, 4773 .
20. Cioslowski, J.; Fleischmann, E.D. *J. Am. Chem. Soc.* **1991**, *113*, 64.
21. Cioslowski, J.; Challacombe, M. *Int. J. Quant. Chem.* **1991**, *25*, 81.
22. Ortiz, J.V.; Cioslowski, J. *Chem. Phys. Lett.* **1991**, *185*, 270.
23. Cioslowski, J.; Surján, P.R. *J. Mol. Struct. (Theochem)* **1992**, *255*, 9.
24. Ponec, R.; Strnad, M. *Collect. Czech. Chem. Commun.* **1990**, *55*, 2583.
25. Ponec, R.; Strnad, M. *J. Phys. Org. Chem.* **1991**, *4*, 701.
26. Ponec, R.; Strnad, M. *Int. J. Quantum Chem.* **1992**, *42*, 501.
27. Ponec, R.; Strnad, M. *Croat. Chem. Acta* **1991**, *66*, 123.
28. Ponec, R. *J. Chem. Inf. Comput. Sci.* **1993**, *33*, 805.
29. Ponec, R.; Strnad, M. *Int. J. Quantum Chem.* **1994**, *50*, 43.
30. *Concepts and Applications of Molecular Similarity*; Johnson, M.A.; Maggiora, G., Eds.; John Wiley & Sons: New York, 1990.
31. Hodgkin, E.E.; Richards, W.G. *Int. J. Quant. Chem.* **1987**, *14*, 105.
32. Good, A.C.; Hodgkin, E.E.; Richards, W.G. *J. Chem. Inf. Comput. Sci.* **1992**, *32*, 188.
33. Good, A.C. *J. Mol. Graphics* **1992**, *10*, 144.
34. Good, A.C.; So, S-S; Richards, W.G. *J. Med. Chem.* **1993**, *36*, 433.
35. Mezey, P. *Shape in Chemistry* VCH: New York, 1993.
36. Martín, M.; Sanz, F.; Campillo, M.; Pardo, L.; Pérez, J.; Turmo, J. *Int. J. Quant. Chem.* **1983**, *23*, 1627.
37. Martín, M.; Sanz, F.; Campillo, M.; Pardo, L.; Pérez, J.; Turmo, J.; Aulló, J.M. *Int. J. Quant. Chem.* **1983**, *23*, 1643.
38. Sanz, F.; Martín, M.; Pérez, J.; Turmo, J.; Mitjana, A.; Moreno, V. *Quantitative Approaches to Drug Design*; Dearden, J.C., Ed.; Elsevier: Amsterdam, 1983.
39. Sanz, F.; Martín, M.; Lapeña, F.; Manaut, F. *Quant. Struct.-Act. Relat.* **1986**, *5*, 54.
40. Sanz, F.; Manaut, F.; José, J.; Segura, J.; Carbó, M.; de la Torre, R. *J. Mol. Struct. (Theochem)* **1988**, *170*,
41. Luque, F.J.; Sanz, F.; Illas, F.; Pouplana, R.; Smeyers, Y.G. *Eur. J. Med. Chem.* **1988**, *23*, 7.
42. *Practical Applications of QSAR in Environmental Chemistry and Toxicology*; Karcher, W.; Devillers, J., Eds.; Kluwer Academic: Dordrecht, 1990.
43. McQuarrie, D.A. *Quantum Chemistry*; University Science Books: Mill Valley, CA, 1983.

44. Born, M.; Oppenheimer, J.R. *Annln. Phys.* **1927**, *84*, 457.
45. Born, M.; Huang, K. *Dynamical Theory of Crystal Lattices*; Clarendon: Oxford, 1954.
46. Longuet-Higgins, H.C. *Adv. in Spectrosc.* **1961**, *2*, 429.
47. Löwdin, P.O. *Phys. Rev.* **1955**, *97*, 1474.
48. Löwdin, P.O. *Phys. Rev.* **1955**, *97*, 1490.
49. Löwdin, P.O. *Phys. Rev.* **1955**, *97*, 1509.
50. McWeeny, R. *Proc. Roy. Soc. A* **1955**, *232*, 114.
51. McWeeny, R. *Proc. Roy. Soc. A* **1956**, *235*, 496.
52. McWeeny, R. *Proc. Roy. Soc. A* **1959**, *253*, 242.
53. Zemanian, A.H. *Generalized Integral Transformations*; Dover: New York, 1987.
54. *Encyclopaedia of Mathematics*; Kluwer Acad.: Dordrecht, 1990.
55. Pople, J.A.; Beveridge, D.L. *Approximate Molecular Orbital Theory*; McGraw-Hill: New York, 1970.
56. Mulliken, R.S. *J. Chem. Phys.* **1955**, *23*, 1833.
57. Mulliken, R.S. *J. Chem. Phys.* **1955**, *23*, 1841.
58. Mulliken, R.S. *J. Chem. Phys.* **1955**, *23*, 2338.
59. Mulliken, R.S. *J. Chem. Phys.* **1955**, *23*, 2343.
60. Tou, J.T.; González, R.C. *Pattern Recognition Principles*; Addison-Wesley Reading, 1974.
61. Petke, J.D. *J. Comput. Chem.* **1991**, *14*, 928.
62. Liotard, D.A.; Healy, E.F.; Ruiz, J.M.; Dewar, M.S.J. AMPAC-version 2.1. Quantum Chemistry Program Exchange, Program 506. *QCPE Bull.*, **1989**, 9.
63. Dewar, M.S.J.; Zoebisch, E.G.; Healy, E.F.; Stewart, J.J.P. *J. Am. Chem. Soc.* **1985**, *107*, 3902.
64. Hehre, W.J.; Stewart, R.F.; Pople, J.A. *J. Chem. Phys.* **1969**, *51*, 2657.
65. Frisch, M.J.; Head-Gordon, M.; Trucks, G.W.; Foresman, J.B.; Schlegel, H.B.; Raghavachari, K.; Binkley, J.S.; Gonzalez, C.; Defrees, D.J.; Fox, D.J.; Whiteside, R.A.; Seeger, R.; Melius, C.F.; Baker, J.; Martin, R.L.; Kahn, L.R.; Stewart, J.J.P.; Topiol, S.; Pople, J.A. (1990) GAUSSIAN 90, Revision H, Gaussian Inc., Pittsburgh, PA.
66. (a) Crum-Brown, A.; Fraser, T. *Trans. Roy. Soc. Edinburgh* **1868–1869**, *25*, 151. (b) Overton, E. *Z. Physikol. Chem.* **1897**, *22*, 189. (c) Meyer, H. *Arch. Exptl. Pathol. Pharmakol.* **1899**, *42*, 109. (d) Traube, T. *Arch. Ges. Physiol.* **1904**, *105*, 541. (e) Moore, W. *Science* **1919**, *49*, 572. (f) Hammet, L.P. *Chem. Rev.* **1935**, *17*, 125. (g) McGowan, J.C. *J. Appl. Chem. (London)* **1954**, *4*, 41. (h) Hansch, C.; Fujita, T. *J. Am. Chem. Soc.* **1964**, *86*, 1616.
67. (a) Gálvez, J.; García-Domenech, R.; de Julián-Ortiz, J.V.; Soler, R. *J. Chem. Inf. Comput. Sci.* **1995**, *35*, 272. (b) Pastor, M.; Alvarez-Bulla, J. *Quant. Struct.-Act. Relat.* **1995**, *14*, 24. (c) Wessel, M.D.; Jurs, P.C. *J. Chem. Inf. Comput. Sci.* **1995**, *35*, 68.
68. Benigni, R.; Cotta-Ramusino, M.; Giorgi, F.; Gallo, G. *J. Med. Chem.* **1995**, *38*, 629.
69. See, for example: (a) Purcell, W.P.; Bass, G.E.; Clayton, J.M. *Strategy of Drug Design*; John Wiley & Sons: New York, 1973. (b) Kier, L.B.; Hall, L.H. *Molecular Connectivity in Chemistry and Drug Research*; Academic: New York, 1976. (c) Richards, W.G. *Quantum Pharmacology*; Butterworths: London, 1977. (d) Martin, Y.C. *Medicinal Research Series*; Marcel Dekker: New York, 1978, Vol. 8. (e) *A Textbook of Drug Design and Development*; Krogsgaard-Larsen, P.; Bundgaard, H., Eds.; Harwood Acad.: Chur (Switzerland), 1991.
(f) *Diseño de Medicamentos*; Mosqueira, A., Ed.; Real Academia de Farmacia: Madrid, 1994.
70. Carbó, R.; Martín, M.; Pons, V. *Afinidad* **1977**, *34*, 348.
71. Bohm, A.; Gadella, M. In *Lecture Notes in Physics*; Springer Verlag: Berlin, 1989, p. 348.
72. Besalú, E.; Carbó, R. *Scientia Gerundensis* **1995**, in press.
73. Montgomery, D.C.; Peck, E.A. *Introduction to Linear Regression Analysis*; John Wiley & Sons: New York, 1992.
74. Tabachnick, B.G.; Fidell, L.S. *Using Multivariate Statistics*; HarperCollins: New York, 1989.
75. Geladi, P.; Kowalski, B.R. *Analytica Chimica Acta* **1986**, *185*, 1–17.
76. *3D QSAR in Drug Design*; Kubinyi, H., Ed.; Escom: Leiden, 1993.

77. Needham, D.E.; Wel, I.C.; Seybold, P.G. *J. Am. Chem. Soc.* **1988**, *110*, 4186–4194.
78. Amoore, J.E. *Molecular Basis of Odor*; C. C. Thomas, 1970.
79. Hopfinger, A.J. *J. Am. Chem. Soc.* **1980**, *102*, 7196.
80. Hadjipavlou-Litina, D.; Hansch, C. *Chem. Rev.* **1994**, *94*, 1483–1505.
81. Clementi, E.; Roetti, C.; *At. Data Nucl. Data Tables* **1974**, *14*, 177.
82. Mestres, J.; Solà, M.; Carbó, R.; Duran, M. *J. Am. Chem. Soc.* **1994**, *116*, 5909–5915.
83. Solà, M.; Mestres, J.; Duran, M.; Carbó, R. *J. Chem. Inf. Comput. Sci.* **1994**, *34*, 1047–1053.

SIMILARITY OF ATOMS IN MOLECULES

Boris B. Stefanov and Jerzy Cioslowski

I. INTRODUCTION

The idea to study the similarity of atoms in molecules has emerged[1] at the interface of the pioneering theories of similarity of quantum mechanical systems[2,3] and of atoms in molecules (AIMs).[4,5] The ability to quantify the extent to which two molecules are similar is of a paramount importance to numerous scientific disciplines such as, to name a few, enzymology, pharmacology, toxicology, and polymer design. The question "How similar is molecule *X* to molecule *Y*?" arises whenever

Advances in Molecular Similarity
Volume 1, pages 43–59

ISBN: 0-7623-0131-7

the phenomenon of molecular recognition is encountered. The recent progress in the real-space analysis of molecular structures has brought about the necessity for the assessment of similarities between atoms in molecules as well.

A reliable measure of similarity between two quantum–chemical systems has to possess certain characteristics, namely:

- general applicability;
- lack of dependence on any information other than that already contained in the electronic wavefunctions of the two systems;
- physical meaningfulness and interpretability;
- symmetry with respect to the interchange of the two systems;
- low computational cost;
- a well-defined dependence on the mutual orientation of the two systems.

A similarity measure satisfying all of the above conditions has been proposed for the first time by Carbó et al.[2] These researchers have quantified the similarity between two molecular structures with an index involving their respective electron densities. A related index has been proposed by Hodgkin et al.[6] A similarity measure based on the overlap of one-electron reduced density matrices has also been put forward.[7] In addition, the extent of similarity between molecules has been quantified by means of various topological shape descriptors applied to the electron density distributions.[8] Carbó and Calabuig[3] have recently developed a more general theory of molecular similarity measures based on a generalization of the overlap between (many-)electron density functions. Their formalism is further elaborated in Section II of this review, in which we provide some general theoretical background on molecular similarity measures.

The theory of AIMs[4,5,9] (outlined in Section III of this review) has bridged the long existing gap between the modern quantum theory and the general concepts of chemistry. It does not only rigorously define AIMs as distinct open quantum–mechanical systems but it also identifies the major interactions within molecules and allows for the partitioning of molecular properties into atomic contributions. The original theory of AIMs has been further extended with the definitions of important chemical concepts such as covalent bond orders,[10] steric crowding,[11] and electronegativities in situ.[12]

An almost perfect transferability of the properties of AIMs has been observed in many chemical systems,[13] implying that new electronic structure methods involving the assembly of large molecules from nearly transferable AIM-based fragments may be feasible.[14] The development of such methods calls for the use of taxonomy of AIMs based on quantitative electronic and geometric criteria.[15] Cioslowski and Nanayakkara[1] have recently proposed a computationally efficient measure of the similarity of atoms in molecules that primarily compares their three-dimensional shapes. This similarity measure and the possible alternatives to it are discussed in some detail in Section IV of this review.

Calculations of the similarity of AIMs require an appropriate representation of the atomic zero-flux surfaces. The original approach to the determination of these surfaces[16] has employed triangulation based upon a numerically determined family of gradient paths passing through the corresponding bond point. The limited accuracy of the triangulation resulted in an insufficient accuracy of the calculated similarities of AIMs. The numerical representation of the atomic zero-flux surfaces degraded the efficiency of the computations and prevented routine archiving of the results. The introduction of the variational approach to the computation of the atomic zero-flux surfaces[17] has substantially improved the speed and the accuracy of AIM similarity calculations. Section V of this review is devoted to some algorithmic and computational aspects of such calculations.

A discussion of the results that emerged from the recent numerical studies of similarities of AIMs in various chemical systems concludes this review. The data presented in Section VI include similarities of carbon atoms in (fluoro)hydrocarbons and those of oxygens in carbonyl compounds.

II. SIMILARITY OF MOLECULES

Let $\Gamma_X(\mathbf{R},\mathbf{R}')$ and $\Gamma_Y(\mathbf{R},\mathbf{R}')$ be the n-th order reduced density matrices describing the molecules X and Y, respectively, and $\mathbf{R} \equiv (\mathbf{r}_1, \mathbf{r}_2, \ldots, \mathbf{r}_n)$ be a position hypervector. The *generalized n-th order density matrix overlap* is defined as the integral,

$$Z^{(n)}_{Z,Y}(\hat{\mathcal{C}};\boldsymbol{\alpha}) = \int\!\!\int\!\!\int\!\!\int \Gamma^*_X(\mathbf{R}_1,\mathbf{R}'_1)\hat{\mathcal{C}}\,[\hat{\mathcal{A}}(\boldsymbol{\alpha})\Gamma_Y(\mathbf{R}_2,\mathbf{R}'_2)]\; d\mathbf{R}_1\, d\mathbf{R}'_1\, d\mathbf{R}_2\, d\mathbf{R}'_2 \tag{1}$$

where integration over the n-fold product of Cartesian spaces $\Re^3 \times \Re^3 \times \ldots \times \Re^3$ is implicitly assumed for each variable.

In Eq. 1, the *alignment operator* $\hat{\mathcal{A}}(\boldsymbol{\alpha})$ defines the mutual orientation of the two coordinate systems in which X and Y are defined. Being parameterized by a six-dimensional vector α whose components correspond to the three components of the translation vector and the three Euler angles, it rotates and translates all the coordinates of $\Gamma_Y(\mathbf{R}_2, \mathbf{R}'_2)$ simultaneously. One should note that,

$$\hat{\mathcal{A}}(0) \equiv \hat{1} \text{ and } \hat{\mathcal{A}}(\boldsymbol{\alpha})\hat{\mathcal{A}}(-\boldsymbol{\alpha}) \neq \hat{1} \tag{2}$$

because of the noncommutativity of elementary Euler rotations.

In order to simplify the equations, the tilde is used in the following to denote the image of a function under the action of the operator $\hat{\mathcal{A}}(\boldsymbol{\alpha})$. For instance, $\tilde{\Gamma}(\mathbf{R}, \mathbf{R}')$ will be used to represent $\hat{\mathcal{A}}(\boldsymbol{\alpha})\Gamma(\mathbf{R}, \mathbf{R}')$.

The symmetrical Hermitian *coupling operator,*

$$\hat{\mathcal{C}} = \hat{\mathcal{C}}(\mathbf{R}_1,\mathbf{R}'_1;\mathbf{R}_2,\mathbf{R}'_2) = \hat{\mathcal{C}}(\mathbf{R}_2,\mathbf{R}'_2;\mathbf{R}_1,\mathbf{R}'_1) \tag{3}$$

describes the coupling between the density matrices of X and Y employed in a particular definition of $Z^{(n)}_{X,Y}$. For example, the choice $\hat{C}_u = \delta(\mathbf{R}_1 - \mathbf{R}'_1)\delta(\mathbf{R}_2 - \mathbf{R}'_2)$, where $\delta(\mathbf{R})$ is a $3n$-dimensional Dirac's delta function, corresponds to a completely decoupled integral,

$$Z^{(n)}_{X,Y}(\hat{C}_u) = \int\int D^{(n)}_X(\mathbf{R}_1)\, \tilde{D}^{(n)}_Y(\mathbf{R}_2)\, d\mathbf{R}_1\, d\mathbf{R}_2 \tag{4}$$

of the n-th order density functions $D^{(n)}(\mathbf{R})$, which is independent of the mutual orientation of the systems X and Y. On the other hand, the choice $\hat{C}_c = \delta(\mathbf{R}_1 - \mathbf{R}_2)\delta(\mathbf{R}'_1 - \mathbf{R}'_2)$, results in a perfect coupling:

$$Z^{(n)}_{X,Y}(\hat{C}_c;\boldsymbol{\alpha}) = \int\int \Gamma^*_X(\mathbf{R}_1,\mathbf{R}'_1)\tilde{\Gamma}_Y(\mathbf{R}_1,\mathbf{R}'_1))\, d\mathbf{R}_1\, d\mathbf{R}'_1 \tag{5}$$

Finally, the choice $\hat{C}_d = \delta(\mathbf{R}_1 - \mathbf{R}'_1)\delta(\mathbf{R}_2 - \mathbf{R}'_2)\,\delta(\mathbf{R}_1 - \mathbf{R}_2)$ transforms $Z^{(n)}_{X,Y}$ into an overlap integral of the n-th order density functions:

$$Z^{(n)}_{X,Y}(\hat{C}_d;\boldsymbol{\alpha}) = \int D^{(n)}_X(\mathbf{R}_1)\, \tilde{D}^{(n)}_Y(\mathbf{R}_1)\, d\mathbf{R}_1 \tag{6}$$

Most of the currently employed molecular similarity measures can be derived from the general form of $Z^{(n)}_{X,Y}$ (Eq. 1) with a suitably chosen operator $\hat{C}$. Let $N_{X,Y}$ be a norm subject to the requirements:

$$N_{X,X} = Z^{(n)}_{X,X}(\hat{C};\mathbf{0}) \text{ and } N_{X,Y} = N_{Y,X} \tag{7}$$

By normalizing the generalized density matrix overlap, one arrives at a general form of a *similarity index*:

$$I^{(n)}_{X,Y}(\hat{C};\boldsymbol{\alpha}) = \frac{1}{N_{X,Y}}\, Z^{(n)}_{X,Y}(\hat{C};\boldsymbol{\alpha}) \tag{8}$$

In the one-electron case the quantities defined in Eqs. 4, 5, and 6 become,

$$Z^{(1)}_{X,Y}(\hat{C}_u) = \int\int \rho_X(\mathbf{r}_1)\tilde{\rho}_Y(\mathbf{r}_2)d\mathbf{r}_1 d\mathbf{r}_2 \tag{9}$$

$$Z^{(1)}_{X,Y}(\hat{C}_c;\boldsymbol{\alpha}) = \int\int \Gamma^*_X(\mathbf{r}_1,\mathbf{r}'_1)\tilde{\Gamma}_Y(\mathbf{r}_1,\mathbf{r}'_1)d\mathbf{r}_1 d\mathbf{r}'_1 \tag{10}$$

and,

$$Z^{(1)}_{X,Y}(\hat{C}_d;\boldsymbol{\alpha}) = \int \rho_X(\mathbf{r})\tilde{\rho}_Y(\mathbf{r})d\mathbf{r} \tag{11}$$

respectively, where $\rho_X(\mathbf{r}) \equiv D^{(1)}_X(\mathbf{r})$ is the electron density. Being equal to the product of the numbers of electrons in X and Y, $Z^{(1)}_{X,Y}(\hat{C}_u)$ is independent of the mutual orientation of the two systems. On the other hand, $Z^{(1)}_{X,Y}(\hat{C}_c;\boldsymbol{\alpha})$ can be readily recognized as the basis of the NOEL similarity measure.[7] Using $Z^{(1)}_{X,Y}(\hat{C}_d;\boldsymbol{\alpha})$ and,

$$N^{C}_{X,Y} = \left[\int \rho^2_X(\mathbf{r}) d\mathbf{r} \int \rho^2_Y(\mathbf{r}) d\mathbf{r} \right]^{1/2} \tag{12}$$

in Eq. 8 yields Carbó's[2] similarity index. Likewise, the substitution of the norm,

$$N^{H}_{X,Y} = \frac{1}{2} \int \left[\rho^2_X(\mathbf{r}) + \rho^2_Y(\mathbf{r}) \right] d\mathbf{r} \tag{13}$$

into Eq. 8 results in Hodgkin's[6] similarity index.

A *normalized measure* $M_{X,Y}(\hat{C})$ *of the similarity* between two molecules X and Y that is invariant to the choice of coordinate system is obtained by maximizing $I^{(n)}_{X,Y}(\hat{C};\boldsymbol{\alpha})$ with respect to $\boldsymbol{\alpha}$:

$$M_{X,Y}(\hat{C}) = \sup_{\boldsymbol{\alpha}} \{I^{(n)}_{X,Y}(\hat{C};\boldsymbol{\alpha})\} = \frac{1}{N_{X,Y}} \sup_{\boldsymbol{\alpha}} \{Z^{(n)}_{X,Y}(\hat{C};\boldsymbol{\alpha})\} \tag{14}$$

If $Z^{(n)}_{X,Y}$ is derived from real-valued wavefunctions, $M_{X,Y}$ itself is real-valued and the norm $N_{X,Y}$ can be chosen in such a way that the convenient inequality $0 < M_{X,Y} \leq 1$ holds. The upper limit $M_{X,Y} = 1$ is attained in the case of a perfect similarity. The unattainable lower limit of $M_{X,Y} = 0$ would indicate a complete dissimilarity.

III. ATOMS IN MOLECULES (AIMS)

The development of the theory of AIMs[5] has been prompted by the experimental observation that the electron densities $\rho(\mathbf{r})$ in molecules exhibit nuclear cusps at which the electron density gradient $\nabla\rho(\mathbf{r})$ is discontinuous. An examination of the density gradient field at points in the neighborhood of the nuclei shows that cusps serve as *attractors* for the *gradient paths*, i.e. the lines of steepest ascent in $\nabla\rho(\mathbf{r})$ (Figure 1). The attractors in a molecule typically coincide with the positions of the nuclei, although nonnuclear attractors are occasionally encountered.[18]

Each attractor (represented by a black dot in Figure 1) constitutes a terminus to a number of gradient paths originating at infinity (thin lines in Figure 1). At the same time, the attractor is the terminus to one or several gradient paths of a finite length (thick lines in Figure 1) known as *bond paths* or, more generally, *attractor interaction lines*. Bond paths originate at *bond critical points* that are characterized by a vanishing $\nabla\rho(\mathbf{r})$ and the electron density Hessian possessing one positive eigenvalue. Each bond critical point is shared by two attractors. This sharing indicates the presence of either chemical bonding or a strong steric interaction between the corresponding atoms.[11]

The Cartesian space can be subdivided into disjoint regions, known as *atomic basins* $\{\Omega_A\}$, each containing an attractor and all the gradient paths that terminate at it. AIM *is defined as the union of a nuclear attractor and its basin*. The boundary of an atom (*atomic surface*) contains all the bond critical points associated with its attractor and all the gradient paths for which those bond critical points serve as

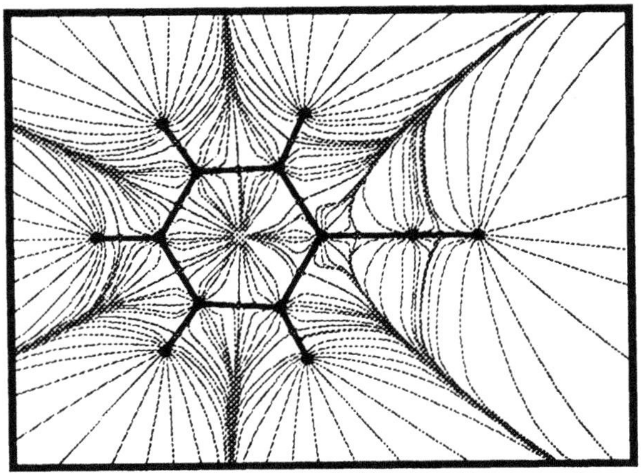

Figure 1. Gradient paths in the molecular plane of the borabenzene·CO complex.

termini. The atomic surface Π_A is therefore tangent everywhere to $\nabla\rho(\mathbf{r})$ and satisfies the zero-flux condition:

$$\int_{\Pi_A} \nabla\rho(\mathbf{r})\cdot d\mathbf{s} = 0 \tag{15}$$

As a consequence of Eq. 15, AIMs conform to all theorems of quantum mechanics.

A one-electron property P_A of an atom A in a molecule is defined as an integral of the corresponding property density $\rho_P(\mathbf{r})$ over the atomic basin Ω_A:

$$P_A = \int_{\Omega_A} \rho_P(\mathbf{r}) d\mathbf{r} \tag{16}$$

Since AIMs are disjoint, yet fill the entire Cartesian space, their properties satisfy the important additivity condition,

$$P_{\text{mol}} = \sum_A P_A \tag{17}$$

where P_{mol} is the respective molecular property.

IV. SIMILARITY OF AIMS: THEORY

The concept of similarity of molecules, described in Section II of this review, can be easily adapted to AIMs by altering the integration limits in Eqs. 9–11. The integration over the entire Cartesian space $\Re^3$ is replaced by integration over the common part,

$$\Omega_{A,B}(\boldsymbol{\alpha}) \equiv \Omega_{A(X),B(Y)}(\boldsymbol{\alpha}) = \Omega_{A(X)} \cap \tilde{\Omega}_{B(Y)} \tag{18}$$

of the atomic basin of atom A in molecule X and that of atom B in molecule Y. Here and in the following, the subscripts A and B are used as a shorthand for $A(X)$ and $B(Y)$, respectively. As before, the tilded quantities refer to a rotated/translated coordinate system. The spatial extent and the shape of $\Omega_{A,B}(\boldsymbol{\alpha})$ vary with the mutual orientation (parameterized by $\boldsymbol{\alpha}$) of A and B. It is important to note that, since the concept of AIMs is based upon the topological properties of the *one-electron* density $\rho(\mathbf{r})$, atomic similarity measures based upon the overlap between many-electron density functions or density matrices are devoid of any physical meaning. Therefore, the most general form of a similarity index $I_{A,B}$ for AIMs reads,

$$I_{A,B}(\hat{C};\boldsymbol{\alpha}) = \frac{1}{N_{A,B}} Z_{A,B}(\hat{C};\boldsymbol{\alpha}) \tag{19}$$

where the generalized overlap integral is given by

$$Z_{A,B}(\hat{C};\boldsymbol{\alpha}) = \int\int_{\Omega_{A,B}(\boldsymbol{\alpha})} \rho_X(\mathbf{r}_1)\hat{C}_{\tilde{\rho}_Y}(\mathbf{r}_2)d\mathbf{r}_1 d\mathbf{r}_2 \tag{20}$$

Within a given AIM, $\rho(\mathbf{r})$ attains its only maximum at the corresponding attractor. Thus, the maximal overlap between the electron densities within the atoms A and B implies coalescence of their attractors. Since it is our ultimate aim to maximize the overlap integral (Eq. 20), it is useful to implicitly assume this coalescence. Such an assumption eliminates the translational degrees of freedom from $\boldsymbol{\alpha}$, leaving it with just three components that correspond to the three Euler angles.

In analogy to $N_{X,Y}$ (Eq. 7), the normalization constant $N_{A,B}$ in Eq. 20 has to satisfy:

$$N_{A,A} = Z_{A,A}(\hat{C};\mathbf{0}) \text{ and } N_{A,B} = N_{B,A} \tag{21}$$

Two meaningful choices are possible for the coupling operator $\hat{C}$. The first choice $\hat{C}_c = \delta(\mathbf{r}_1 - \mathbf{r}_2)$ results in a full coupling and gives rise to similarity measures of the Carbó–Hodgkin type:

$$M_{A,B}(\hat{C}_c) = \frac{1}{N_{A,B}} \sup_{\boldsymbol{\alpha}} \left\{ \int_{\Omega_{A,B}(\boldsymbol{\alpha})} \rho_X(\mathbf{r})\tilde{\rho}_Y(\mathbf{r})d\mathbf{r} \right\} \tag{22}$$

The norms,

$$N^C_{A,B} = \left[\int_{\Omega_A} \rho^2_X(\mathbf{r})\, d\mathbf{r}_1 \int_{\Omega_B} \rho^2_Y(\mathbf{r})\, d\mathbf{r}_1 \right]^{1/2} \tag{23}$$

and,

$$N^{H}_{A,B} = \frac{1}{2}\left[\int_{\Omega_A} \rho_X^2(\mathbf{r})\, d\mathbf{r}_1 + \int_{\Omega_B} \rho_Y^2(\mathbf{r})\, d\mathbf{r}_1\right] \tag{24}$$

which are analogous to those appearing in Eqs. 12 and 13, can be substituted into Eq. 22 to form the corresponding similarity measures $M^{C}_{A,B}$ and $M^{H}_{A,B}$. The following scaling analysis can be used to demonstrate that the choice of norm in Eq. 22 is not as minor a matter as it might seem.

Let $\rho(\mathbf{r})$ be the electron density within a given atomic basin Ω_A, and $\nabla\rho(\mathbf{r})$ be the corresponding electron density gradient. $\rho(\mathbf{r})$ uniquely determines the atomic zero-flux surface Π_A. Let α be an arbitrary positive constant different from one and $\rho'(\mathbf{r}) = \alpha\, \rho(\mathbf{r})$ be the electron density within the basin $\Omega_{A'}$ of a hypothetical atom A'. As the field of $\nabla\rho'(\mathbf{r})$ is collinear with that of $\nabla\rho(\mathbf{r})$:

$$\Pi_{A'} \equiv \Pi_A \text{ and } \Omega_{A'} \equiv \Omega_A \tag{25}$$

The similarity between the atom A and its hypothetical counterpart A' as measured by $M^{C}_{A,A'}$ would equal 1, while the $M^{H}_{A,A'}$ similarity measure would assume the value of $2\alpha(1 + \alpha^2)^{-1} < 1$. This result shows that the norm $N^{C}_{A,B}$ emphasizes the similarity of shapes of the atoms in comparison, while $N^{H}_{A,B}$ is more sensitive to the similarity of the electron density distributions within their basins.

The choice $\hat{C} = \hat{C}_u = \hat{1}$ in Eq. 20 produces a completely decoupled similarity measure:

$$M_{A,B}(\hat{C}_u) = \frac{1}{N_{A,B}} \sup_{\boldsymbol{\alpha}} \left\{ \iint_{\Omega_{A,B}(\boldsymbol{\alpha})} \rho_X(\mathbf{r}_1)\, \tilde{\rho}_Y(\mathbf{r}_2) d\mathbf{r}_1 d\mathbf{r}_2 \right\} \tag{26}$$

Substitution of the norm,

$$N_{A,B} = \int_{\Omega_A} \rho_X(\mathbf{r})\, d\mathbf{r} \int_{\Omega_B} \rho_Y(\mathbf{r})\, d\mathbf{r} = N_A N_B \tag{27}$$

where N_A and N_B are the numbers of electrons in atoms A and B, respectively, into Eq. 26 produces Cioslowski's similarity measure:[1]

$$S_{A,B} = \sup_{\boldsymbol{\alpha}} \ \{s_{A,B}(\boldsymbol{\alpha})\}$$

$$s_{A,B}(\boldsymbol{\alpha}) = \frac{1}{N_A N_B} \int_{\Omega_{A,B}(\boldsymbol{\alpha})} \rho_X(\mathbf{r}) d\mathbf{r} \int_{\Omega_{B,A}(\boldsymbol{\alpha})} \rho_Y(\mathbf{r}) d\mathbf{r} \tag{28}$$

A scaling analysis similar to that performed for $M^{C}_{A,B}$ and $M^{H}_{A,B}$, produces $S_{A,A'} = 1$, demonstrating that $S_{A,B}$ measures mostly the similarity of shapes of AIMs.

In contrast to $M^{C}_{A,B}$ and $M^{H}_{A,B}$, the computation of $S_{A,B}$ involves integrals linear in $\rho(\mathbf{r})$ and requires only N_A and N_B (which are routinely calculated) for the computation of the norm $N_{A,B}$ It is primarily due to its computational simplicity that $S_{A,B}$ is the only measure of the similarity of AIMs that has been employed in practical calculations thus far.[1,15,19]

V. SIMILARITY OF AIMS: COMPUTATIONS

The calculation of $S_{A,B}$ involves the following three steps: First, the electron densities $\rho_X(\mathbf{r})$ and $\rho_Y(\mathbf{r})$ of the molecules containing the atoms in comparison are obtained. Second, the atomic zero-flux surfaces of the atom A in the molecule X and the atom B in the molecule Y are determined, and the respective numbers of electrons, N_A and N_B, in their atomic basins, Ω_A and Ω_B, are calculated. Third, given the initial relative orientation of A and B [parameterized by the vector of Euler angles $\boldsymbol{\alpha} \equiv (\alpha_1,\alpha_2,\alpha_3)$], the similarity index $s_{A,B}(\boldsymbol{\alpha})$ is globally optimized. Each iteration of the optimization procedure requires the calculation of the common part $\Omega_{A,B} = \Omega_{B,A} \equiv \Omega_{A,B}(\boldsymbol{\alpha})$ of the two atomic basins, as well as its derivatives with respect to α_1, α_2, and α_3.

Superior efficiency and accuracy in the calculations of the similarity measure $S_{A,B}$ are achieved by employing the recently developed variational approach to the determination of the atomic zero-flux surfaces[17] in conjunction with the semi-analytical integration algorithm.[20] The former provides atomic boundaries of excellent accuracy in an analytically differentiable representation; the latter offers accurate integrations with considerable computational savings. The atomic zero-flux surfaces are constructed from the atomic zero-flux surface sheets.[17] Each of these sheets intersects its respective attractor interaction line at the corresponding bond critical point. The j-th zero-flux surface sheet of the atom A is explicitly given by ,

$$\eta = H_{A,j}(\xi,\varphi) \tag{29}$$

where $H_{A,j}$ is an analytical function and (ξ,φ,η) are suitable curvilinear coordinates.

A convenient curvilinear coordinate system (ξ,φ,η) can be constructed in the following manner (Figure 2): The Hessian of the electron density at the bond critical point C has one negative and two positive eigenvalues. A local Cartesian coordinate system (x_o,y_o,z_o) has the z_o coordinate axis collinear with the eigenvector $\mathbf{e}_z$ of the Hessian that corresponds to its negative eigenvalue. The x_o axis is chosen to be parallel to the eigenvector that corresponds to the larger of the two positive eigenvalues of the Hessian [note that the (x_o,y_o,z_o) coordinates are different from the Cartesian coordinates $\mathbf{r} \equiv (x,y,z)$ in which the densities are defined and the integrations are performed; the transformation $(x,y,z) \rightarrow (x_o,y_o,z_o)$ involves a rotation of the axes and a translation of the origin]. If A_1' and A_2' are the orthogonal projections of the two attractors A_1 and A_2 onto the z_o axis, then the midpoint O between A_1' and A_2' is the origin of the (x_o,y_o,z_o) coordinate system. The set of equations,

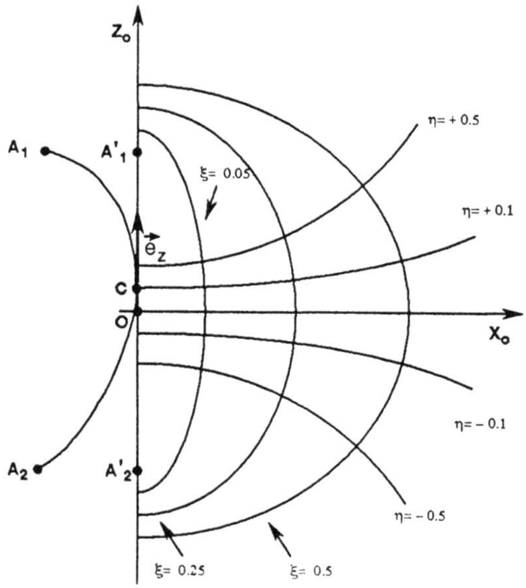

Figure 2. Elements of the curvilinear coordinate system (ξ,φ,η) including the intersections of several ξ and η isosurfaces with the $\varphi = 0$ half-plane.

$$x_o = \tau\sqrt{\frac{\xi}{1-\xi}}\,\sqrt{1-\eta^2}\,\cos\phi$$

$$y_o = \tau\sqrt{\frac{\xi}{1-\xi}}\,\sqrt{1-\eta^2}\,\sin\phi$$

$$z_o = \tau\sqrt{\frac{1}{1-\xi}}\,\eta \tag{30}$$

$$\eta \in [-1,1],\quad \xi \in [0,1),\quad \varphi \in [0,2\pi)$$

where τ is one half of the distance between A_1' and A_2', define the prolate spheroidal coordinate system (ξ,φ,η).

The function $H_{A,j}(\xi,\varphi)$ (Eq. 29) is such that $H_{A,j}(0,\varphi) = \eta_{0,j}$, where $(0,\varphi,\eta_{0,j})$ are the prolate spheroidal coordinates of the j-th bond critical point (note that φ is indefinite at that point). It is convenient to define $H_{A,j}$ as,

$$H_{A,j} = \frac{h_{A,j}}{\sqrt{1+h_{A,j}^2}} \tag{31}$$

where $h_{A,j}$ is expanded in a basis of orthogonal functions $\Phi_k(\xi,\varphi)$,

$$h_{A,j}(\xi,\varphi) = C_{A,j,0} + \xi \sum_{k=1}^{N} C_{A,j,k}\, \Phi_k(\xi,\varphi) \tag{32}$$

and $C_{A,j,0} = \eta_{0,j}(1 - \eta_{0,j}^2)^{-1/2}$. It is preferable to expand $h_{A,j}$, which can take arbitrary real values, instead of $H_{A,j}$, which is allowed to vary only within the $[-1,1]$ interval. The coefficients $\{C_{A,j,k}, k = 1,N\}$ are optimized subject to the requirement that the surface given by Eq. 29 satisfies an approximate zero-flux condition that everywhere on a grid $\{(\xi_m,\varphi_m)\}$,

$$\mathbf{n}_{A,j,m} \cdot \mathbf{g}_{A,j,m} = 0 \tag{33}$$

i.e. the normal to the surface,

$$\mathbf{n}_{A,j,m} = \nabla[\eta - H_{A,j}(\xi_m,\varphi_m)] \tag{34}$$

is orthogonal to the electron density gradient,

$$\mathbf{g}_{A,j,m} = \nabla\rho\{\mathbf{r}[\xi_m,\varphi_m,H_{A,j}(\xi_m,\varphi_m)]\} \tag{35}$$

All the integrals involved in the calculation of the similarity index $s_{A,B}(\boldsymbol{\alpha})$ (Eq. 28) are approximated by sums of radial integrals with weights $W_{A,m}$ stemming from numerical angular integration. For example, the number of electrons in atom A is given by:

$$N_A = \int_{\Omega_A} \rho_X(\mathbf{r})d\mathbf{r} \approx \sum_i W_{A,i} \int_{\omega_{A,i}} \rho_X(\mathbf{r}_A + R\,\mathbf{u}_{A,i})R^2 dR \tag{36}$$

In Eq. 36, $\mathbf{r}_A$ denotes the position of the attractor of A and $\mathbf{u}_{A,i}$ is the i-th radial unit vector. The range of integration, defined symbolically by $\omega_{A,i} \equiv \bigcup_k [R_{A,i,2k-1}, R_{A,i,2k}]$, comprises a union of the intervals $[R_{A,i,2k-1}, R_{A,i,2k}]$ along the direction of $\mathbf{u}_{A,i}$ that belong to the basin Ω_A of the atom A. The end-points of these intervals correspond to the set of intersections of the atomic zero-flux surface with the i-th ray,

$$\mathbf{r}_i(R) = \mathbf{r}_A + R\,\mathbf{u}_{A,i} \tag{37}$$

that emanates from $\mathbf{r}_A$ along $\mathbf{u}_{A,i}$. These intersections are obtained by solving simultaneously Eqs. 29, 30, and 37 or, equivalently, by finding the roots R of:

$$\eta_{A,i}(R) - H_{A,j}[\xi_{A,i}(R),\varphi_{A,i}(R)] =$$

$$\eta(\mathbf{r}_A + R\,\mathbf{u}_{A,i}) - H_{A,j}\,[\xi(\mathbf{r}_A + R\,\mathbf{u}_{A,i}),\, \varphi(\mathbf{r}_A + R\,\mathbf{u}_{A,i})] = 0 \tag{38}$$

The weights $W_{A,i}$ and the sets $\omega_{A,i}$ are precomputed with the adaptive integration scheme that is employed in the calculation of atomic charges.[17]

It is possible to compute the integrals,

$$I_A = \int_{\Omega_{A,B}} \rho_X(\mathbf{r})d\mathbf{r} \quad \text{and} \quad I_B = \int_{\Omega_{B,A}} \rho_Y(\mathbf{r})d\mathbf{r} \tag{39}$$

that enter Eq. 28 in the same manner as N_A and N_B, i.e.,

$$I_A \approx \sum_i W_{A,i} \int_{\omega_{AB,i}} \rho_X(\mathbf{r}_A + R\,\mathbf{u}_{A,i})\, R^2 dR \tag{40}$$

and,

$$I_B \approx \sum_i W_{B,i} \int_{\omega_{BA,i}} \rho_Y(\mathbf{r}_B + R\,\mathbf{u}_{B,i})\, R^2 dR \tag{41}$$

The observation that many of the sets $\varpi_{AB,i} \equiv \omega_{A,i} \backslash \omega_{AB,i}$ are empty when A is similar to B leads to the conclusion that the calculation of the integrals can be significantly accelerated by evaluating them as,

$$I_A = N_A - \bar{I}_A \quad \text{and} \quad I_B = N_B - \bar{I}_B \tag{42}$$

where,

$$\bar{I}_A = \int_{\bar{\Omega}_{A,B}} \rho_X(\mathbf{r})d\mathbf{r} = \int_{\Omega_A \backslash \Omega_{A,B}} \rho_X(\mathbf{r})d\mathbf{r} \approx \sum_i W_{A,i} \int_{\varpi_{AB,i}} \rho_X(\mathbf{r}_A + R\,\mathbf{u}_{A,i})\, R^2 dR \tag{43}$$

and,

$$\bar{I}_B = \int_{\bar{\Omega}_{B,A}} \rho_Y(\mathbf{r})d\mathbf{r} = \int_{\Omega_B \backslash \Omega_{B,A}} \rho_Y(\mathbf{r})d\mathbf{r} \approx \sum_i W_{B,i} \int_{\varpi_{BA,i}} \rho_X(\mathbf{r}_B + R\,\mathbf{u}_{B,i})\, R^2 dR \tag{44}$$

The new ranges of integration $\{\varpi_{AB,i}\}$ require the calculation of the roots of the equations (compare with Eq. 38),

$$\begin{aligned} \eta_{AB,i}(R) - H_{B,j}[\xi_{AB,i}(R), \varphi_{AB,i}(R)] &= \eta(\mathbf{r}_B + R\hat{T}_{AB}\,\mathbf{u}_{A,i}) \\ &\quad - H_{B,j}[\xi(\mathbf{r}_B + R\hat{T}_{AB}\,\mathbf{u}_{A,i}), \varphi(\mathbf{r}_B + R\hat{T}_{AB}\,\mathbf{u}_{A,i})] = 0 \end{aligned} \tag{45}$$

where the orthogonal matrix $\hat{T}_{AB} = \hat{T}_{AB}(\alpha_1, \alpha_2, \alpha_3)$ depends on the three Euler angles that determine the mutual orientation of A and B. The solution $R \equiv R_{AB,ijl}$ of Eq.(45) corresponds to the l-th intersection of the i-th ray in the numerical integrations for atom A with the j-th surface sheet of atom B. Similarly to Eq. 38, Eq. 45 constitutes a one-dimensional nonlinear problem that can be readily solved with a linear search algorithm.[17] A permutation of the subscripts A and B in Eq. 45 leads to the equations,

$$\eta(\mathbf{r}_A + R\hat{T}_{BA}\,\mathbf{u}_{B,i}) - H_{A,j}[\xi(\mathbf{r}_A + R\hat{T}_{BA}\,\mathbf{u}_{B,i}),\varphi(\mathbf{r}_A + R\hat{T}_{BA}\,\mathbf{u}_{B,i})] = 0 \qquad (46)$$

for the intersections that determine the ranges $\{\varpi_{BA,i}\} \equiv \{\omega_{B,i}\backslash\omega_{BA,i}\}$ of the radial integrals involved in the calculation of $\bar{I}_B$. In Eq. 46:

$$\hat{T}_{BA} = \hat{T}_{AB}^{-1} = \hat{T}_{AB}^{T} \qquad (47)$$

The maximization of $s_{A,B}(\boldsymbol{\alpha})$ requires its derivatives with respect to the Euler angles to be computed. The derivation starts with

$$\frac{\partial s_{A,B}}{\partial \alpha_k} = -\frac{1}{N_A}\left[1 - \frac{\bar{I}_B}{N_B}\right]\frac{\partial \bar{I}_A}{\partial \alpha_k} - \frac{1}{N_B}\left[1 - \frac{\bar{I}_A}{N_A}\right]\frac{\partial \bar{I}_B}{\partial \alpha_k},\; k = 1,2,3 \qquad (48)$$

Since all the dependence of $\bar{I}_A$ and $\bar{I}_B$ on $\boldsymbol{\alpha}$ is contained in $\varpi_{BA,i}$ and $\varpi_{AB,i}$, the only terms that have to be calculated are the derivatives $\partial R_{ijl}/\partial\alpha_k$ of the intersections $R_{ijl} \in \{R_{A,ijl}, R_{B,ijl}, R_{AB,ijl}, R_{BA,ijl}\}$ [solutions of Eqs. 38, 45, and 46] with respect to the Euler angles. For example, in order to obtain $\partial R_{AB,ijl}/\partial\alpha_k$, one differentiates Eq. 45 with respect to α_k. This results in,

$$\left[\vec{S}^T \cdot \hat{T}_{AB} \cdot \mathbf{u}_{A,i}\right]\frac{\partial R_{ijl}}{\partial \alpha_k} + \left[\vec{S}^T \cdot \frac{\partial \hat{T}_{AB}}{\partial \alpha_k} \cdot \mathbf{u}_{A,i}\right] R_{ijl} = 0 \qquad (49)$$

with,

$$\vec{S}^T = \nabla\eta(\mathbf{r}) - \frac{\partial H_{B,j}}{\partial \xi}\nabla\xi(\mathbf{r}) - \frac{\partial H_{B,j}}{\partial \varphi}\nabla\varphi(\mathbf{r}) \qquad (50)$$

where $\mathbf{r} = \mathbf{r}_B + R_{ijl}\,\hat{T}_{AB}\cdot\mathbf{u}_{A,i}$. The second derivatives $\{\partial^2 R/\partial\alpha_k\partial\alpha_m\}$ required for the calculation of the Hessian of $s_{A,B}$ are evaluated in an analogous manner.

Many procedures for the maximization of $s_{A,B}(\boldsymbol{\alpha})$ are possible in principle. In practice, a modification of the variable metric method has been found to perform well in actual calculations of $S_{A,B}$.[1,15] The gradient, Eq. 48, is calculated at each step, while the Hessian $\{\partial^2 R/\partial\alpha_k\partial\alpha_m\}$ is computed during the first step of the optimization and updated with the BFGS[21] formula in each subsequent iteration.

In many cases multiple maxima of $s_{A,B}(\boldsymbol{\alpha})$ are encountered. The safest approach in such cases is to locate and compare all the maxima in order to determine the global one. The use of geometrical and heuristic considerations in the selection of the initial orientation often results in substantial computational savings. In many instances, such considerations can also be successfully employed in the determination of the anticipated number of maxima in $s_{A,B}(\boldsymbol{\alpha})$ and the estimation of their relative magnitudes.

VI. SIMILARITIES OF AIMS: APPLICATIONS

The similarity measure $S_{A,B}$ (Eq. 28) has been invoked in the original work[1] in order to compare the carbon atoms in some simple hydrocarbons and the carbon, hydrogen, and fluorine atoms in fluoro-substituted methanes. Two convenient definitions have been introduced. The *ligands* of a given atom have been defined as the atoms whose attractors share bond paths with the attractor of the atom in question. Atoms with the same nuclei and the same total numbers of ligands have been termed *formally identical.*

The comparison of carbon atoms in fluoro-substituted methanes[1] has shown that the similarity of formally identical atoms connected to different ligands can be as low as 62% (CH_4 vs. CF_4 in Table 1), exposing the nature of the ligands as the primary factor that affects the similarity of formally identical atoms. On the other hand, the comparison of carbon atoms in simple hydrocarbons[1] has demonstrated that, surprisingly, atoms that are not formally identical (C_2H_4 vs. C_2H_2 in Table 2) can possess similarities as high as 87%.

The effect of the hybridization of the ligands has been observed in the example of the formally identical hydrogen atoms in simple hydrocarbons (Table 3). In the cases of different carbon hybridizations, similarities between hydrogens as low as 92% have been obtained. In the case of formally identical ligands (CH_4 vs. C_2H_6) the similarity of the hydrogens has exceeded 99%.

The comparison of the four formally identical hydrogen atoms in the acrolein molecule[15] (Figure 3) illustrates the usefulness of the similarity measure $S_{A,B}$ in the detection and quantification of steric interactions among AIMs. The four zero-flux

Table 1. Similarity of Carbon Atoms in Fluoro-Substituted Methanes

$S_{A,B}$	CH_4	CH_3F	CH_2F_2	CHF_3
CH_3F	0.909			
CH_2F_2	0.816	0.890		
CHF_3	0.711	0.776	0.865	
CF_4	0.623	0.667	0.737	0.838

Table 2. Similarity of Carbon Atoms in Simple Hydrocarbons

$S_{A,B}$	C_2H_2	C_2H_4	C_2H_6
C_2H_4	0.873		
C_2H_6	0.803	0.861	
CH_4	0.821	0.857	0.963

Table 3. Similarity of Hydrogen Atoms in Simple Hydrocarbons

$S_{A,B}$	C_2H_2	C_2H_4	C_2H_6
C_2H_4	0.934		
C_2H_6	0.923	0.979	
CH_4	0.927	0.985	0.993

surface sheets, which are associated with the bonds C_2–H_4, C_2–C_3, C_3–C_6, and C_6–H_7 and pass through the relatively narrow opening between the atoms H_4 and H_7, are severely distorted. The resulting changes in the shapes of atoms H_4 and H_7 relative to the shapes of the congestion-free hydrogens H_5 and H_8 are revealed by visual inspection and also reflected in the calculated similarities (Table 4). The atomic similarity between the "undistorted" hydrogens H_5 and H_8 amounts to 99.33%, whereas the "distorted"–"undistorted" pairs H_4–H_5 and H_4–H_8 exhibit the lowest similarities of 95.25% and 95.47%, respectively. The hydrogen H_7 is significantly less distorted than H_4, as indicated by its 98.31% similarity with H_5 and 98.86% similarity with H_8. The significant additional distortion of H_4 by its second-neighbor O_1 is also reflected in the similarity of only 96.42% between H_4 and H_7.

An extensive study of carbonyl oxygens[19] in diverse molecular systems has employed the similarity measure $S_{A,B}$ to quantify the variability in atomic shapes (Figure 4). The possibility of a correlation between shapes of AIMs and their one-electron properties has been investigated. The concept of *similarity graphs* has been invoked to provide a visual representation of similarity patterns among formally identical atoms. The study, which involved a set of 21 molecules with

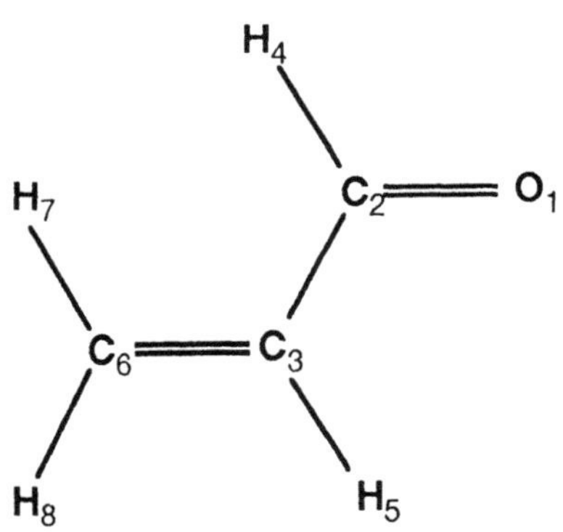

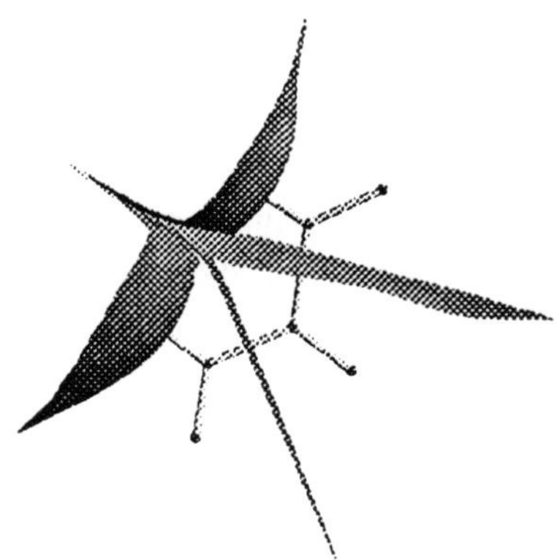

Figure 3. The numbering of atoms in the acrolein molecule (*left*) and the four zero-flux surface sheets that pass between the H_4 and H_7 hydrogens (*right*).

Table 4. Similarities of Hydrogen Atoms in the Acrolein Molecule

$S_{A,B}$	H_4	H_5	H_7
H_5	0.9525		
H_7	0.9642	0.9831	
H_8	0.9547	0.9933	0.9886

$$CH_3COCN \leftrightarrow CH_3CONH_2$$

$$CO(NH_2)_2 \leftrightarrow CH_3COOH$$

$$HCHO \leftrightarrow CH_3CHO \rightarrow CH_3COCH_3$$

$$HCOCN \leftrightarrow NH_2CHO \rightarrow HCOOH$$

$$NH_2COCl \leftrightarrow ClCOOH \rightarrow COCl_2$$

$$ClCOCN \leftarrow CNCOOH \leftrightarrow NH_2COCN \rightarrow CO(CN)_2$$

$$NH_2COOH \leftrightarrow CO(OH)_2 \rightarrow CH_3COCl \rightarrow HCOCl$$

Figure 4. Relations of maximal similarity between the carbonyl oxygen atoms in various molecular systems. The notation X←Y denotes the carbonyl oxygen of Y being the most similar to that in X from among all the systems under study other than X.

general structure R_1COR_2, where R_1, R_2 = H, CH_3, NH_2, Cl, CN, or OH has produced several important findings.

It has been shown that the shapes of atoms in molecules are primarily affected by the size of their neighbors. Effects due to the electron-withdrawing or electron-donating properties of the second neighbors have not been observed. Unlike the atomic shapes, the computed atomic charges have been found to reflect the ability of the second neighbors to donate or withdraw electrons. Most importantly, the study has unequivocally demonstrated that no correlation exists between the shapes and the electronic properties of AIMs.

VII. SUMMARY

The aforementioned investigations have illustrated the usefulness of the atomic similarity index $S_{A,B}$ (Eq. 28) in the research on the taxonomy of atomic shapes. Other potentially important applications of $S_{A,B}$ include the evaluation of the degree of transferability of atoms in molecules and the detection and quantification of steric interactions within molecules. These applications hold the promise to make the atomic similarity measures indispensable tools of quantum chemistry.

ACKNOWLEDGMENT

This work was partially supported by the National Science Foundation under the grant CHE-9224806.

REFERENCES

1. Cioslowski, J.; Nanayakkara, A. *J. Am. Chem. Soc.* **1993**, *115*, 11213.
2. Carbó, R.; Leyda, L.; Arnau, M. *Int. J. Quantum Chem.* **1980**, *17*, 1185.
3. Carbó, R.; Calabuig, B. *Int. J. Quantum Chem.* **1992**, *42*, 1681.
4. Bader, R.W.F.; Tal, Y.; Anderson, S.G.; Nguyen-Dang, T.T. *Israel J. Chem.* **1979**, *19*, 8.
5. Bader, R.W.F. *Atoms in Molecules: A Quantum Theory*; Clarendon Press: Oxford, 1990.
6. Hodgkin, E.E.; Richards, W.G. *Int. J. Quantum Chem., Quantum Biol. Symp.* **1987**, *14*, 105.
7. Cioslowski, J.; Fleischmann, E.D. *J. Am. Chem. Soc.* **1991**, *113*, 64.
8. Mezey, P.G. *Shape in Chemistry: An Introduction to Molecular Shapes and Topology*; VCH Publishers: New York, 1993.
9. Bader, R.W.F. *Chem. Rev.* **1991**, *91*, 893.
10. Cioslowski, J.; Mixon, S.T. *J. Am. Chem. Soc.* **1991**, *113*, 4142.
11. Cioslowski, J.; Mixon, S.T. *J. Am. Chem. Soc.* **1992**, *114*, 4382.
12. Cioslowski, J.; Mixon, S.T. *J. Am. Chem. Soc.* **1993**, *115*, 1084.
13. Bader, R.W.F., Becker, P. *Chem. Phys. Lett.* **1988**, *148*, 452; Bader, R.W.F.; Larouche, A.; Gatti, C.; Carroll, M.T.; MacDougall, P.J.; Wiberg, K.B. *J. Chem. Phys.* **1987**, *87*, 1142; Bader, R.W.F.; Carroll, M.T.; Cheeseman, J.R.; Chang, C. *J. Am. Chem. Soc.* **1987**, *109*, 7968; Bader, R.W.F., *Can. J. Chem.* **1986**, *64*, 1036; Bader, R.W.F.; Keith, T.A.; Gough, K.M.; Laidig, K.E., *Mol. Phys.* **1992**, *75*, 1167; Bader, R.W.F.; Keith, T.A. *J. Chem. Phys.* **1993**, *99*, 3693.
14. Chang, C.; Bader, R.W.F. *J. Phys. Chem.* **1992**, *96*, 1654.
15. Cioslowski, J.; Stefanov, B.B.; Constans, P. *J. Comp. Chem.*, in press.
16. Biegler-König, F.W.; Bader, R.W.F.; Tang, T.H. *J. Comp. Chem.* **1982**, *3*, 317.
17. Cioslowski, J.; Stefanov, B.B. *Mol. Phys.* **1995**, *84*, 707; Stefanov, B.B.; Cioslowski, J. *J. Comp. Chem.* **1995**, *16*, 1394.
18. Gatti, C.; Fantucci, P.; Pacchioni, G. *Theor. Chim. Acta* **1987**, *72*, 433; Cao, W.L.; Gatti, C.; MacDougall, P.J.; Bader, R.W.F. *Chem. Phys. Lett.* **1987**, *141*, 380; Cioslowski, J., *J. Phys. Chem.* **1990**, *94*, 5497.
19. Stefanov, B.B.; Cioslowski, J. *Can. J. Chem.*, in press.
20. Cioslowski, J.; Nanayakkara, A.; Challacombe, M. *Chem. Phys. Lett.* **1993**, *203*, 137.
21. Broyden, C.G. *Math. Comput.* **1967**, *21*, 368; Fletcher, R. *Comput. J.* **1970**, *13*, 317; Goldfarb, D., *Math. Comput.* **1970**, *24*, 23; Shanno, D.F. *Math. Comput.* **1970**, *24*, 647.

MOMENTUM–SPACE SIMILARITY:
SOME RECENT APPLICATIONS

Peter T. Measures, Neil L. Allan, and David L. Cooper

ABSTRACT

We describe three applications of momentum–space quantum similarity indices, each linking features of the electron distribution to observed activity. The three applications are: (1) the molecular hyperpolarizabilities of conjugated systems, such as disubstituted benzenes, styrenes, stilbenes and diphenylacetylenes; (2) the use of clustering techniques to analyze momentum–space similarity matrices, taking a range of anti-HIV1 phospholipids as a test case; and (3) the HIV1 inhibition of a series of

Advances in Molecular Similarity
Volume 1, pages 61–87

ISBN: 0-7623-0131-7

nucleotides, introducing a new dissimilarity index which, unlike our previous distance-like measures, emphasizes the shape rather than the magnitude of the electron densities being compared.

I. INTRODUCTION

In recent years we have investigated the use of quantum similarity indices based on momentum–space concepts. These are a valuable addition to other techniques of molecular similarity, such as graph theoretical methods and database searching,[1,2] the comparison of position–space electron densities,[3–7] and electrostatic potentials,[8–10] and the topological analysis of the three-dimensional shapes of charge densities.[11] In previous reviews,[12,13] we have discussed in detail the underlying methodology, including the form of momentum–space electron densities, the indices used to quantify similarity using these densities, and some applications. We concentrate here on applications of our techniques and present three case studies involving large molecules and situations for which it is difficult to rationalize the observed physical or biological behavior with conventional chemical intuition. First, we extend our previous studies[14] of molecular hyperpolarizabilities of conjugated systems, such as disubstituted benzenes, styrenes, stilbenes, and diphenylacetylenes. Secondly, we investigate the use of two different clustering techniques to analyze momentum–space similarity matrices, taking as our example the diverse biological behavior of a range of phospholipids. Finally, we examine a series of nucleotide HIV1 inhibitors, introducing a new dissimilarity index that is largely size independent, unlike our previous distance-like measures.

II. MOMENTUM–SPACE MOLECULAR SIMILARITY

We start with molecular orbitals, $\psi(\mathbf{r})$, of the form,

$$\psi(\mathbf{r}) = \sum_i c_i \phi_i^\alpha(\mathbf{r} - \mathbf{R}_\alpha) \tag{1}$$

where the index i sums over the position–space atomic basis functions, ϕ_i^α, centered on nuclei with positions vectors $\mathbf{R}_\alpha$. The momentum–space wavefunction, $\Psi(\mathbf{p})$, is obtained by a Fourier transform of this position–space wavefunction, so that,

$$\Psi(\mathbf{p}) = \sum_i c_i \Phi_i^\alpha(\mathbf{p}) \exp(-i\mathbf{p}.\mathbf{R}_\alpha) \tag{2}$$

in which the $\Phi_i^\alpha(\mathbf{p})$ are the Fourier transforms of the respective $\phi_i^\alpha(\mathbf{r})$. The relationship in momentum space between the wavefunction and the electron density is exactly the same as in position space, i.e. the momentum–space density, $\rho(\mathbf{p})$, for this molecular orbital is given by the product $\Psi^*(\mathbf{p})\Psi(\mathbf{p})$. The momentum–space basis functions, $\Phi_i^\alpha(\mathbf{p})$, fall off sharply with $p = |\mathbf{p}|$ and so the corresponding electron density emphasizes the slowest moving valence electrons, whereas position–space electron densities tend to be dominated by the regions close to the nuclei.

The basic approach used to quantify the momentum–space similarity is the analogue of the scheme first proposed for position–space densities by Carbó et al.[3] In the present case, the generalized overlap between momentum–space densities ρ_A and ρ_B takes the form:

$$I_{AB}(n) = \int p^n \, \rho_A(\mathbf{p})\rho_B(\mathbf{p})d\mathbf{p} \tag{3}$$

The momentum–space densities can be total electron densities, total valence densities, or those associated with one or more orbitals of interest or with particular molecular fragments. The function p^n is included in the integrand to emphasize particular regions of the density. For example, a value of n of -1 focuses on the slowest moving electron density. This corresponds in turn to emphasizing the long-range valence density in position space. In this review we extend earlier work,[13,14] which considered only $n = -1$, 0, 1, and 2 by investigating also the use of noninteger values of n.

It is often useful to scale $I_{AB}(n)$ into the range 0–100%. This can, of course, be achieved in many ways and a number of these have been employed in our previous work. In the studies described here, we concentrate on just two families of scaled indices: $R_{AB}(n)$ and $T_{AB}(n)$. The index $T_{AB}(n)$, which takes the form,

$$T_{AB}(n) = 100 \frac{I_{AB}(n)}{(I_{AA}(n) + I_{BB}(n) - I_{AB}(n))} \tag{4}$$

has often turned out to be the most discriminating of our scaled similarity indices. The index $R_{AB}(n)$, defined according to,

$$R_{AB}(n) = 100 \frac{I_{AB}(n)}{(I_{AA}(n)I_{BB}(n))^{1/2}} \tag{5}$$

is particularly sensitive to the shape of the momentum densities and has turned out to be especially useful in certain applications.

For cases with extremely high similarities ($\approx$ 100%), the distance-like dissimilarity index $D_{AB}(n)$, can be more informative. This index takes the form,

$$D_{AB}(n) = 100 \, \{I_{AA}(n) + I_{BB}(n) - 2I_{AB}(n)\} \tag{6}$$

and can take values from zero (total similarity), with no upper limit. We introduce later a further dissimilarity index, $P_{AB}(n)$, which is more shape-dependent than is $D_{AB}(n)$.

III. HYPERPOLARIZABILITIES

Nonlinear optics (NLO) deals with the interaction of applied electromagnetic fields with materials to generate new electromagnetic fields, altered in frequency or phase. Materials able to manipulate photonic signals efficiently are important in laser physics, optical communication, optical computing and dynamic image processing.[15–17] The development of actual devices has been limited by the lack of readily processed materials with sufficiently large NLO responses and with other desirable properties, and so there is considerable current interest in the synthesis of more efficient materials.[18,19]

Light incident on a medium can induce an oscillating dipole moment in that medium and the induced polarization generates a second optical field that can interfere with the incident field. The magnitude of this field-induced polarization, P_I, can be expressed as a Taylor series,

$$P_I = \sum_{J} \chi^{(1)}_{IJ} F_J + \tfrac{1}{2} \sum_{J,K} \chi^{(2)}_{IJK} F_J F_K + \tfrac{1}{6} \sum_{J,K,L} \chi^{(3)}_{IJKL} F_J F_K F_L + \ldots \tag{7}$$

in which the labels I, J, K, and L denote the macroscopic axes of the material, $\mathbf{F}$ is the applied field, and the coefficients $\chi^{(1)}_{IJ}$, $\chi^{(2)}_{IJK}$ and $\chi^{(3)}_{IJKL}$ are the first-, second-, and third-order responses of the material, respectively. The first-order term can only give rise to an emitted field of the same frequency as the incident radiation, whereas the higher order terms allow the secondary field to possess frequencies different from that of the applied field. These new frequencies correspond to various NLO effects, such as second-harmonic generation.

For a material to be suitable for a practical NLO application, it must of course have the desired chemical and physical properties. In particular, new materials must possess a crystal structure of the correct symmetry, have suitable mechanical properties, and consist of molecules with large NLO coefficients. The ability to control the alignment of the chromophores is relatively unrefined,[20] so that most effort, both experimental[21,22] and theoretical,[23] has been directed at improving the molecular hyperpolarizabilities. The molecular polarization, p_i, is given by,

$$p_i = \sum_{j} \alpha_{ij} F_j + \tfrac{1}{2} \sum_{j,k} \beta_{ijk} F_j F_k + \tfrac{1}{6} \sum_{j,k,l} \gamma_{ijkl} F_j F_k F_l + \ldots \tag{8}$$

where α_{ij}, β_{ijk}, and γ_{ijkl} are the polarizability (linear response), first-order hyperpolarizability and second-order hyperpolarizability (nonlinear responses), respectively, and the subscripts i, j, k, and l label the molecular Cartesian axes. The macroscopic susceptibilities ($\chi^{(1)}_{IJ}$, $\chi^{(2)}_{IJK}$, and $\chi^{(3)}_{IJKL}$) are related to the corresponding molecular coefficients (α_{ij}, β_{ijk}, and γ_{ijkl}) by local correction fields, the number density, and cosines of the angles between the macroscopic and molecular axes.

A number of experimental techniques are available for the determination of molecular NLO coefficients. Of particular relevance to the systems examined in the present study is electric-field-induced second-harmonic (EFISH) generation. The EFISH experiment can be used to determine β, the vector component of the first-order hyperpolarizability tensor β_{ijk} along the direction of the ground state dipole moment ($\boldsymbol{\mu}$):

$$\beta = \frac{\mu_x\beta_x + \mu_y\beta_y + \mu_z\beta_z}{|\boldsymbol{\mu}|}$$

$$\beta_i = \beta_{iii} + \tfrac{1}{3}\sum_{j \neq i} (\beta_{ijj} + \beta_{jij} + \beta_{jji}) \tag{9}$$

Our principal concern here is with values of β for a range of molecules with asymmetric electron distributions, arising from conjugated organic frameworks separating electron-donor and electron-acceptor groups. Examples of these types of molecules, for which β has been determined using EFISH,[21] include 1,4-disubstituted benzenes, 1,β-disubstituted styrenes, 4,4′-disubstituted stilbenes, and 4,4′-disubstituted diphenylacetylenes,

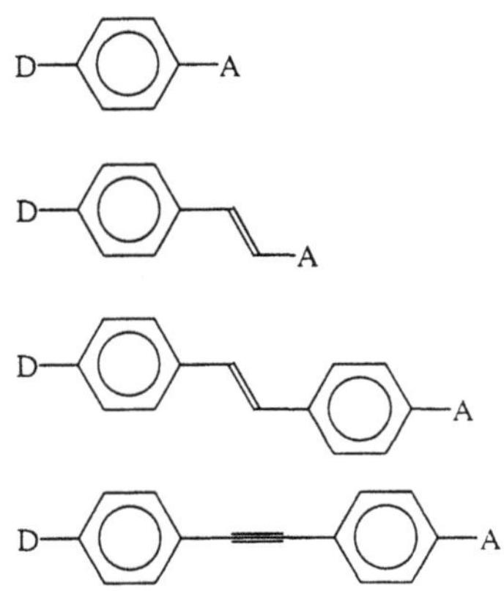

where A and D denote donor and acceptor groups, respectively.

Synthesis of these systems can often be difficult and the experimental determination of the second-order response is far from straightforward. Accordingly, theoretical approaches to β are of considerable interest. Most first-principles evaluations of β involve either finite-field or sum-over-states methodologies. However, when using ab initio quality wavefunctions, the application of these techniques even to small systems is computationally expensive and the results are dependent on the quality of the basis set. These approaches are more tractable when used with semiempirical wavefunctions, but then provide only semiquantitative measures of β.

The alternative to direct evaluation of β is to try to find a correlation between β and a quantity that is easy and cheap to evaluate as well as relatively insensitive to the quality of the wavefunction. In looking for such a structure–activity relationship, we have been guided by the two-state model, which is often applicable when the molecule shows a strong charge-transfer interaction.

In such a case the sum-over-states is likely to be dominated by the first excited state, such that,

$$\beta = 12\frac{\langle g|\mu_i|e\rangle^2\{\langle e|\mu_i|e\rangle - \langle g|\mu_i|g\rangle\}}{(E_e - E_g)^2} \tag{10}$$

in which E_e and E_g are the energies of the excited state e and the ground state g, respectively.

In the simplest treatment, the excited state arises from the excitation of an electron from the highest occupied molecular orbital (HOMO) to the lowest unoccupied molecular orbital (LUMO). As a consequence, we chose to compare the HOMO with the LUMO in each of the molecules of interest. In earlier work[12] we presented a correlation between β and $R_{\mathrm{HL}}(-1)$ for 1,4-benzene derivatives, considering only the contributions to these frontier orbitals from basis functions associated with the benzene ring. No such correlation was found for disubstituted styrenes, stilbenes, or diphenylacetylenes. More recently,[14] again prompted by the form of the two-state model, we have established correlations for all four series of derivatives between β and the quantity Ω, where,

$$\Omega = \frac{R_{HL}(-1)}{(E_H - E_L)^2} \tag{11}$$

and $E_{\mathrm{H}} - E_{\mathrm{L}}$ is the HOMO/LUMO energy separation.

Our previous work used wavefunctions generated using semiempirical MNDO geometry optimizations.[24] We have now also calculated values of $R_{HL}(-1)$ and $E_H - E_L$ using the AM1 scheme,[24] and these are listed in Table 1, together with experimentally determined values of β.[21] In Figure 1, Ω_{MNDO}[14] is plotted vs. Ω_{AM1}. There is a good linear relationship between these, which indicates that our empirical correlation is not sensitive to the choice of semiempirical parameterization for the wavefunction.

In Figure 2 we plot Ω_{AM1} vs. β for 1,4-disubstituted benzenes, 1,β-disubstituted styrenes, 4,4′-disubstituted stilbenes, and 4,4′-disubstituted diphenylacetylenes. Clearly (nonlinear) correlations exist between Ω and β for each series. The curves fitted in Figure 2 are quadratic, of the form,

$$\beta = A_1\,\Omega^2 + A_2\,\Omega + A_3 \tag{12}$$

Table 1. Experimental Values of β[a] and Calculated Values of $R_{HL}(-1)$, $(E_H - E_L)^2$ Using the Semiempirical AM1 Parameterization

Donor	*Acceptor*	β (10^{-30}esu)	$R_{HL}(-1)$	$(E_H - E_L)^2$
		1,4-Disubstituted Benzenes		
CN	Cl	0.8	43.1	84.55
CN	Me	0.7	44.4	86.78
CN	NH_2	3.1	48.2	74.84
CN	NMe_2	5.0	50.3	71.16
CN	OMe	1.9	45.4	81.95
CN	OPh	1.2	44.5	78.01
COH	Me	1.7	49.1	85.98
COH	NMe_2	6.3	56.6	69.56
COH	OMe	2.2	50.7	80.84
COH	OPh	1.9	49.5	76.96
NO_2	Me	2.1	48.4	85.76
NO_2	NH_2	9.2	57.2	71.66
NO_2	NMe_2	12.0	59.9	67.16
NO_2	OH	3.0	50.5	81.13
NO_2	OMe	5.1	51.3	79.81
NO_2	OPh	4.0	50.5	75.30
		4,4′-Disubstituted Stilbenes		
CN	NMe_2	36	54.6	52.70
CN	OH	13	52.0	58.14
CN	OMe	19	52.3	57.74
NO_2	Br	14	53.1	57.90
NO_2	OMe	28	55.7	56.01
NO_2	Me	15	53.8	56.81
NO_2	NH_2	40	56.3	50.76
NO_2	NMe_2	73	57.3	48.74
NO_2	OH	17	54.4	55.26
NO_2	OPh	18	40.2	38.57
		4,β-Substituted Styrenes		
CN	NMe_2	23	55.6	60.26
CN	OMe	7.0	51.6	68.54
COH	Br	6.5	51.8	70.34
COH	OMe	11	53.7	67.70
COH	NMe_2	30	57.8	59.10
NO_2	NMe_2	50	60.3	56.32
NO_2	OH	18	54.9	66.55
NO_2	OMe	17	55.4	65.74

(continued)

Table 1. (Continued)

Donor	*Acceptor*	β (10^{-30}esu)	$R_{HL}(-1)$	$(E_H - E_L)^2$
		4,4′-Disubstituted Diphenylacetylenes		
CN	NH_2	20	65.0	56.23
CN	NHMe	27	65.4	54.56
CN	NMe_2	29	65.8	53.67
NO_2	OMe	14	65.0	57.34
NO_2	Br	10	64.9	61.18
NO_2	NH_2	40, 24[b]	65.6	51.97
NO_2	NHMe	46	66.1	50.11
NO_2	NMe_2	46	66.4	49.14

Notes: [a] Cheng et al., 1991

[b] Two distinct experimental results are obtained for the diphenylacetylene A = NO_2, D = NH_2 when measured in different solvents.

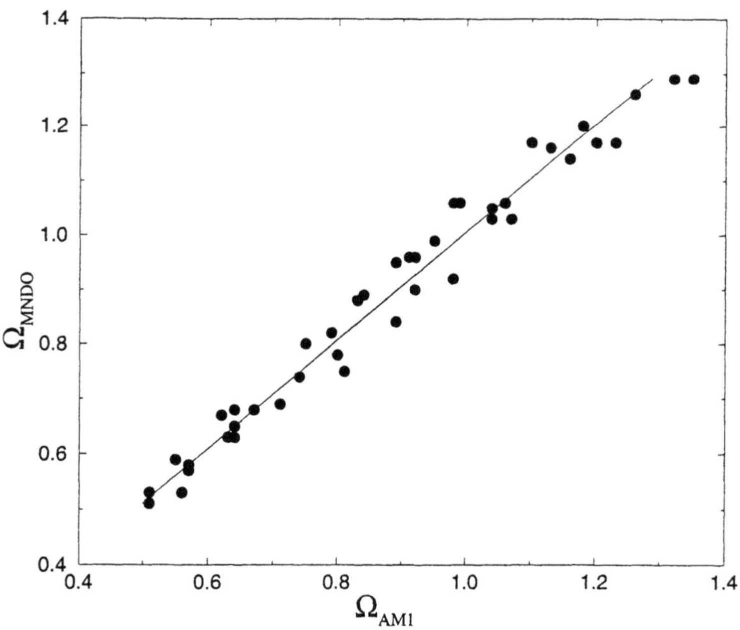

Figure 1. Ω_{MNDO} (values taken from Measures et al., 1995) plotted against Ω_{AM1} which was calculated according to Eq. 11 using the values listed in Table 1.

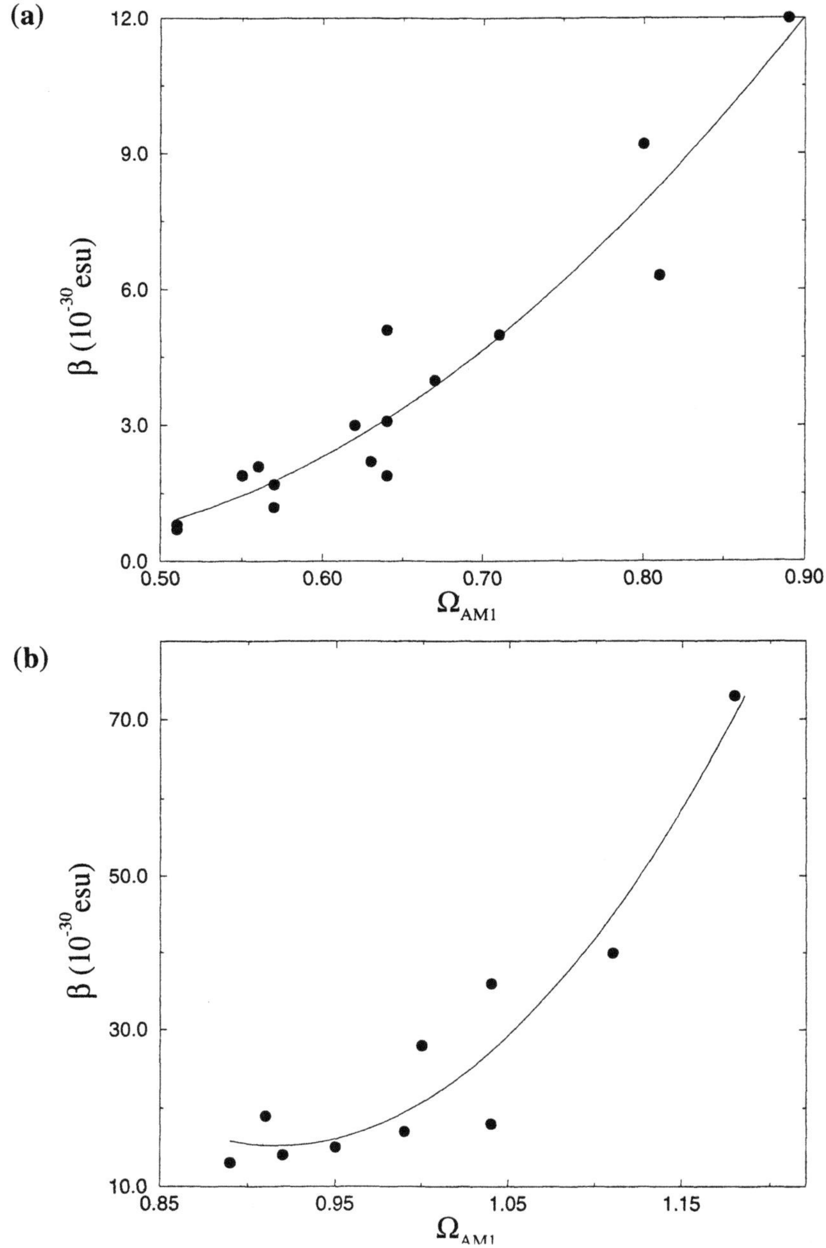

(continued)

Figure 2. Experimentally determined β (in 10^{-30} esu) (Cheng et al., 1991) versus calculated values of Ω_{AM1} (defined in Eq. 11) for (**a**) 1,4-disubstituted benzenes, (**b**) 4,4′-disubstituted stilbenes, (**c**) 4,β-substituted styrenes and (**d**) 4,4′-disubstituted diphenylacetylenes. The two point marked * in (**d**) are for A = NO_2 and D = NH_2 in different solvents. Details of the fitted curves are given in Table 2.

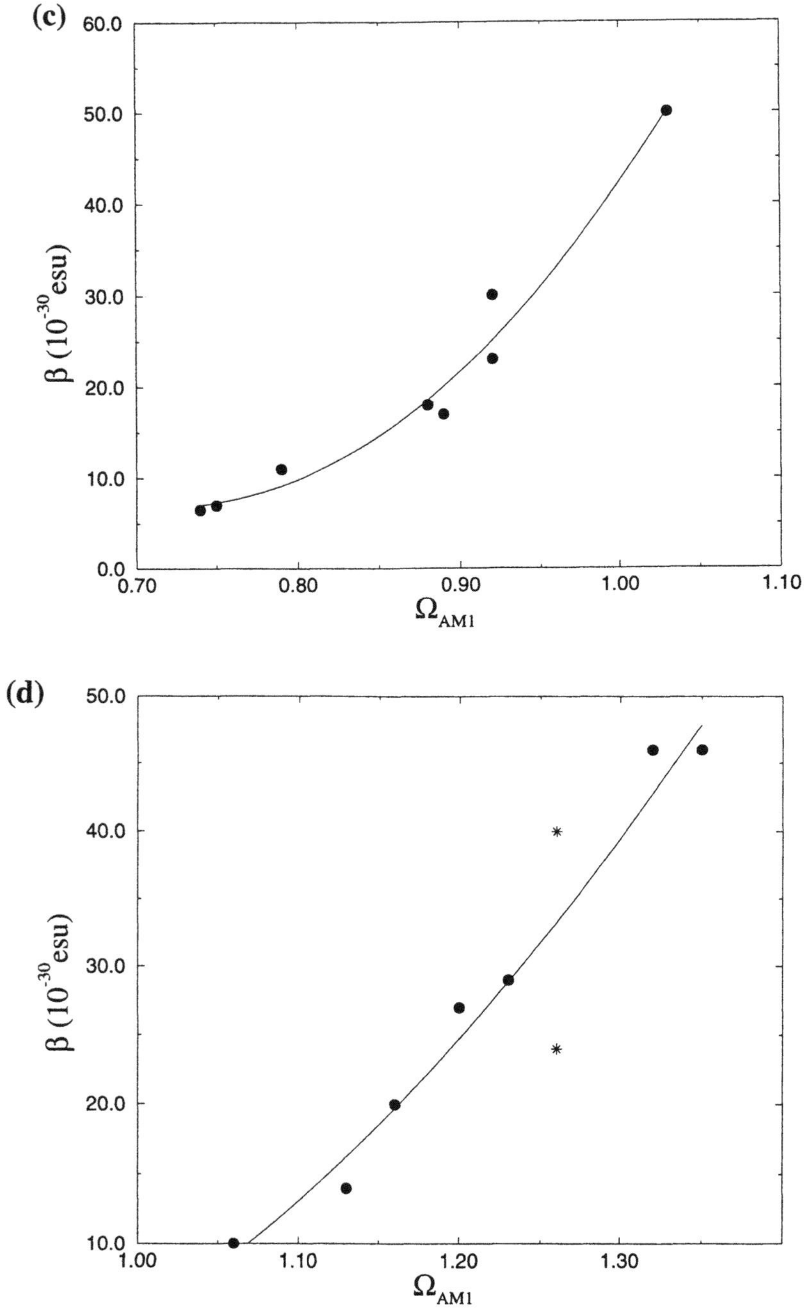

Figure 2. (Continued)

Table 2. Coefficients A_1, A_2 and A_3 and RMS Deviations for Quadratic Fits of the Form $\beta = A_1X^2 + A_2X + A_3$ for $X = \Omega_{AM1}$ or $(E_H - E_L)^{-2}$

X	A_1	A_2	A_3	RMS
		1,4-Disubstituted Benzenes		
$(E_H - E_L)^{-2}$	135.7	−22735.5	963944	1.41
Ω_{AM1}	6.5	−33.3	43.8	0.89
		4,4′-Disubstituted Stilbenes		
$(E_H - E_L)^{-2}$	−1090.1	105831	-2.43×10^6	7.81
Ω_{AM1}	689.4	−1470.3	801.6	5.26
		4,β-Substituted Styrenes		
$(E_H - E_L)^{-2}$	448.5	−64196.7	2.34×10^6	3.46
Ω_{AM1}	233.0	−631.7	441.0	2.30
		4,4′-Disubstituted Diphenylacetylenes		
$(E_H - E_L)^{-2}$	124.5	−20272.2	809315	4.27
Ω_{AM1}	79.5	223.2	148.0	4.16

and the coefficients $A_1, A_2,$ and A_3 are listed in Table 2. For all the series of molecules these correlations are more successful than the analogous quadratic fits between β and $(E_H - E_L)^{-2}$ (see Table 2).

The effects of substituting donor and acceptor groups at the two ends of a two-state nondipolar model system can be treated to a first approximation using perturbation theory, as is common in the frontier orbital approach. The two states of the new system, H' and L', can be expressed as linear combinations of the unperturbed states, H and L, with mixing coefficient C. Within this model, the hyperpolarizability of the new system can be expressed as a function of C, of matrix elements involving wavefunctions of the unperturbed states, and of the difference in energy between H' and L'. The difference, $R_{H'L'} - R_{HL}$, is also a function of C, and so it seems reasonable to seek relationships of the general form:

$$\beta = \frac{f(R_{H'L'} - R_{HL})}{(E_{H'} - E_{L'})^2} \tag{13}$$

This type of argument has prompted us to investigate relationships of the form of Eq. 13 for each of our series of molecules. The acceptor and donor groups are viewed, crudely, as a perturbation to the bridging molecule (i.e. the bridging framework plus hydrogen atoms at each end). $R_{H'L'}(-1)$ is calculated exactly as before, in the spirit of the two-state model, considering only contributions from basis functions associated with the bridging framework, and $R_{HL}(-1)$ is calculated using the frontier orbitals of the bridging molecule. In Figure 3 we plot

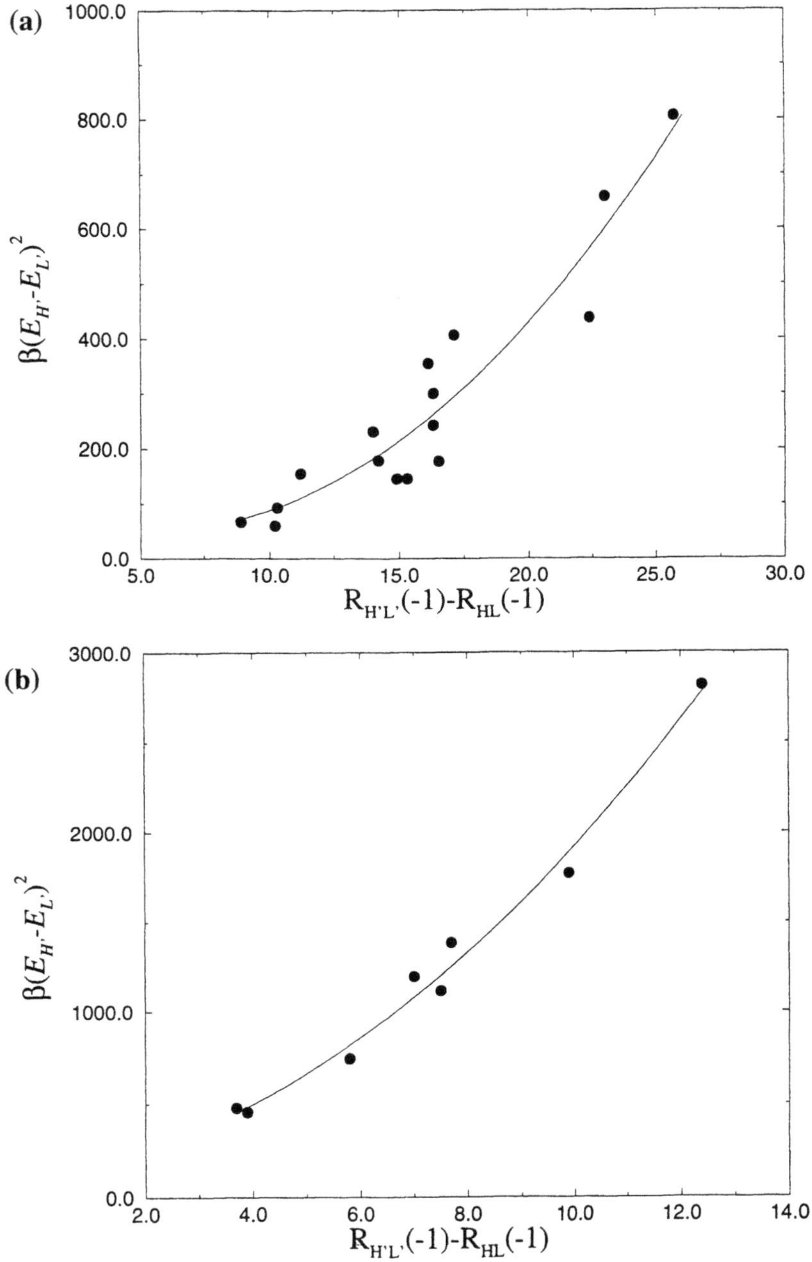

Figure 3. Plots of $\beta(E_{H'} - E_{L'})^2$ versus $R_{H'L'}(-1) - R_{HL}(-1)$ for (**a**) 1,4-disubstituted benzenes and (**b**) 4,β-substituted styrenes. β, $(E_{H'} - E_{L'})^2$ and $R_{H'L'}(-1)$ all correspond to the values given in Table 1; the different $R_{HL}(-1)$ are presented in Table 3.

Table 3. Coefficients A_1, A_2 and A_3 and RMS Deviations for Cubic Fits of the Form:[a]

$$\beta = \frac{A_1X + A_2X^2 + A_3X^3}{(E_{H'} - E_{L'})^2} \qquad X = R_{H'L'}(-1) - R_{HL}(-1)$$

$R_{HL}(-1)$	A_1	A_2	A_3	*RMS*
		1,4-Disubstituted Benzenes		
34.2	4.26	0.21	0.32	0.87
		4,4′-Disubstituted Stilbenes		
49.1	63.01	27.69	1.33	6.78
		4,β-Substituted Styrenes		
47.9	104.49	3.55	0.49	1.34
		4,4′-Disubstituted Diphenylacetylenes		
62.4	1434.01	-1048.8	266.26	5.64

Note: [a] The values of $R_{H'L'}(-1)$ and $(E_{H'} - E_{L'})^2$ are the corresponding quantities reported in Table 1 and the $R_{HL}(-1)$ are as listed below.

$\beta(E_{H'} - E_{L'})^2$ vs. $R_{H'L'} - R_{HL}$ for the benzene and styrene series using AM1 densities. The fitted curves shown are for a cubic polynomial in $R_{H'L'} - R_{HL}$ restricted to pass through the origin. The RMS deviations in β listed in Table 3 suggest that these fits are an improvement over those given earlier, based on Eq. 12. Results for the stilbene and diphenylacetylene series are slightly worse than those presented earlier (Table 2). We note that Cheng et al.[21] have concluded from a comparison of the experimental values of β and the positions of peaks in the UV spectra that the two-state model is more applicable to benzene derivatives than to stilbene derivatives. Our results are relatively insensitive to the type of semiempirical wavefunction used (AM1 or MNDO).

IV. CLUSTER ANALYSIS

If the similarity of each pair of a set of N molecules, S_{ij} ($i = 1...N, j = 1...N$) is calculated, an $N \times N$ *similarity matrix* is obtained. In general, we wish to analyze this matrix in such a way that molecules are grouped into clusters according to their similarity. The overall aim, of course, is that the molecules in each cluster should exhibit like behavior. In simple cases, where the divisions between similar and dissimilar species are clear-cut, the clustering of molecules into groups can be carried out by eye. However, in many other cases, such an approach is far from straightforward. Many different methods have been developed for scrutinizing similarity matrices, under the general title of cluster analysis.[25] Such techniques include mapping methods, similarity trees, hierarchical clustering methods, partitioning schemes, density search techniques, and clumping procedures.

We have now applied two methods of cluster analysis to a momentum–similarity matrix. In the proceedings of the First Girona Seminar[13] we presented a similarity–activity relationship for a series of phospholipid HIV1 inhibitors, all of which possess the general formula,

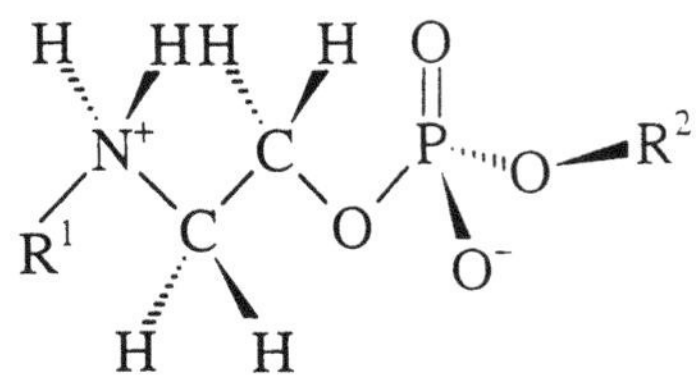

where the molecules possessing different R^1 and R^2 are listed in Table 4 together with their respective mnemonics and activities (ED_{50} values from Cooper et al.[26]) A low ED_{50} value indicates high activity. The similarity matrix obtained using momentum–space total densities and the index $T_{AB}(-1)$ is given in Table 5.

The first method of cluster analysis that we investigate here is a clumping technique. In this procedure we select two clusters, A and B, which can overlap, allowing some molecules to reside in both. The preferred outcome is that the molecules belonging to *both* clusters should exhibit intermediate activity, whereas those species only in A should be active and those only in B inactive, or vice versa. The optimum clustering is determined using a variation on the method proposed by Needham.[27] Given two clusters, A and B, the quantities Γ_{AA}, Γ_{BB}, and Γ_{AB} are calculated according to:

Table 4. Experimental ED_{50} Values for the Inhibition of HIV1 in C8166 T-Lymphoblastoid Cells for a Series of Phospholipids

Mnemonic	R^1	R^2	ED_{50} (μM)
HX1	methyl	*n*–hexyl	> 200
DD1	methyl	*n*–dodecyl	> 200
OD1	methyl	*n*–octadecyl	25
EG1	methyl	ethyl glycolate	110
OL1	methyl	oleyl	10
HX2	*t*–butyl	*n*–hexyl	40
DD2	*t*–butyl	*n*–dodecyl	10
OD2	*t*–butyl	*n*–octadecyl	3
EG2	*t*–butyl	ethyl glycolate	200
OL2	*t*–butyl	oleyl	3
HX3	hydrogen	*n*–hexyl	> 200
DD3	hydrogen	*n*–dodecyl	4
OD3	hydrogen	*n*–octadecyl	3.5
OL3	hydrogen	oleyl	0.5

Table 5. Similarities for Each Pair of Phospholipids Calculated Using Momentum-Space Total Densities and the Index $T_{AB}(-1)$

	DD1	*DD2*	*DD3*	*OD1*	*OD2*	*OD3*	*OL1*	*OL2*	*OL3*	*HX1*	*HX2*	*HX3*	*EG1*
DD2	97.5												
DD3	99.9	95.5											
OD1	92.5	98.0	89.3										
OD2	85.7	94.1	82.1	98.3									
OD3	94.6	99.5	91.7	99.7	97.1								
OL1	93.3	98.4	90.3	99.9	97.9	99.9							
OL2	86.8	94.7	83.2	98.7	99.8	97.6	98.4						
OL3	95.2	99.4	92.6	99.4	96.7	99.6	99.6	97.4					
HX1	86.4	76.5	89.7	67.7	60.6	70.7	68.9	61.0	71.8				
HX2	96.8	90.0	98.4	82.0	74.4	84.8	83.1	75.5	86.0	95.4			
HX3	80.6	70.3	84.3	61.7	54.4	64.6	62.8	55.3	62.8	99.3	91.2		
EG1	76.0	65.8	79.9	57.4	50.4	60.1	58.4	51.4	61.3	97.5	87.4	99.4	
EG2	91.3	82.5	94.1	73.7	66.0	76.7	74.9	67.0	77.9	99.0	98.3	96.6	94.0

$$\Gamma_{XY}=\sum_{i\in X}\sum_{j\in Y} S_{ij} \qquad XY = AA,\ BB, \text{ or } AB \tag{14}$$

Varying the members of A and B, but forbidding their total union, we search for the global minimum of $G(\kappa)$, where:

$$G(\kappa) = \frac{\Gamma_{AB}}{(\Gamma_{AA}\Gamma_{BB})^{\kappa}} \tag{15}$$

The power κ, which lies in the range $\frac{1}{2} \le \kappa \le 1$, is included to influence the size of the intersection. If κ is large, $G(\kappa)$ is dominated by the value of $\Gamma_{AA}\Gamma_{BB}$, favoring large intersections.

The second method that we investigate here is a "density search" technique, as proposed by Carmicheal et al.[28,29] A cluster is initiated by finding the two most similar molecules, x and y. A third molecule, z, is then selected by finding the maximum value of S_{xz} or S_{yz}. A decision is now made as to whether z really belongs to this cluster:

- The average similarity of the cluster containing x and y is subtracted from twice the average similarity of the proposed cluster containing all three molecules.
- If this value is greater than a specified tolerance τ, the molecule z is accepted into the cluster and a new molecule, i, is then chosen by finding the maximum of S_{ix}, S_{iy}, or S_{iz}, and it is then judged for suitability by the same criterion.
- If a molecule is not accepted into an existing cluster, then a new cluster is started by finding the highest similarity between molecules *not* already assigned to clusters.

The process continues until all the molecules have been assigned. Unlike the clumping technique, the number of clusters is not fixed beforehand and it is a function of the tolerance τ.

In previous work,[14] we clustered the similarity matrix for the phospholipids by eye, having first replaced the numerical values of the index $T_{AB}(-1)$ by different

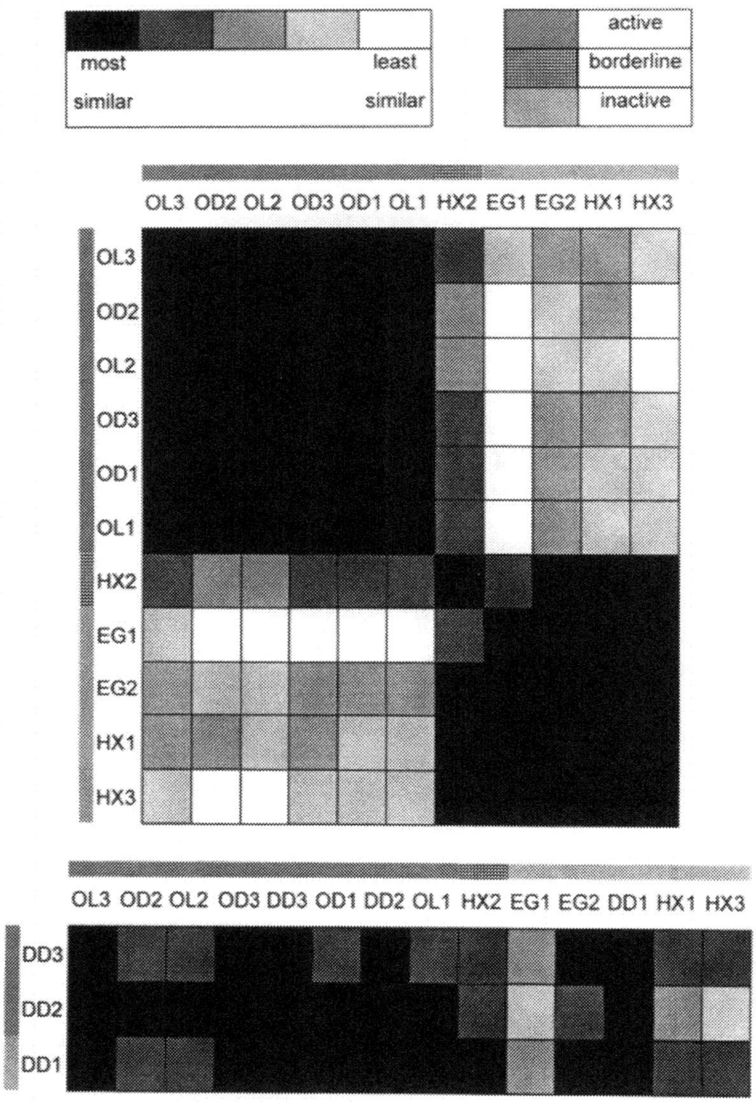

Figure 4. Visually clustered phospholipid similarity matrix with $T_{AB}(-1)$ values replaced by varying degrees of shading (see text).

degrees of shading, as shown in Figure 4. Black denotes very high similarity ($\geq 91\%$) and white denotes low momentum–space similarity ($\leq 60\%$). The bars next to the labels indicate that the biological activity is high ($ED_{50} \leq 10$ μM), intermediate, or low ($ED_{50} \geq 110$ μM). The most active molecules are located in the top left-hand corner of the figure. Clearly, the active molecules are very similar to each other and they are dissimilar to the inactive molecules, and vice versa. HX2 shows intermediate similarity and activity. However, the DD molecules (DD1 inactive, DD2, and DD3 active), shown separately in Figure 4, appear to be very similar to all of the active species *and* to some of the inactive species. We suggested previously that an experimental redetermination of the activities of the DD molecules could be worthwhile. Our purpose here is to examine whether our two numerical clustering techniques produce the same results as those produced by visual clustering.

With the clumping technique, the two clusters formed when $\kappa \geq 0.62$ are:

Cluster A: DD1, DD2, DD3, OD1, OD3, OL1, OL2, OL3, HX1, HX2, HX3, EG1, EG2

Cluster B: DD1, DD2, DD3, OD1, OD2, OD3, OL1, OL2, OL3, HX1, HX2, HX3, EG2

All the molecules belong to the intersection of the two clusters except EG1 ($ED_{50} = 110$ μM) and OD2 ($ED_{50} = 3$ μM). This suggests that these two molecules (EG1 and OD2) would show the most extreme behavior in the series, with the other molecules displaying intermediate activity. This hypothesis is clearly false, given the actual activities of these species. Analyzing the similarity matrix using a lower value of $\kappa \leq 0.61$, produces clusters that relate much more straightforwardly to the activities. Clusters A and B now consist of the following molecules:

Cluster A: DD1, DD2, DD3, OD1, OD2, OD3, OL1, OL2, OL3

Cluster B: HX1, HX2, HX3, EG1, EG2

and turn out to be mutually exclusive. It is clear that cluster B consists of molecules which display ED_{50} values ≥ 40 μM while cluster A, with the exception of DD1, contains molecules with $ED_{50} \leq 25$ μM.

Applying the "density search" technique to the phospholipid similarity matrix at $\tau = 0.8$ produced two clusters:

Cluster A: DD1, DD3, HX2, EG2, HX1, HX3, EG1

Cluster B: OD1, OL1, OD3, OL3, DD2, OL2, OD2

These two clusters model adequately the actual activities, given that the second cluster is composed of species with ED_{50} values ≤ 25 μM and the first cluster contains molecules which have $ED_{50} \geq 40$ μM, with the exception of DD3. Increasing τ to 0.95, the first cluster splits into two to give a total of three clusters:

Cluster A: DD1, DD3, HX2

Cluster B: OD1, OL1, OD3, OL3, DD2, OL2, OD2

Cluster C: HX3, EG1, HX2, EG2

As before, the second cluster consists of the active species ($ED_{50} \leq 25$ μM). The third cluster collects mostly inactive species ($ED_{50} \geq 110$ μM). However, the first cluster consists of an active molecule, DD3, an inactive molecule, DD1, and HX2 which has an ED_{50} value of 40 μM, suggesting that DD1 and DD3 might display intermediate activity (between 25 and 110 μM).

For certain input parameters (κ for the clumping technique and τ for the "density search" technique), both procedures give results in broad agreement with the visual approach we employed previously. DD2 is correctly predicted to be active in these cases. However, there is no consistency in the results for DD1 and DD3, about which we were able to make no conclusions from visual clustering. Definitive experimental values for the DD molecules would be very useful in assessing the merits of the different approaches.

V. NUCLEOTIDES

In this section we consider a further set of molecules that inhibit the HIV1 virus, namely a series of nine nucleotides with general formula:

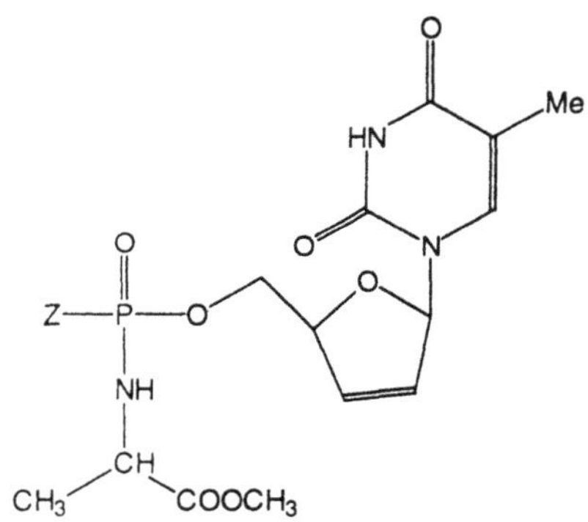

The different Z groups are listed in Table 6, together with their individual molecular labels and, for molecules 1–7, biological activities.[30] The activities of molecules 8 and 9 had not yet been determined when we received the data (ED_{50} values).

Molecular similarity concepts are particularly helpful in situations such as these where the inhibition mechanism is not completely understood. In view of the size of the molecules we generated computationally inexpensive semiempirical MNDO wavefunctions. As was the case in our previous work on phospholipids, no search for the global minimum conformation was carried out, but full geometry optimizations were performed starting from a consistent geometry for the common framework.

Comparing the total densities for the complete molecules, the values of $R_{AB}(n)$ and $T_{AB}(n)$ are very high (>96%) for all pairs of molecules. In such situations, it

Table 6. Z Groups of the Nine Nucleotides and their Respective ED_{50} Values

Molecular Label	*Z*	ED_{50} (μM)
1	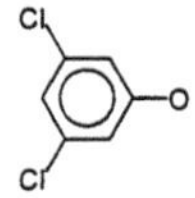	0.06
2	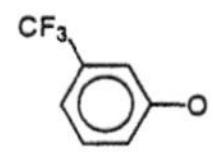	0.08
3	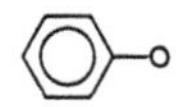	0.085
4	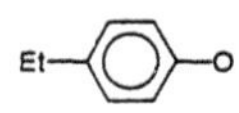	0.2
5	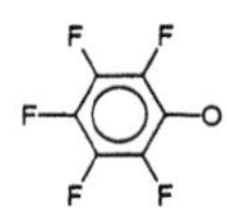	2.5
6	Et—O	20
7	Pr—O	100
8	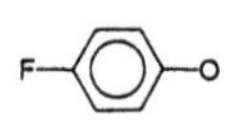	—
9	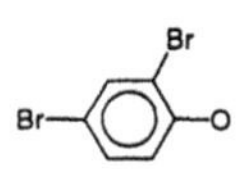	—

may be more useful to examine momentum–space dissimilarity indices such as $D_{AB}(n)$ (Eq. 6).

First of all, we introduce a new family of dissimilarity indices which should be appropriate for situations in which the shape of the densities is particularly important. This index, which we denote $P_{AB}(n)$, is defined according to,

$$P_{AB}(n) = 100\,\frac{(I_{AA}(n)I_{BB}(n))^{\frac{1}{2}} - I_{AB}(n)}{N_A N_B} \tag{16}$$

in which:

$$N_X = \int \rho_X(\mathbf{p})\, d\mathbf{p} \qquad X = A, B \tag{17}$$

As in the case of $D_{AB}(n)$, the index $P_{AB}(n)$ takes values from zero upwards. In the special case that $\rho_A(\mathbf{p}) = m \times \rho_B(\mathbf{p})$, $P_{AB}(n)$ is invariant to the choice of nonzero m, whereas $D_{AB}(n)$ is not. It is in this sense that values of $P_{AB}(n)$ are determined more by the shape of the momentum–space electron densities than are values of $D_{AB}(n)$.

In the present work, we evaluate $D_{1X}(n)$ and $P_{1X}(n)$ (*for* $n = 0,-1$) for comparisons of the most active molecule (molecule 1) with each of the other nucleotides, matching as closely as possible the positions of the nuclei in the thyamine group and the position of the phosphorous atom. The results of these calculations are listed in Table 7. Clearly, $P_{1X}(-1)$ provides the best relationship between dissimilarity and the ED_{50} values, although the activity of molecule 6 is predicted to be too high, relative to those of molecules 4 and 5. Figure 5 shows separately molecules 4 and 6 superimposed on molecule 1. The overlay between the amino groups in molecules 1 and 4 is noticeably poorer than that between molecules 1 and 6. The same is true if molecule 4 is replaced by molecule 5. This appears to suggest that the variation of $P_{1X}(-1)$ in the comparisons of 1 with 4, of 1 with 5, and of 1 with 6 is dominated by conformational differences rather than the chemical composition of the group Z. These conformational differences might not be important in determining the biological activity.

An alternative is to compare *only* the fragments Z. With this in mind, we replaced the P atom and its substituents by H. We denote the resulting alcohols derived from molecules 1 . . . 9 with the corresponding letters of the alphabet, "a . . . i" (see Table 8). MNDO wavefunctions were used to investigate the similarities between these alcohols. The momentum–space dissimilarity measures, $P_{ax}(n)$, for n values of -1, $-\frac{3}{4}$, $-\frac{1}{2}$, $-\frac{1}{4}$, and 0 were calculated using the total electron density for alcohol "a" (derived from molecule 1) and each of the other alcohols (x = b . . . i). These

Table 7. Dissimilarity Indices $D_{1X}(n)$ and $P_{1X}(n)$ ($n = 0,-1$) Calculated Using Total Momentum–Space Electron Densities for Molecule 1 and for Each of the Other Eight Molecules

X	$D_{1X}(-1)$	$D_{1X}(0)$	$P_{1X}(-1) \times 10^2$	$P_{1X}(0) \times 10^2$
1	0	0	0	0
2	88.1	104.2	0.12	0.14
3	529.0	538.4	0.19	0.10
4	375.8	158.7	0.28	0.18
5	356.1	405.1	0.30	0.49
6	3664.2	3188.0	0.22	0.14
7	314.9	349.5	0.37	0.26
8	425.9	404.3	0.12	0.07
9	139.4	102.5	0.07	0.08

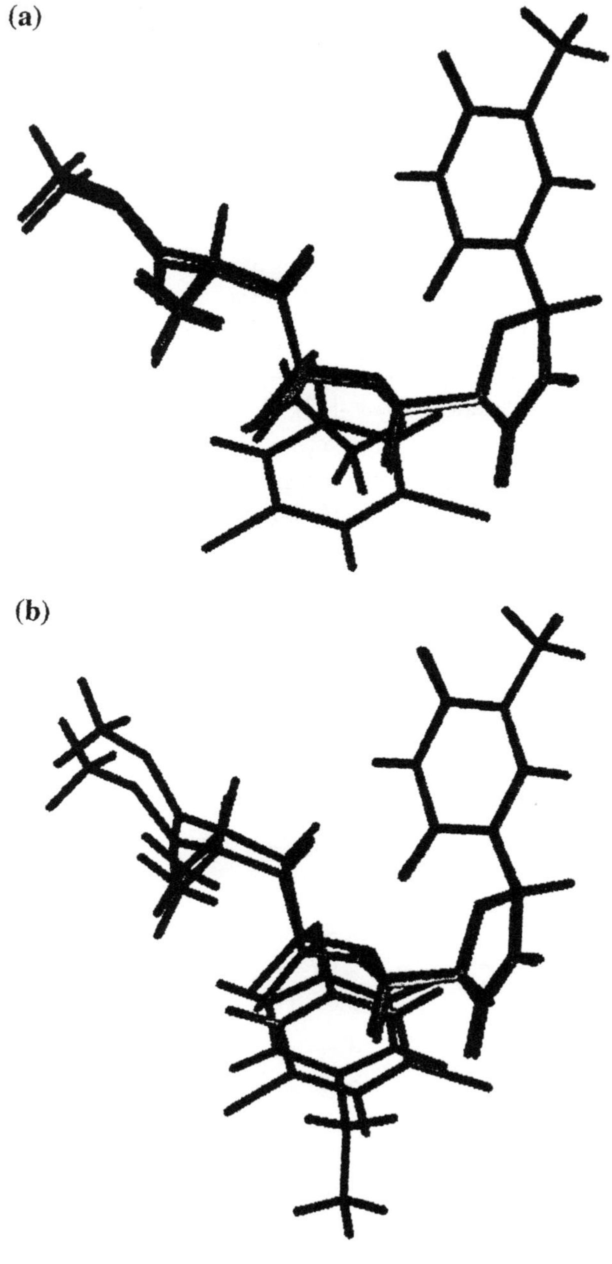

Figure 5. Nucleotide molecule 1 (grey) superimposed (**a**) on molecule 6 (black) and (**b**) on molecule 4 (black).

Table 8. $P_{ax}(n)$ (× 10^2) Values Calculated Using Total Densities for Alcohol a and Each of the Other Alcohols (x = b . . . i)[a]

Molecule–Alcohol	−1	−¾	−½	−¼	0
1 – a	0	0	0	0	0
2 – b	1.44	1.42	1.46	1.56	1.72
3 – c	3.40	2.85	2.42	2.07	1.80
4 – d	3.91	3.40	3.00	2.67	2.42
5 – e	3.61	4.06	4.57	5.16	5.82
6 – f	5.93	5.24	4.68	4.25	3.91
7 – g	7.49	6.62	5.96	5.46	5.08
8 – h	1.79	1.49	1.26	1.11	1.01
9 – i	0.84	0.88	0.92	0.97	1.03

Note: [a] Dissimilarities are evaluated for *n* values of −1, −¾, −½, −¼, and 0.

molecules were superimposed by overlaying the C–O–H groups in each molecule, matching the positions of the C, O, and H atoms as closely as possible. The resulting dissimilarity indices are listed in Table 8 and they are shown graphically in Figure 6, where $P_{ax}(n)$ is plotted for different values of n. The effect here of altering the value of n in the p^n term in the generalized overlap (Eq. 3) is larger than that noted

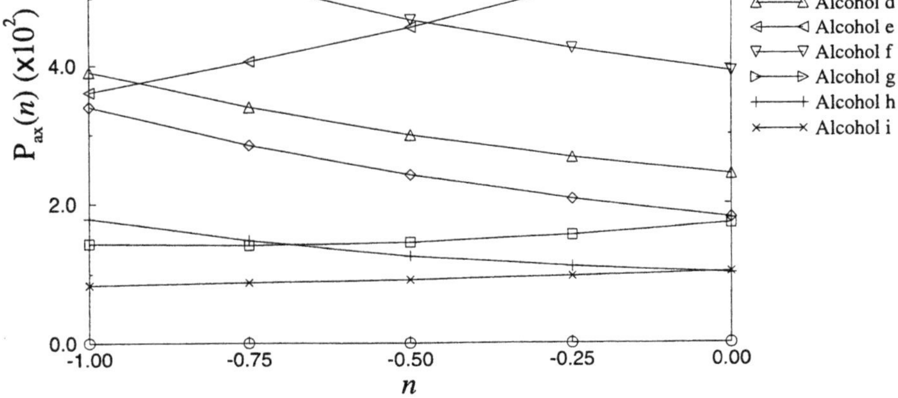

Figure 6. $P_{ax}(n)$ (× 10^2) values for $n = -1, -¾, -½, -¼$ and 0 for comparisons of alcohol a and the other eight alcohols (x = b . . . i).

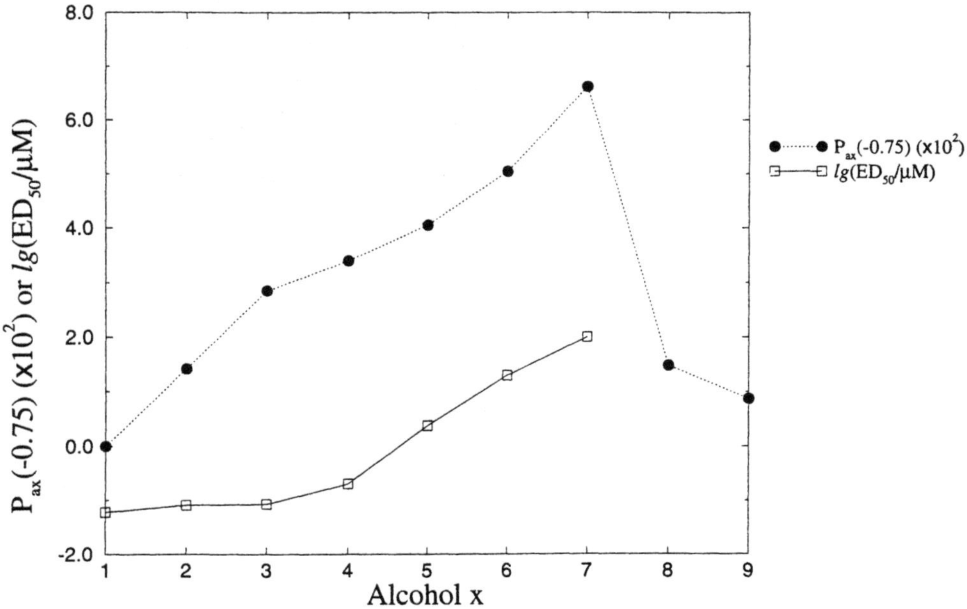

Figure 7. Values of $P_{ax}(-\frac{3}{4})$ ($\times$ 10^2) and $lg(ED_{50}/\mu M)$ for the different alcohols x.

in any of the examples in our previous work.[12,14] A good structure–activity relationship between the ED_{50} values and $P_{1X}(n)$ is found only when $-1 < n < -\frac{1}{2}$. Figure 7 plots $P_{ax}(-\frac{3}{4})$ together with $lg(ED_{50}/\mu M)$ for each alcohol. Considering the molecules with unknown activity (8 and 9), we predict the biological behavior of molecule 8 to be similar to that of molecule 2. Molecule 9 should show comparable activity to molecule 1.

To investigate this dissimilarity–activity relationship further, we have constructed the entire dissimilarity matrix using the $P_{AB}(-\frac{3}{4})$ values for each pair of fragments. These are given in Table 9 and are shown diagrammatically in Figure 8, in which the dissimilarities have been replaced by varying degrees of shading. Although the shading appears to follow the ED_{50} values reasonably well, it is difficult to cluster this matrix visually. Alcohol e is different from most of the other molecules. Furthermore, alcohols c, d, f, and g are all rather similar to each other, in spite of very different ED_{50} values.

To help with the analysis of the whole matrix, we used the clumping method and the density search technique, discussed in the previous section. Distance-like values of $P_{AB}(-\frac{3}{4})$ were first converted to similarity measures by selecting the maximum dissimilarity from Table 9, subtracting each dissimilarity index from this value, and then scaling these new quantities so that they lie in the range 0 to 100%.

Table 9. Dissimilarities for Each Pair of Alcohols Calculated Using Momentum–Space Total Densities and the Index $P_{AB}(-3/4)$ ($\times 10^2$)

	a	*b*	*c*	*d*	*e*	*f*	*g*	*h*
b	1.42							
c	2.85	2.21						
d	3.40	3.07	0.20					
e	4.06	1.72	7.89	9.80				
f	5.24	3.59	1.19	0.80	9.37			
g	6.62	5.08	1.66	0.99	11.91	0.48		
h	1.48	0.60	0.55	1.07	4.25	1.88	2.90	
i	0.88	3.14	2.48	2.46	10.31	3.10	5.33	2.13

Applying the clumping procedure, for $\kappa < 0.72$, alcohol e is separated from the other eight species; when $\kappa > 0.72$, only molecules g and e are not in the intersection. To gain further insight from this matrix, we chose to exclude alcohol e from the clustering procedure. We find three different domains:

$\kappa \leq 0.52$	$0.53 \leq \kappa \leq 0.75$	$\kappa \geq 0.76$
Cluster A: b, c, d, f, g, h	**Cluster A:** a, b, h, i	**Cluster A:** b, c, d, f, g, h, i
Cluster B: a, i	**Cluster B:** c, d, f, g	**Cluster B:** a, b, c, d, f, h, i

These various results suggest that molecules 1 and 7 differ most in activity, with molecules 2, 8, and particularly 9 showing comparable activity to molecule 1. Rather disappointingly, the alcohols c and d are never separated from f and g, and alcohols b and c (ED_{50} values of 0.08 and 0.085 μM) do not always cluster together. The scheme does, however, predict activities for molecules 8 and 9 that are consistent with those predicted earlier by considering the values of $P_{ax}(-3/4)$ (the first row of the matrix).

When the density search technique is used to analyze the matrix, the result most consistent with the biological data is obtained with a tolerance $\tau = 0.8$. This yields the following clustering:

Cluster A: c, d, h, b

Cluster B: f, g

Cluster C: a, i

Cluster D: e

Again alcohol e is in its own separate cluster and molecule 9 is predicted to behave in a similar fashion to molecule 1. Molecule 8 is predicted to show comparable activity to molecule 2 and, in this case, also to species 3 and 4.

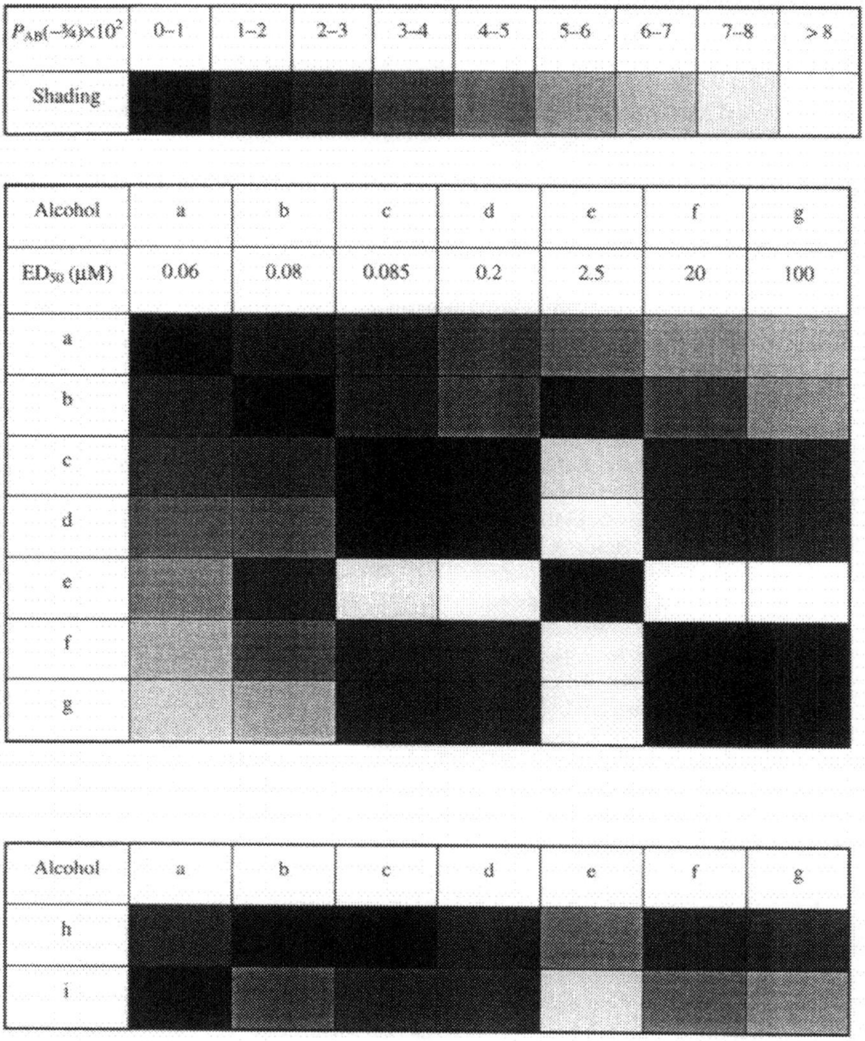

Figure 8. Visually clustered nucleotide derivatives dissimilarity matrix with values of $P_{AB}(-^3\!/_4)$ ($\times\ 10^2$) replaced by varying degrees of shading.

All of our methods of analysis suggest that molecule 9 should have high activity. Subsequent to our work, molecule 9 was shown experimentally to have an extremely high activity ($ED_{50} = 0.04\ \mu M$). This success suggests that our momentum–space approach can be effective even in situations where the data is relatively sparse and/or where the molecules appear to be very similar indeed.

VI. CONCLUSIONS

Momentum–space similarity techniques allow us to rationalize physical properties and biological activities. In this chapter we have presented several examples of structure–activity relationships based on momentum–space quantities for the molecular hyperpolarizabilities of series of conjugated systems and for the HIV1 inhibition of series of both phospholipids and nucleotides. Momentum–space indices can be particularly useful when the property or activity appears to have no obvious dependence on the bonding topology of the molecules, or the nature of the atomic backbone, but is more sensitive instead to the variation of the long-range valence electron density.

REFERENCES

1. Johnson, M.A.; Maggiora, G.M., Eds. *Concepts and Applications of Molecular Similarity*; Wiley: New York, 1990.
2. Johnson, M.A.; Maggiora, G.M. *J. Chem. Inf. Comput. Sci.* **1992**, *32*, 577.
3. Carbó, R.; Leyda, L.; Arnau, M. *Int. J. Quantum Chem.* **1980**, *17*, 1185.
4. Carbó, R.; Domingo, Ll. *Int. J. Quantum Chem.* **1987**, *32*, 517.
5. Carbó, R.; Calabuig, B. *Int. J. Quantum Chem.* **1992**, *42*, 1681.
6. Ponec, R.; Strnad, M. *J. Phys. Org. Chem.* **1991**, *4*, 701.
7. Ponec, R.; Strnad, M. *Int. J. Quantum Chem.* **1992**, *42*, 501.
8. Hodgkin, E.E.; Richards, W.G. *Int. J. Quantum Chem., Quantum Biol. Symp.* **1987**, *14*, 105.
9. Richards, W.G.; Hodgkin, E.E. *Chem. Br.* **1988**, *24*, 1141.
10. Burt, C.; Richards, W.G. *J. Comput.-Aided Mol. Design* **1990**, *4*, 231.
11. Walker, P.D.; Arteca, G.A.; Mezey, P.G. *J. Comput. Chem.* **1991**, *12*, 220.
12. Allan, N.L.; Cooper, D.L. In *Molecular Similarity*; Sen, K.D., Ed.; *Topics in Current Chemistry* **1995**, *173*, 85.
13. Cooper, D.L.; Allan, N.L. In *Molecular Similarity and Reactivity: From Quantum Chemical to Phenomenological Approaches*; Carbó, R., Ed.; Kluwer Academic Publishers: Netherlands, 1995, p. 31.
14. Measures, P.T.; Mort, K.A.; Allan, N.L.; Cooper, D.L. *J. Comput.-Aided Mol. Design* **1995**, *9*, 331.
15. Bloembergen, N. *Nonlinear Optics*; W.A. Benjamin: New York, 1965.
16. Shen, Y.R. *The Principles of NonLinear Optics*; Wiley: New York, 1984.
17. Boyd, R.W. *Nonlinear Optics*; Academic Press: New York, 1992.
18. Prasad, N.P.; Williams, D.J. *Introduction to Nonlinear Optics in Molecules and Polymers*; Wiley: New York, 1991.
19. Marder, S.R.; Sohn, J.E.; Stucky, G.D., Eds.; *Materials for Nonlinear Optics: Chemical Perspectives*; ACS Symposium Series 455, American Chemical Society: Washington DC, 1991.
20. Marks, T.J.; Ratner, M.A. *Ange. Chemie* **1995**, *34*, 155.
21. Cheng, L.; Tam, W.; Stevenson, S.H.; Meredith, G.R.; Rikken, G.; Marder, S.R. *J. Phys. Chem.* **1991**, *95*, 10631.
22. Steigman, A.E.; Graham, E.; Perry, K.J.; Khundkar, L.R.; Cheng, L.; Perry, J.W. *J. Am. Chem. Soc.* **1991**, *113*, 7658.
23. Kanis, D.R.; Ratner, M.A.; Marks, T.J. *Chem. Rev.* **1994**, *94*, 195.
24. Stewart, J.J.P. *J. Comput.-Aided Mol. Design* **1990**, *4*, 1.
25. Everitt, B. *Cluster Analysis*; Heinemann Educational Books: London, 1974.
26. Cooper, D.L.; Mort, K.A.; Allan, N.L.; Kinchington, D.; McGuigan, C. *J. Am. Chem. Soc.* **1993**, *115*, 12615.

27. Needham, R.M. *The Statistician* **1967**, *19*, 45.
28. Carmicheal, J.W.; George, J.A.; Julius, R.S. *Syst. Zool.* **1968**, *17*, 144.
29. Carmicheal, J.W.; Sneath, P.H.A. *Syst. Zool.* **1969**, *18*, 402.
30. Kinchington, D.; McGuigan, C., private communication (1995).

MOLECULAR SIMILARITY MEASURES OF CONFORMATIONAL CHANGES AND ELECTRON DENSITY DEFORMATIONS

Paul G. Mezey

Advances in Molecular Similarity
Volume 1, pages 89–120

ISBN: 0-7623-0131-7

I. INTRODUCTION

From the fundamental, quantum mechanical description of similarity[1–8] to applied similarity studies[9–39] of special importance in pharmacological drug design, molecular similarity involves a diverse array of disciplines and methodologies. Two aspects of molecular similarity are of special importance: the similarity of nuclear arrangements,[40] and the similarity of electron density distributions.[29] A molecule can be regarded as an electron density distribution superimposed on a nuclear distribution, where these interacting distributions are dependent on each other. In this review special aspects of these distributions are discussed. The fundamental roles of additive fuzzy density fragmentation methods as tools for similarity analysis are described. Two similarity measures, one based on nuclear distributions, and another based on a generalization of the Hausdorff metric to fuzzy electron densities are discussed, and some properties of two families of tools of similarity analysis, T-hulls and interior T-aggregates, are described.

II. THE CONFORMATION OF NUCLEAR ARRANGEMENT AND THE SHAPE OF ELECTRON DENSITY

In conventional stereochemistry, the nuclear arrangement defines the molecular conformation. This arrangement can be specified using a set of internal coordinates of the N nuclei of the molecule; for example, in the simplest version of the Born–Oppenheimer approximation, $3N$-6 internal coordinates are used to specify the molecular conformation of a polyatomic ($N > 2$) molecule.

In general, nuclear arrangements can be described within a ($3N$-6)-dimensional, reduced nuclear configuration space M provided with an appropriate metric.[40] Whereas this space M is not a vector space, many of the familiar concepts of the ordinary 3D Euclidean space apply. The distance $d(K,K')$ between two nuclear configurations represented by points K and K' of the space M is a valid measure of the dissimilarity of the corresponding two nuclear arrangements.[40]

Whereas nuclear arrangements and the associated stereochemical bond structure are the more commonly used concepts for the identification and comparison of molecules, such "skeletal" models of molecules represent only a simplistic description of the wealth of molecular shape features. The actual fuzzy molecular body, as represented by the electronic density charge cloud surrounding the nuclei, is primarily responsible for bond creation and bond breaking in any chemical reaction. Also, the changes in the electronic density are important components in any conformational process. Since the electronic density fully reflects the nuclear distribution, and since in a molecule there is nothing else than nuclei and an electronic density cloud, the electronic density and its changes contain all the relevant chemical information about the molecule. In this context, the analysis of the shape of the electronic density is of fundamental importance in the study, comparison, and eventual understanding of molecular properties.

For a molecule of some fixed conformation K, the SCF LCAO ab initio electronic density $\rho(\mathbf{r})$ can be expressed in terms of the set of atomic orbitals, $\varphi_i(\mathbf{r})$ ($i = 1,2,\ldots, n$), serving as the basis for the expansion of the molecular wavefunction, where n is the number of orbitals and $\mathbf{r}$ is the three-dimensional position vector variable. Using the notation $\mathbf{P}$ for the $n \times n$ dimensional density matrix determined in the course of SCF calculation, the electronic density $\rho(\mathbf{r})$ is computed as:

$$\rho(\mathbf{r}) = \sum_{i=1}^{n} \sum_{j=1}^{n} \mathbf{P}_{ij}\, \varphi_i(\mathbf{r})\, \varphi_j(\mathbf{r}) \qquad (1)$$

The electronic density $\rho(\mathbf{r})$ is the fuzzy "body" of the electronic charge cloud, fully describing the shape of the molecule.

Detailed and quantum chemically rigorous shape analysis of electron density clouds is possible using "shape group methods", based on algebraic topological properties of molecular isodensity contour surfaces (MIDCOs). For a review of the shape group methods and the associated algebraic–topological computational techniques the reader is referred to a recent review (Ref. 29). For more details, the reader may consult the original Refs. 41–44.

III. ADDITIVE FUZZY ELECTRON DENSITY FRAGMENTATION (AFDF) METHODS

It is advantageous if electron density fragmentation schemes fulfill two, natural criteria; namely that the fragment densities are:

1. *additive*, and
2. boundaryless, *fuzzy* charge clouds analogous to those of complete molecules.

Condition 1 is a natural requirement if a fragmentation scheme is used to build electron densities for complete molecules. Condition 2 allows one to use shape analysis and other techniques developed for complete molecules, and it also eliminates the accumulation of local errors occurring if fragments with boundaries are transferred and used for building electron densities for large molecules. Some of the often severe problems with nonadditive schemes or schemes involving density fragments with boundaries have been discussed before,[45–48] and here we shall focus on schemes fulfilling both criteria.

Following the notations introduced earlier,[45] a generalized form of the additive, fuzzy Mulliken–Mezey scheme[34,39] can be given in terms of a subdivision of the set of nuclei of the molecule into mutually exclusive families. The simplest version of these schemes, the original Mulliken–Mezey scheme, is the basis of the MEDLA method of Walker and Mezey.[46,49–52] Although any subset of the nuclei can be declared as a nuclear family, it is advantageous if nuclei of a given family are located within a common region of the space.

If the molecule contains m mutually exclusive nuclear families,

$$f_1, f_2, \ldots, f_k, \ldots f_m \tag{2}$$

then the corresponding fragment density functions,

$$\rho^1(\mathbf{r}), \rho^2(\mathbf{r}), \ldots, \rho^k(\mathbf{r}), \ldots \rho^m(\mathbf{r}) \tag{3}$$

of m additive, fuzzy density fragments,

$$F_1, F_2, \ldots, F_k, \ldots F_m \tag{4}$$

can be computed using the $m_k(i)$ the membership functions of the AO basis functions $\varphi_i(\mathbf{r})$ in the set of AOs centered on a nucleus of family f_k:

$$m_k(i) = 1 \text{ if AO } \varphi_i(\mathbf{r}) \text{ is centered on one of the nuclei of set } f_k,$$

$$0 \text{ otherwise} \tag{5}$$

Using the Mulliken–Mezey scheme, the elements P^k_{ij} of the $n \times n$ fragment density matrix $\mathbf{P}^k$ for the k-th additive, fuzzy density fragment F_k is written as

$$\mathbf{P}^k_{ij} = 0.5\ [m_k(i) + m_k(j)]\ P_{ij} \tag{6}$$

Any of the generalized additive fuzzy density fragmentation schemes proposed[34,39,45] can also be formulated in terms of the membership functions $m_k(i)$, by taking Mezey's fragment density matrix as,

$$P^k_{ij} = [m_k(i)\ w_{ij} + m_k(j)\ w_{ji}]\ P_{ij} \tag{7}$$

where for the generalized weighting factors w_{ij} and w_{ji} the following condition holds:

$$w_{ij} + w_{ji} = 1 \tag{8}$$

The Mulliken–Mezey scheme corresponds to the choice of $w_{ij} = w_{ji} = 0.5$, and can be regarded as Mulliken's population analysis[53,54] without integration.

If this general scheme is applied for the construction of the fragment density matrix $\mathbf{P}^k$ for the k-th fragment, then the *fuzzy density fragment* $\rho^k(\mathbf{r})$ of the molecule can be calculated as:

$$\rho^k(\mathbf{r}) = \sum_{i=1}^{n} \sum_{j=1}^{n} P^k_{ij}\ \varphi_i(\mathbf{r})\ \varphi_j(\mathbf{r}) \tag{9}$$

As it is easily verified by substitution, the sum of Mezey's fragment density matrices $\mathbf{P}^k$ is equal to the density matrix $\mathbf{P}$ of the molecule,

$$\mathbf{P} = \sum_{k=1}^{m} \mathbf{P}^k \tag{10}$$

since for each element,

$$P_{ij} = \sum_{k=1}^{m} P_{ij}^k \tag{11}$$

holds.

Both the fragment density, Eq. 9, and the full molecular density, Eq. 1 are linear in the respective density matrices, consequently, the sum of fragment densities $\rho^k(\mathbf{r})$ is equal to the density $\rho(\mathbf{r})$ of the molecule;

$$\rho(\mathbf{r}) = \sum_{k=1}^{m} \rho^k(\mathbf{r}) \tag{12}$$

that is, an *additive, fuzzy electron density fragmentation* (AFDF) scheme is obtained.

Whereas a fuzzy fragmentation and subsequent reconstruction of electron densities of molecules is of interest in quantum chemical studies on functional groups and local molecular shape analysis, another important application of the AFDF schemes is based on the computation of fragment densities from small molecules and using them to construct electron densities for different molecules. Using this latter approach, the AFDF scheme has been used to build ab initio quality electron densities for large molecules, such as the HIV-1 protease of more than a thousand atoms,[46] utilizing electron density functions $\rho^1(\mathbf{r}), \rho^2(\mathbf{r}), \ldots, \rho^k(\mathbf{r}), \ldots \rho^m(\mathbf{r})$ of density fragments $F_1, F_2, \ldots, F_k, \ldots F_m$ calculated and taken from small "parent" molecules,

$$M_1, M_2, \ldots, M_k, \ldots M_m \tag{13}$$

where the local nuclear geometry and the local surroundings of the fragment match those found within the large "target" molecule. These calculations are based on a numerical electron density database and on a simple superposition of the additive density fragments, referred to as the molecular electron density "lego" assembler (MEDLA) technique.[46,49–52] Test calculations for smaller molecules have indicated that the resulting MEDLA electron densities are of better quality than densities obtained by conventional Hartree–Fock ab initio techniques using smaller basis sets, and are virtually indistinguishable from densities obtained using standard Hartree–Fock ab initio techniques with a 6-31G** basis set.

IV. MACROMOLECULAR DENSITY MATRIX METHODS BASED ON THE AFDF PRINCIPLE

A more advanced technique, the adjustable density matrix assembler (ADMA) method employs the AFDF scheme (in its simplest form, the Mulliken–Mezey scheme) for a direct, algebraic generation of ab initio quality approximate *density matrices* of macromolecules.[45,55–57] Besides the two conditions of additivity and fuzziness, the fragmentation scheme of the ADMA method must fulfill an additional condition of mutual compatibility of parent and target density fragments. The overall, mutually compatible AFDF scheme is referred to as the MC-AFDF scheme. The advantage of the ADMA method over the numerical MEDLA approach is a direct link to mainstream quantum chemical techniques for property and energy calculations based on density matrices.

The ADMA macromolecular density matrices constructed from mutually compatible fragment density matrices correspond to the same level of accuracy as the ideal, infinite resolution numerical MEDLA densities. The ADMA method describes interactions between local fragments to the same level of accuracy as the MEDLA technique; however, the ADMA density matrices can also be readjusted for small nuclear geometry changes of the macromolecule, a feature advantageous in biochemical applications.

The mutual compatibility of the ADMA macromolecular density matrix $\mathbf{P}$ and the family of additive fragment density matrices (MC-AFDM) $\mathbf{P}^k$ used for its construction involves two conditions:[55–57]

1. constraints on AO basis set orientation,
2. compatible target - parent fragmentation condition.

The basis set orientation condition (1) requires that all the fragment density matrices should refer to local coordinate systems where the coordinate axes are oriented the same way as the reference axes of a common, macromolecular coordinate system.

If required, then local coordinate transformations can be carried out on each fragment density matrix $\mathbf{P}^k$, changing the orientations of atomic orbitals to those in the common, macromolecular coordinate system. Take the k-th fragment density matrix $\mathbf{P}^k$ obtained from an ab initio calculation for the parent molecule M_k, where the vector $\varphi^{(k)}(\mathbf{r})$ represents the set of atomic orbitals for the parent M_k where the orientations of AOs are given in a local coordinate system. The k-th fragment density matrix $\mathbf{P}^k(\varphi)$ is defined with respect to this local coordinate system. If vector $\psi^{(k)}(\mathbf{r})$ represents the same sequence of atomic orbitals at the same nuclear centers in local coordinate systems of axes aligned with the axes of the common, macromolecular coordinate system, then these two representations are interrelated by an orthogonal matrix transformation $\mathbf{T}^{(k)}$:

$$\psi^{(k)}(\mathbf{r}) = \mathbf{T}^{(k)}\varphi^{(k)}(\mathbf{r}) \tag{14}$$

Matrix $\mathbf{T}^{(k)}$ is block-diagonal, a direct sum of the one-dimensional identity matrix for each of the *s*-orbitals, a three-dimensional rotation matrix for each triple of *p*-orbitals, the standard five-dimensional conversion matrix for each set of five orthonormalized *d*-orbitals, the seven-dimensional conversion matrix for each set of seven orthonormalized *f*-orbitals, and so on. (For non-orthonormal AOs an appropriately modified transformation matrix $\mathbf{T}^{(k)}$ should be used).

The actual fragment density matrix $\mathbf{P}^k$ used in the construction of the macromolecular ADMA density matrix $\mathbf{P}$ is obtained by the following similarity transformation:

$$\mathbf{P}^k = \mathbf{T}^{(k)}\,\mathbf{P}^k(\varphi)\mathbf{T}'^{(k)} \tag{15}$$

Using such transformation, fragment density matrix $\mathbf{P}^k$ fulfills the basis set compatibility condition (1) above.

Condition (2) on the compatibility of target and parent molecules is essential for the proper combination of fragment density matrix contributions $\mathbf{P}^k$ when building the macromolecular target density matrix $\mathbf{P}$. This condition can be summarized[55–57] as follows: If the nuclei of the target molecule M are classified into m families, then each parent molecule M_k may contain only *complete* nuclear families $f_{k'}$ from the target molecule M.

The parent molecule M_k, the source of the fragment density matrix $\mathbf{P}^k$ of the nuclear family f_k, either contains another complete nuclear family $f_{k'}$ as part of the surroundings of nuclear set f_k, or M_k does not contain any part of this nuclear family $f_{k'}$, with the possible exception of some peripheral H nuclei (or, possibly other nuclei) used to tie off dangling bonds in parent molecule M_k. These extra nuclei are at large distances from the actual nuclear set f_k of the fragment density matrix $\mathbf{P}^k$, hence they are assumed to have negligible influence on the actual fragment density matrix based on nuclear set f_k. By coincidence, a peripheral nucleus might occur at the same location as a nucleus of another nuclear family $f_{k''}$.

A natural restriction on the fragment AO basis sets apply: the AO basis functions with centers at nuclear locations of any family f_k are the same in all parent molecules where the nuclear family f_k occurs, either in the role of the central family (as in M_k), or as a part of the surrounding "coordination shell" for a fragment based on a different nuclear family $f_{k'}$ in a parent molecule $M_{k'}$.

Only those density matrix elements P^k_{ij} of each parent molecule M_k are involved in the construction of the final, macromolecular density matrix $\mathbf{P}$ of the ADMA method which fulfill the following conditions:

1. the selection conditions of the defining Eqs. 6 or 7 of any of the alternative, generalized additive fuzzy density fragmentation schemes proposed[34,39,45] for the fragment density matrix $\mathbf{P}^k$; and
2. no element of the fragment density matrix $\mathbf{P}^k$ involves the peripheral extra H (or other) nuclei of the parent molecules used to tie off dangling bonds.

Nuclear families f_k and appropriate parent molecules M_k fulfilling the above conditions can always be obtained for any macromolecule M.

In the target macromolecule M, the integers $n_1, n_2, \ldots, n_k, \ldots$, and n_m denote the number of AOs in the nuclear families $f_1, f_2, \ldots, f_k, \ldots$, and f_m, respectively.

For each pair (k,k') of nuclear families, $k,k' = 1, 2, \ldots, m$, define:

$$c_{k'k} = \begin{cases} 1, \text{ if nuclear family } f_{k'} \text{ contributes parent molecule } M_k, \\ 0 \text{ otherwise} \end{cases} \tag{16}$$

Each AO $\varphi(\mathbf{r})$ is assigned three indices, depending on the context. The notation $\varphi_{b,k'}(\mathbf{r})$ is used to indicate that this basis orbital is the b-th AO within the set,

$$\left\{\varphi_{a,k'}(\mathbf{r})\right\}_{a=1}^{n_{k'}} \tag{17}$$

of AOs associated with the nuclear family $f_{k'}$. The notation $\varphi_j^k(\mathbf{r})$ is used to indicate that the same basis orbital is the j-th AO within the basis set,

$$\left\{\varphi_i^k(\mathbf{r})\right\}_{i=1}^{n_{\mathrm{P}^k}} \tag{18}$$

involved in the definition of the k-th fragment density matrix $\mathbf{P}^k$, where the number of such AOs is calculated as:

$$n_{\mathrm{P}^k} = \sum_{k'=1}^{m} c_{k'k}\, n_{k'}. \tag{19}$$

The notation $\varphi_y(\mathbf{r})$ is used to indicate that y is the serial index of the same AO within the basis set,

$$\left\{\varphi_x(\mathbf{r})\right\}_{x=1}^{n} \tag{20}$$

involved in the definition of the macromolecular density matrix $\mathbf{P}$.

There are simple relations among these indices. For the AO basis function,

$$\varphi_{a,k'}(\mathbf{r}) = \varphi_i^k(\mathbf{r}) = \varphi_x(\mathbf{r}) \tag{21}$$

index x can be determined from index a within nuclear family k' using the relation,

$$x = x(k',a,f) = a + \sum_{b=1}^{k'-1} n_b \tag{22}$$

where symbol f in the argument of index function $x(k',a,f)$ indicates that indices k' and a originate from a nuclear family.

For each index k of fragment density matrices $\mathbf{P}^k$, index x can be determined from indices i and k by a simple procedure. One defines,

$$a'_k(k'',i) = i + \sum_{b=1}^{k''} n_b c_{bk} \tag{23}$$

$$k' = k'(i,k) = \min\{k'' : a'_k(k'',i) \leq 0\} \tag{24}$$

and,

$$a_k(i) = a'_k(k',i) + n_{k'} \tag{25}$$

for each nuclear family $f_{k''}$ for which:

$$c_{k''k} \neq 0 \tag{26}$$

In terms of the index function $x(k',a,f)$ of Eq. 22 and index k' given in Eq. 24, the actual AO index $x = x(k,i,P)$ in the macromolecular density matrix $\mathbf{P}$ can be calculated from indices i and k using the relation,

$$x = x(k,i,P) = x(k',a_k(i),f) \tag{27}$$

where symbol P in the argument of the index function $x(k,i,P)$ indicates that indices k and i refer to a fragment density matrix.

Using these index relations, the macromolecular density matrix $\mathbf{P}$ is calculated by identifying each nonzero matrix element P^k_{ij} of each fragment density matrix $\mathbf{P}^k$ and by setting:

$$P_{x(k,i,P),y(k,j,P)} = P_{x(k,i,P),y(k,j,P)} + P^k_{ij} \tag{28}$$

If the fragment density matrices $\mathbf{P}^1, \mathbf{P}^2, \ldots, \mathbf{P}^k, \ldots \mathbf{P}^m$ for nuclear families $f_1, f_2, \ldots, f_k, \ldots f_m$ are calculated from the series of parent molecules $M_1, M_2, \ldots, M_k, \ldots M_m$ fulfilling the compatibility conditions with one another and with the macromolecule M, then this algorithm[55–57] generates the ADMA macromolecular density matrix $\mathbf{P}$. This density matrix $\mathbf{P}$ and the macromolecular AO basis set $\{\varphi_x(\mathbf{r})\}_{x=1,..n}$ give a detailed ab initio quality quantum chemical description of macromolecule M. By taking large enough parent molecules M_k, the ADMA macromolecular density matrix $\mathbf{P}$ approximates the exact macromolecular density matrix of the same basis set as accurately as desired. For practical purposes, a "coordination shell" of approximately 4–5 Å thickness surrounding the "central" nuclear family f_k in each parent molecule M_k appears sufficient to represent the macromolecular interfragment interactions of each fragment.

There are practical limitations on the size of the AO basis set used in the ab initio calculation for the parent molecule M_k. Consequently, the computer time needed for the index reassignment for elements of each fragment density matrix is bounded by a constant. This implies that the overall computer time for the ADMA computation scales linearly with the number of fragments that is proportional with the size of the macromolecule.

The macromolecular electron density $\rho(\mathbf{r})$ is computed from the ADMA density matrix **P** using Eq. 1. Using the ADMA method, ab initio quality density matrices can be calculated for large molecules without first determining a molecular wavefunction. Within the Hartree–Fock framework, all higher order density matrices are determined by the first-order density matrix **P**; furthermore, the expectation values of one-electron and two-electron operators can be expressed in terms of the first-order and second-order density matrices. Several molecular properties can be computed using standard methodologies based on density matrices.[58,59] The ADMA method can be used to calculate approximate expectation values for many macromolecular properties, including energy, further extending the applicability of quantum chemistry to macromolecules.

If the size of the coordination shells used in the parent molecules is small, then the neglect of the density matrix contributions from the atomic orbitals of the peripheral H atoms of the "dangling" bonds in the parent molecules may result in small deviations from perfect charge conservation and the condition of idempotency for the macromolecular density matrix **P**. Charge conservation can be restored using the scaling method described earlier.[46]

If a product operation $*$ for density matrices is defined in terms of the matrix product **PSP** where **S** is the overlap matrix for a given nonorthogonal AO basis, then the idempotency condition can be written as:

$$\mathbf{P} * \mathbf{P} = \mathbf{P} \tag{29}$$

If an approximate macromolecular density matrix does not fulfill the idempotency condition to the desired level of accuracy, then by a small modification of **P** idempotency can be restored using standard methods.[58,59]

For a simple first approximation to a macromolecular density matrix $\mathbf{P}(K')$ of a nuclear arrangement K' slightly distorted with respect to a nuclear geometry K used for the construction of the original macromolecular density matrix $\mathbf{P}(K)$, one may use the same matrix with respect to the new basis orbitals located at the displaced nuclei. This crude approach gives useful approximate electron densities[56] for the new nuclear geometry K'; however, for larger displacements or if higher accuracy is needed, alternative approximations provide better results. One such approach involves a pair of Löwdin's transforms,[58,60–62] applied to the macromolecular density matrix $\mathbf{P}(K)$. These transformations are based on orthonormalization.

If the macromolecular density matrix is defined in terms of an orthonormal basis set, then the overlap matrix becomes the unit matrix and the idempotency condition takes the usual, simpler form. Another additional advantage of orthonormal basis sets is the fact that the transformations interconverting such bases and the corresponding density matrices are also simpler.

Löwdin's symmetric orthogonalization method[58,60–62] for the generation of orthonormal molecular basis sets is a technique used in many algorithms, including most implementations of the molecular Hartree–Fock method. Löwdin's transformation is especially suitable for converting density matrices of different bases into one another. Analogous transformations are used in quantum crystallography,[63] generating N-representable "experimental" density matrices based on experimental electronic densities obtained from crystallographic diffraction data, fulfilling the N-representability condition.[64–82]

If the overlap matrix of the basis set located at nuclei of arrangement K is denoted by $\mathbf{S}(K)$, then the Löwdin's transform of the density matrix $\mathbf{P}(K)$ involves multiplications from both left and right by the matrix $\mathbf{S}(K)^{1/2}$, leading to the matrix,

$$\mathbf{S}(K)^{1/2}\,\mathbf{P}(K)\,\mathbf{S}(K)^{1/2} \tag{30}$$

that is idempotent with respect to ordinary matrix multiplication.

If in a subsequent step, the inverse Löwdin's transform based on the appropriate power $\mathbf{S}(K')^{-1/2}$ of the new overlap matrix $\mathbf{S}(K')$ at the new nuclear configuration K' is applied, then an idempotent, improved approximation $\mathbf{P}(K',[K])$ of the density matrix $\mathbf{P}(K')$ is obtained:

$$\mathbf{P}(K',[K]) = \mathbf{S}(K')^{-1/2}\,\mathbf{S}(K)^{1/2}\,\mathbf{P}(K)\,\mathbf{S}(K)^{1/2}\,\mathbf{S}(K')^{-1/2} \tag{31}$$

Idempotency of $\mathbf{P}(K',[K])$ with respect to $*$ multiplication can be easily verified by substitution.

The two Löwdin-type transformations involve only the relatively inexpensive macromolecular overlap matrices $\mathbf{S}(K)$ and $\mathbf{S}(K')$ for two, slightly different nuclear geometries, K, and K'. The overall transformation can be regarded as "orthonormalization–deorthonormalization". The approximation $\mathbf{P}(K',[K])$ of the density matrix $\mathbf{P}(K')$ obtained in terms of the density matrix $\mathbf{P}(K)$ and the transformed overlap matrices is referred to as the SADMA approximation, where the name refers to the involvement of overlap matrices $\mathbf{S}$, as well as the ordinary ADMA approach.[83]

One approach based on ADMA and SADMA density matrices is approximate macromolecular force calculations using ADMA and SADMA electron densities and the electrostatic Hellmann–Feyman theorem.[84–86]

If $\rho(\mathbf{r})$ is the ADMA or SADMA macromolecular electron density, $\mathbf{R}_a$ is the position vector of nucleus a of nuclear charge z_a, and if $\mathbf{F}_a$ is the force operator representing the force acting on nucleus a, then, according to the electrostatic Hellmann–Feynman theorem, the expectation value of this force is:

$$\langle \mathbf{F}_a \rangle = -z_a \int \rho(\mathbf{r})(\mathbf{R}_a - \mathbf{r})|\mathbf{R}_a - \mathbf{r}|^{-3}\,d\mathbf{r} + z_a \sum_{a \neq b}^{N} z_b(\mathbf{R}_a - \mathbf{R}_b)|\mathbf{R}_a - \mathbf{R}_b|^{-3} \tag{32}$$

This expectation value is a simple sum of a classical contribution from the electronic charge density and the nuclear repulsion term. If ADMA or SADMA macromolecular density matrices are available, then the 3D integral in the first term of the expectation value can be computed efficiently; the summation in the second term is trivial. Whereas the calculated Hellmann–Feynman forces are sensitive to the quality of the quantum chemical representation of the electronic density,[87,88] this approach provides the basis of an approximate technique for macromolecular geometry optimization.

The study of small amplitude vibrations and other, restricted geometry changes, minor conformational motions in protein folding processes, as well as applications in the structure refinement process of X-ray structure determination are the areas where the adjustability of the SADMA macromolecular density matrices and the calculated electronic densities appear advantageous.

V. MOLECULAR FRAGMENTS AND CHEMICAL FUNCTIONAL GROUPS

The concept of similarity plays a profound role in the chemistry of functional groups: A functional group is usually perceived as a collection of nuclei and the associated electron density which occur with a similar arrangement in a variety of molecules. Furthermore, a functional group typically exhibits similar reactivities in most molecules; it has *similar function* in chemical reactions, hence the name, "functional group." Using the tools of the AFDF methods and shape analysis, a systematic, approach to the quantum chemistry of functional groups has been proposed.[47]

This treatment of functional groups is based on the *density domain* (DD) approach to chemical bonding.[20,32] A density domain DD(a,K) is a formal body enclosed by a molecular isodensity contour (MIDCO) $G(a,K)$, where some fixed nuclear configuration K and some electron density threshold a are indicated in the argument,

$$G(a,K) = \{\mathbf{r} : \rho(\mathbf{r},K) = a\} \tag{33}$$

and;

$$\mathrm{DD}(a,K) = \{\mathbf{r} : \rho(\mathbf{r},K) \geq a\} \tag{34}$$

Density domains are used in molecular shape analysis and in the computation of various molecular similarity measures.[29]

A formal molecular body at an electronic density threshold a and nuclear configuration K is represented by the density domain DD(a,K). In general, such a body DD(a,K) is either a single piece or it may be composed from several disconnected pieces, called the maximum connected components $\mathrm{DD}_i(a,K)$ of DD(a,K):

$$\mathrm{DD}(a,K) = \bigcup_i \mathrm{DD}_i(a,K) \tag{35}$$

(Note that the present usage of the term "domain" does not follow the usual mathematical terminology.)

Based on the connectedness properties of these bodies, a natural density domain condition has been proposed for a functional group. If within a given molecule of conformation K there exists a threshold a such that a corresponding connected density domain contains a subset of nuclei while separating them from the rest of the nuclei of the molecule, then this subset of nuclei is the *nuclear family* of a functional group. The existence of a separate density domain indicates that the part of the electronic density cloud dominated by this subset of nuclei is an entity with some limited "autonomy" within the complete molecule.

In general, the collection of all nuclei within a maximum connected density domain component $DD_i(a,K)$, together with $DD_i(a,K)$ is regarded as a *functional group* of the molecule[20,32] at the density threshold a.

This quantum chemical model of functional groups is consistent with the essentially geometrical framework discussed earlier[47] where an algebraic structure—a mathematical lattice—has been proposed for the description of the interrelations between families of functional groups.

Within the AFDF schemes, molecular fragment electron densities have short- and long-range properties analogous to those of complete molecules. This analogy allows one to apply a common fuzzy set approach for the description of molecular density fragments and functional groups using the same technique that has been introduced for families of complete molecules.[47]

It is natural to use fuzzy set methods[89–99]—in particular, fuzzy membership functions—to treat the fuzzy electron density contributions from a molecular assembly to the combined electron density of the resulting interacting system. If a family L of several molecules $X_1, X_2, \ldots X_i, \ldots X_m$ is located within a common spatial domain D, then it is of some interest to determine the extent various points $\mathbf{r}$ of the space can be assigned to individual molecules. The individual electron density contributions,

$$\rho_{X_1}(\mathbf{r}), \rho_{X_2}(\mathbf{r}), \ldots \rho_{X_i}(\mathbf{r}), \ldots \rho_{X_m}(\mathbf{r}) \qquad (36)$$

respectively, represent the "share" of each molecule in the total electron density of the molecular family L. Each "share" $\rho_{X_i}(\mathbf{r})$ is regarded as a separate, individual object in the absence of all other molecules of the family.

The electron density $\rho_{X_i}(\mathbf{r})$ takes its maximum value $\rho_{\max,i}$ within a spatial domain D_{X_i} containing all the nuclei of molecule X_i:

$$\rho_{\max,i} = \max \{\rho_{X_i}(\mathbf{r}), \mathbf{r} \in D_{X_i}\} \qquad (37)$$

The (not necessarily unique) point $\mathbf{r}_{\max i}$ where this maximum density value $\rho_{\max i}$ is realized,

$$\rho_{X_i}(\mathbf{r}_{\max,i}) = \rho_{\max,i} \tag{38}$$

is of special importance.

The total, composite electron density of the spatially "fixed" molecular family $X_1, X_2, \ldots X_j, \ldots X_m$ is denoted by $\rho_L(\mathbf{r})$, and is defined at any point $\mathbf{r}$ by:

$$\rho_L(\mathbf{r}) = \sum_j \rho_{X_j}(\mathbf{r}) \tag{39}$$

Using $\rho_L(\mathbf{r})$ as a reference, a fuzzy membership function $\mu_{X_i,L}(\mathbf{r})$ can be defined that expresses the extent of how much each point $\mathbf{r}$ of the space belongs to molecule X_i of the molecular family L. A consistent model is obtained if one takes,

$$\mu_{X_i,L}(\mathbf{r}) = \rho_{X_i}(\mathbf{r})/\rho_L(\mathbf{r}_{\max,i}) \tag{40}$$

for each molecule X_i.

The fuzzy electron density membership functions $\mu_{X_i,L}(\mathbf{r})$ express the relative contributions of the fuzzy, three-dimensional charge clouds of individual molecules to the total electronic density of molecular family L.

Whereas for complete molecules the MIDCOs are commonly used for shape and similarity analysis, for molecular fragments and functional groups obtained within the density domain approach, the analogous constructions are the fragment isodensity contours (FIDCOs). For a collection L of molecules, their relative contributions to the overall electronic density can be treated using the membership functions given by Eq. 40. A similar method can be used in order to decide what contribution of the electronic charge density cloud of a single molecule belongs to which functional group.

The fuzzy electron density membership function formalism used for molecular families can be used for a family of functional groups within a molecule X. The functional groups which appear as separate density domains,

$$\mathrm{DD}_1(a,K), \mathrm{DD}_2(a,K), \ldots, \mathrm{DD}_k(a,K), \ldots, \mathrm{DD}_m(a,K) \tag{41}$$

at some density threshold a are denoted by:

$$F_1, F_2, \ldots, F_k, \ldots, F_m \tag{42}$$

The actual density threshold value a identifies some of the possible functional groups of molecule X. If a different threshold value a' is chosen, a different set of density domains and a different assignment of nuclei to individual density domains may be obtained that may identify a different set of functional groups within the same molecule X. Clearly, the identity of functional groups depends on the density threshold; for example, at high-density thresholds for the density domains, the ultimate density domains are individual nuclear neighborhoods, hence the ultimate functional groups are individual atoms.

The nuclear set k for each fuzzy fragment density can be chosen as the nuclear set embedded in the corresponding density domain $DD_k(a,K)$ representing functional group F_k. The AFDF scheme determines the electron density contribution $\rho^k(\mathbf{r})$ of each functional group F_k to the molecular density $\rho_X(\mathbf{r})$.

The corresponding fuzzy electron density fragment contributions,

$$\rho_{F_1}(\mathbf{r}), \rho_{F_2}(\mathbf{r}), \ldots \rho_{F_k}(\mathbf{r}), \ldots \rho_{F_m}(\mathbf{r}) \tag{43}$$

respectively, represent the "share" of each functional group F_k in the total electron density $\rho_X(\mathbf{r})$ of molecule X. That is, the fuzzy functional group electron density membership functions measure the relative contributions of the fuzzy electron density charge clouds of the functional groups to the total electronic density of molecule X.

The fragment electron density $\rho_{F_k}(\mathbf{r})$ takes its maximum value $\rho_{\max,k}$ within some spatial domain D_{F_k} containing all the nuclei of functional group F_k:

$$\rho_{\max,k} = \max\,\{\rho_{F_k}(\mathbf{r}), \mathrm{r} \in D_{F_k}\}. \tag{44}$$

There must exist a (not necessarily unique) point $\mathbf{r}_{\max k}$ where this maximum density value $\rho_{\max k}$ is realized for the given functional group F_k:

$$\rho_{F_k}(\mathbf{r}_{\max,k}) = \rho_{\max,k} \tag{45}$$

According to the AFDF principle, the exact additivity property of Mezey's fragmentation scheme implies that at each point $\mathbf{r}$ the total, composite electron density of the spatially "fixed" family $F_1, F_2, \ldots, F_k, \ldots, F_m$ of molecule X determines the total electronic density $\rho_X(\mathbf{r})$ of molecule X, and is given as the sum of the individual functional group electron densities:

$$\rho_X(\mathbf{r}) = \sum_k \rho_{F_k}(\mathbf{r}) \tag{46}$$

If the density $\rho_X(\mathbf{r})$ is used as a reference, then a fuzzy membership function is defined for each functional group F_k as,

$$\mu_{F_k,}X(\mathbf{r}) = \rho_{F_k}(\mathbf{r})/\rho_X(\mathbf{r}_{\max,k}) \tag{47}$$

expressing the extent how much each point $\mathbf{r}$ of the space belongs to functional group F_k of molecule X.

The fuzzy membership functions $\mu_{F_k,X}(\mathbf{r})$ describe the relative influence of various functional groups $F_1, F_2, \ldots F_k, \ldots F_m$ of molecule X at each point $\mathbf{r}$ of the three-dimensional space.

The local shapes of various functional groups can be analyzed using the AFDF schemes. In the simplest version of this approach, the shape analysis is carried out on a molecular density fragment directly, where the interactions with the rest of the molecule are taken into account only in a limited sense: these interactions are used only to truncate the fragment density to restrict it to ranges where it is the dominant fragment within the molecule. This approach, where the density thresholds a are given for the fragment electron density $\rho_F(\mathbf{r})$, is referred to as the local shape approach of *noninteracting* functional groups.

If the local shape of functional group or molecular fragment F is studied, and M' represents the rest of the molecule M, where M' is possibly composed from several fragments, $F_1, F_2, \ldots, F_{m-1}$, then a noninteracting FIDCO for a fragment F in a molecule $M = FM'$ is defined as follows:

$$G_{F\setminus M'}(a) = \{\mathbf{r} : \rho_F(\mathbf{r}) = a,\ \rho_F(\mathbf{r}) \geq \rho_{F_k}(\mathbf{r}),\ k = 1, \ldots m-1\} \tag{48}$$

This definition is equivalent to,

$$G_{F\setminus M'}(a) = G_F(a) \cap \{\mathbf{r} : \rho_F(\mathbf{r}) \geq \rho_{F_k}(\mathbf{r}),\ k = 1, \ldots m-1\} \tag{49}$$

and to:

$$G_{F\setminus M'}(a) = G_F(a) \setminus \{\mathbf{r} : \exists\, k \in \{1, \ldots m-1\} : \rho_F(\mathbf{r}) < \rho_{F_k}(\mathbf{r})\} \tag{50}$$

The noninteracting FIDCO $G_{F\setminus M'}(a)$ of fragment F in molecule FM' is the collection of all those points of the FIDCO $G_F(a)$ where the electron density contribution of fragment F is dominant if regarded within the molecule FM'.

An alternative, also noninteracting FIDCO $G_{F\setminus \Sigma F_i}(a)$ of fragment F in molecule $FM'$is obtained if the composite electron density,

$$\rho_{M'}(\mathbf{r}) = \rho_{F_1}(\mathbf{r}) = \rho_{F_2}(\mathbf{r}) + \ldots + \rho_{F_{m-1}}(\mathbf{r}) \tag{51}$$

of all other fragments is used in the definition:

$$G_{F\setminus \Sigma_{F_i}}(a) = \{\mathbf{r} : \rho_F(\mathbf{r}) = a,\ \ \rho_F(\mathbf{r}) \geq \rho_{M'}(\mathbf{r})\}. \tag{52}$$

The usual shape group analysis of MIDCOs is based on the topological pattern and the resulting homology groups obtained when the surface is subdivided into various curvature domains of types $D_0(G_M(a))$, $D_1(G_M(a))$, and $D_2(G_M(a))$, with respect to some reference curvature b. (For details of the notations, terminology, and methodology, the reader should consult Ref. 29.) If FIDCOs in a molecule M are defined by Eq. 48 or by Eq. 52, then additional domain types arise, corresponding to those ranges on $G_F(a)$ where the electron density $\rho_F(\mathbf{r})$ of the given fragment F is not dominant.

In the case of $G_{F\setminus M'}(a)$, these new domain types are defined by,

$$D_{-1}(G_{F\setminus M'}(a)) = \{\mathbf{r} : \mathbf{r} \in G_F(a),\ \exists\, k \in \{1, \ldots m-1\} : \rho_F(\mathbf{r}) < \rho_{Mk}(\mathbf{r})\} \tag{53}$$

where the actual domain $D_{-1}(G_{F\setminus M'}(a))$ exists only on the original $G_A(a)$ contour.

In the case of $G_{F\setminus\Sigma F_i}(a)$, the new type of domain is defined as:

$$D_{-1}(G_{F\setminus\Sigma F_i}(a)) = \{\mathbf{r} : \mathbf{r} \in G_F(a),\ \rho_F(\mathbf{r}) \le \rho_{M'}(\mathbf{r})\} \tag{54}$$

In an alternative approach, the density thresholds a are given for the electron density $\rho_M(\mathbf{r})$ of the entire molecule M, and the local shape features of a functional group are described with respect to contour surfaces derived for the complete molecule, involving all interfragment interactions. This approach is referred to as the local shape approach of *interacting* functional groups. In this case, a new contour calculation is needed for a detailed description of the interactions between fragments, leading to the interactive FIDCO $G_{F(M')}(a)$ in molecule $M = FM'$. Here $G_{F(M')}(a)$ is defined in terms of a density threshold a for the actual, complete molecule:

$$G_{F(M')}(a) = \{\mathbf{r} : \rho_F(\mathbf{r}) + \rho_{M'}(\mathbf{r}) = a,\ \rho_F(\mathbf{r}) \ge \rho_{M'}(\mathbf{r})\}. \tag{55}$$

Interactive FIDCOs $G_{F(M')}(a)$ often have holes, with boundaries $\Delta D_{-1}(G_{F(M')}(a))$ where the constraints in the defining equation are fulfilled with the weak inequality becoming an equality:

$$\Delta D_{-1}(G_{F(M')}(a)) = \{\mathbf{r} : \mathbf{r} \in G_{F(M')}(a),\ \rho_F(\mathbf{r}) = \rho_{M'}(\mathbf{r})\}. \tag{55}$$

No actual $D_{-1}(G_{F(M')}(a))$ domain exists on the interactive FIDCO $G_{F(M')}(a)$ in the molecule M, and the reference to the fictitious $D_{-1}(G_{F(M')}(a))$ domain in the boundary expression $\Delta D_{-1}(G_{F(M')}(a))$ serves only for notational convenience.

The study of interactive FIDCO surfaces $G_{F(M')}(a)$ for local shape analysis involves additional contour calculations for the complete molecule that is more expensive than the study of noninteractive FIDCO surfaces $G_{F\setminus M'}(a)$ and $G_{F\setminus\Sigma F_i}(a)$; however, interactive FIDCO surfaces $G_{F(M')}(a)$ provide a better representation of physical reality.

The original techniques of the "shape group" methods[29] of electron density shape analysis are applicable to both types of FIDCO surfaces, provided that the domains $D_{-1}(G_{F\setminus M'}(a))$, and $D_{-1}(G_{F\setminus\Sigma F_i}(a))$ on the individual FIDCO $G_F(a)$, as well as the "phantom" domains $D_{-1}(G_{F(M')}(a))$, associated, respectively, with the additional formal domain boundaries $\Delta D_{-1}(G_{F\setminus M'}(a))$, $\Delta D_{-1}(G_{F\setminus\Sigma F_i}(a))$, and $\Delta D_{-1}(G_{F(M')}(a))$, are characterized by one additional index -1. This new index and domain type are treated the same way as the indices of various relative curvature domains. The shape groups of FIDCO surfaces are the one-dimensional homology groups obtained by truncations using all possible index combinations. The corresponding (a,b)-parameter maps and shape codes for similarity analysis are computed by the same algorithm as that introduced for complete molecules.[29]

VI. A SIMILARITY MEASURE BASED ON THE LÖWDIN TRANSFORM

A special similarity measure, motivated in part by the quantum similarity measures of Carbó,[1–8] is obtained if the density matrix comparisons are expressed in terms of the Löwdin transforms involved in nuclear geometry readjustments.

The similarity measure based on Löwdin's transform is suitable for assessing the similarities of electron densities of two nuclear configurations, K and K', slightly distorted with respect to each other. The two corresponding overlap matrices are $\mathbf{S}(K)$ and $\mathbf{S}(K')$, respectively. For macromolecules, these overlap matrices contain many negligible elements, and by setting all elements with absolute value below some suitable threshold equal to zero, both $\mathbf{S}(K)$ and $\mathbf{S}(K')$ become sparse matrices. For such sparse matrices, efficient numerical methods are available for the computation of the powers $\mathbf{S}(K)^{1/2}$ and $\mathbf{S}(K')^{-1/2}$, required for the Löwdin transform and the inverse Löwdin transform.

If the two nuclear configurations were to agree, then the product of the matrices $\mathbf{S}(K)^{1/2}$ and $\mathbf{S}(K')^{-1/2}$ would be the unit matrix $\mathbf{I}$ of appropriate dimension. On the other hand, for two different nuclear configurations, the deviation of the product from the unit matrix $\mathbf{I}$ provides a measure of dissimilarity of the corresponding electron densities with respect to the basis set ψ associated with the two overlap matrices $\mathbf{S}(K)$ and $\mathbf{S}(K')$:

$$\mathbf{D}_{L,\psi}(K,K') = \mathbf{I} - \mathbf{S}(K')^{-1/2}\, \mathbf{S}(K)^{1/2} \tag{56}$$

The trace of the product of the difference matrix $\mathbf{D}_{L,\psi}(K,K')$ with its transpose provides a numerical dissimilarity measure:

$$\mathbf{d}_{L,\psi}(K,K') = (1/n)\ \mathrm{trace}(\mathbf{D}_{L,\psi}(K,K')\mathbf{D}'_{L,\psi}(K,K')) \tag{57}$$

Somewhat simpler to calculate is another dissimilarity measure defined as,

$$\mathbf{d}_{S,\psi}(K,K') = (1/n)\ \mathrm{trace}(\mathbf{D}_{S,\psi}(K,K')\ \mathbf{D}'_{S,\psi}(K,K')) \tag{58}$$

where:

$$\mathbf{D}_{S,\psi}(K,K') = \mathbf{I} - \mathbf{S}(K')^{-1}\ \mathbf{S}(K). \tag{59}$$

This latter dissimilarity measure, however, does not have the same direct link to the actual transformation between the two density matrices $\mathbf{P}(K)$ and the approximation $\mathbf{P}(K',[K])$ of the density matrix $\mathbf{P}(K')$, as given by the "orthonormalization–deorthonormalization" step using $\mathbf{S}(K)^{1/2}$ and $\mathbf{S}(K')^{-1/2}$.

The actual similarity measures obtained from the dissimilarity measures $\mathbf{d}_{L,\psi}(K,K')$ and $\mathbf{d}_{S,\psi}(K,K')$ are defined as,

$$\mathbf{s}_{L,\psi}(K,K') = 1 - |\mathbf{d}_{L,\psi}(K,K')| \tag{60}$$

and,

$$\mathbf{s}_{S,\psi}(K,K') = 1 - |\mathbf{d}_{S,\psi}(K,K')| \tag{61}$$

respectively.

These similarity measures depend on the actual basis set representation and provide a numerical characterization of the similarities of electron densities of two, not drastically different nuclear configurations K and K', for example, of two molecular arrangements slightly distorted with respect to each other along a conformational path.

VII. A SIMILARITY MEASURE BASED ON A FUZZY HAUSDORFF METRIC FOR ELECTRON DENSITIES

The concept of α-cut[98,99] facilitates the description of an electron density similarity measure based on a fuzzy Hausdorff metric. The α-cut of a fuzzy subset A of a set X is defined as the crisp set of all those points x of X where the membership function $\mu_A(x)$ is equal to the value α:

$$G_A(\alpha) = \{x : \mu_A(x) = \alpha\}. \tag{62}$$

One can easily recognize the level set interpretation of the α-cut.

For two ordinary subsets A and B of a metric space X, the ordinary Hausdorff distance[100] $h(A,B)$ is the smallest value r such that each ball of radius r centered at any point of either set contains at least one point of the other set.

If the set X is provided with a metric $d(\mathbf{x},\mathbf{x}')$ for every point $\mathbf{x},\mathbf{x}' \in X$, then the distance between a point $\mathbf{x} \in X$ and a subset $A \subset X$ is usually defined by,

$$d(\mathbf{x}, A) = \inf_{\mathbf{a}\in A} \{d(\mathbf{x},\mathbf{a})\} \tag{63}$$

as the greatest lower bound of distances between points $\mathbf{a}$ of A and the point $\mathbf{x}$. If the distance d is continuous, then for a closed set A, the infimum becomes minimum.

The formal definition of the ordinary Hausdorff distance $h(A,B)$ between two subsets A and B of X can be given as,

$$h(A,B) = \sup_{\substack{\mathbf{a}\in A \\ \mathbf{b}\in B}} \{d(\mathbf{a}, B), d(\mathbf{b}, A)\} \tag{64}$$

the lowest upper bound of distances between points $\mathbf{a}$ of A and the set B and distances between points $\mathbf{b}$ of B and the set A. If the distance function $d(\mathbf{a},\mathbf{b})$ is continuous, then for closed sets A and B the supremum in the definition becomes maximum.

Molecular isodensity contour surfaces are closed sets; the Hausdorff distance between two such superimposed contours is the minimum r value satisfying the condition that any point on either contour surface has at least one point of the other contour surface within a distance r.

The Hausdorff distance $h(A,B)$ itself is a proper metric within any family of compact sets. In particular, the Hausdorff distance $h(A,B)$ is zero if and only if the two sets are the same, $A = B$.

For a generalization of the Hausdorff distance to fuzzy sets, the α-cuts provide a useful link to ordinary sets. If A and B are two fuzzy sets, then take their α-cuts $G_A(\alpha)$, and $G_B(\alpha)$, respectively, for each membership function value α. In terms of the ordinary Hausdorff distances $h(G_A(\alpha), G_B(\alpha))$ for each pair of α-cuts, one can define a function $g(A,B)$,

$$g(A,B) = \sup_{\alpha \in [0,1]} \{h(G_A(\alpha), G_B(\alpha))\} \tag{65}$$

that is a fuzzy set generalization of the Hausdorff metric, equivalent to the fuzzy Hausdorff distance suggested earlier.[96]

In chemical applications, the energetically most important spatial ranges of the molecule are enclosed by those level sets of the fuzzy electronic density where the density threshold a is high. Within a fuzzy set context, the α-cuts with large α values are of special importance. For emphasis of this importance, it is useful to consider a similarity metric for electron density fuzzy sets where the differences for α-cuts with large α values are weighted by the α values—in fact, emphasizing the "more committed points" of the fuzzy sets. For such a measure, if the membership function is positive, then the 0-cut $G_F(0)$ of the fuzzy set F is the empty set.

By scaling the fuzzy Hausdorff distance in Eq. 65 by the α value, a new fuzzy, "commitment-weighted" Hausdorff-type metric $f(A,B)$ is obtained:

$$f(A,B) = \sup_{\alpha \in [0,1]} \{\alpha h(G_A(\alpha), G_B(\alpha))\} \tag{66}$$

A proof is given below showing that the scaled fuzzy Hausdorff distance defined by Eq. 66 is also a metric in the space of fuzzy subsets of the underlying set X.

1. First we show that function $f(A,B)$ is non-negative:

$$f(A,B) \geq 0 \tag{67}$$

   Since each element in the set $\{\alpha h(G_A(\alpha), G_B(\alpha))\}$ in Eq. 66 is non-negative, the supremum over this set is necessarily non-negative.
2. The second metric property we prove is $f(A,B) = 0$ iff $A = B$. If $f(A,B) = 0$, then the fact that according to Eq. 66 $f(A,B)$ is a supremum in the set $\{\alpha h(G_A(\alpha), G_B(\alpha))\}$ implies that for each $\alpha > 0$, the α-scaled ordinary Hausdorff distance $\alpha h(G_A(\alpha'), G_B(\alpha'))$ of α-cuts is zero, hence:

$$h(G_A(\alpha'), G_B(\alpha')) = 0 \tag{68}$$

Consequently, for each value $\alpha > 0$, the pair of α-cuts for A and B agree:

$$G_A(\alpha) = G_B(\alpha) \tag{69}$$

Since all pairs of these α-cuts coincide, we conclude that there must exist a one-to-one and onto correspondence between the points of the two fuzzy sets A and B that preserves membership function, $\mu_A(x) = \mu_B(x)$, for every point $x \in X$ where this membership function is positive, $\mu_A(x) = \mu_B(x) = \alpha > 0$. Specifically, for any point $x' \in X$, $\mu_A(x') = 0$ and $\mu_B(x') = \alpha' > 0$ is impossible since then $x' \in G_A(\alpha')$ but $x' \notin G_B(\alpha)$; that contradicts Eq. 69 for the choice of $\alpha = \alpha'$. This implies that $\mu_A(x) = 0$ if and only if $\mu_B(x) = 0$ also holds. We conclude that the two fuzzy sets A and B are identical, $A = B$.

On the other hand, if $A = B$, then for each choice of α,

$$G_A(\alpha) = G_B(\alpha) \tag{70}$$

holds, consequently,

$$\alpha h(G_A(\alpha), G_B(\alpha)) = 0 \tag{71}$$

also holds for each α value. Consequently:

$$\sup_{\alpha\in[0,1]} \{\alpha h(G_A(\alpha), G_B(\alpha))\} = 0 \tag{72}$$

By combining these results, the second condition for metric follows:

$$f(A,B) = 0 \quad \textit{iff} \quad A = B \tag{73}$$

3. The third metric property we prove is symmetry, $f(A,B) = f(B,A)$.

   We know that the ordinary Hausdorff distance $h(G_A(\alpha'), G_B(\alpha'))$ of each α-cut in the set $\{\alpha h(G_A(\alpha), G_B(\alpha))\}$ is symmetric with respect to interchange of sets A and B,

$$h(G_A(\alpha'), G_B(\alpha')) = h(G_B(\alpha'), G_A(\alpha')) \tag{74}$$

   implying that the supremum $f(A,B)$ in Eq. 66 is also necessarily symmetric:

$$f(A,B) = f(B,A) \tag{75}$$

4. We prove the fourth metric property: the "commitment-weighted" fuzzy Hausdorff-type distance $f(A,B)$ satisfies the triangle inequality.

If continuity is understood within the metric topology of the underlying space X, then we assume that the α-cuts, $G_A(\alpha)$, $G_B(\alpha)$, and $G_C(\alpha)$ of three fuzzy subsets, A, B, and C, respectively, depend at least piecewise continuously on the α parameter from the unit interval [0,1]. On the closed interval [0,1], the proposed function

$\alpha h(G_A(\alpha), G_B(\alpha))$ is at least piecewise continuous in α, and either attains its maximum $h(G_A(\alpha'), G_B(\alpha'))$ at some value α' within [0,1], or it converges to the supremum value,

$$\lim_{\alpha \to \alpha'} \alpha h(G_A(\alpha), G_B(\alpha)) = \sup_{\alpha \in [0,1]} \{\alpha h(G_A(\alpha), G_B(\alpha))\} \tag{76}$$

as α converges within [0,1] to some value α' at a discontinuity of function $\alpha h(G_A(\alpha), G_B(\alpha))$. Equation 76 also holds when $\alpha h(G_A(\alpha), G_B(\alpha))$ attains its maximum at some value α', that is, when:

$$f(A,B) = \sup_{\alpha \in [0,1]} \{\alpha h(G_A(\alpha), G_B(\alpha))\} = \lim_{\alpha \to \alpha'} \alpha h(G_A(\alpha), G_B(\alpha)) \tag{77}$$

For the other two pairs of fuzzy sets, (B,C) and (A,C), there exist threshold values α'' and α''' within the interval [0,1], such that the equations,

$$f(B,C) = \sup_{\alpha \in [0,1]} \{\alpha h(G_B(\alpha), G_C(\alpha))\} = \lim_{\alpha \to \alpha''} \alpha h(G_B(\alpha), G_C(\alpha)) \tag{78}$$

and,

$$f(A,C) = \sup_{\alpha \in [0,1]} \{\alpha h(G_A(\alpha), G_C(\alpha))\} = \lim_{\alpha \to \alpha'''} \alpha h(G_A(\alpha), G_C(\alpha)) \tag{79}$$

hold.

Since the function is defined as a supremum, for limits of convergence to any other threshold value α''', the constraints,

$$\begin{aligned} \sup_{\alpha \in [0,1]} \{\alpha h(G_A(\alpha), G_B(\alpha))\} &= \lim_{\alpha \to \alpha'} \alpha h(G_A(\alpha), G_B(\alpha)) \\ &\geq \lim_{\alpha \to \alpha'''} \alpha h(G_A(\alpha), G_B(\alpha)) \end{aligned} \tag{80}$$

and,

$$\begin{aligned} \sup_{\alpha \in [0,1]} \{\alpha h(G_B(\alpha), G_C(\alpha))\} &= \lim_{\alpha \to \alpha''} \alpha h(G_B(\alpha), G_C(\alpha)) \\ &\geq \lim_{\alpha \to \alpha'''} \alpha h(G_B(\alpha), G_C(\alpha)) \end{aligned} \tag{81}$$

apply.

The triangle inequality holds for the α - scaled ordinary Hausdorff distances for each set of α-cuts taken for each individual α value as $\alpha \to \alpha'''$,

$$\alpha h(G_A(\alpha), G_B(\alpha)) + \alpha h(G_B(\alpha), G_C(\alpha)) \geq \alpha h(G_A(\alpha), G_C(\alpha)) \tag{82}$$

consequently:

$$\lim_{\alpha \to \alpha'''} \alpha h(G_A(\alpha), G_B(\alpha)) + \lim_{\alpha \to \alpha'''} \alpha h(G_B(\alpha), G_C(\alpha)) \geq \lim_{\alpha \to \alpha'''} \alpha h(G_A(\alpha), G_C(\alpha)). \quad (83)$$

This inequality is only strengthened if in the first and second terms on the left hand side the limits $\alpha \to \alpha'''$ are replaced by the optimum limits of $\alpha \to \alpha'$ and $\alpha \to \alpha''$, respectively, that cannot decrease the left hand side, as implied by inequalities, Eqs. 80 and 81:

$$\lim_{\alpha \to \alpha'} \alpha h(G_A(\alpha), G_B(\alpha)) + \lim_{\alpha \to \alpha''} \alpha h(G_B(\alpha), G_C(\alpha)) \geq \lim_{\alpha \to \alpha'''} \alpha h(G_A(\alpha), G_C(\alpha)) \quad (84)$$

Substitutions using Eqs. 77, 78, and 79 proves the triangle inequality:

$$f(A,B) + f(B,C) \geq f(A,C) \quad (85)$$

The four proven properties imply that the "commitment-weighted" fuzzy Hausdorff-type distance $f(A,B)$ is a metric.

If $G_A(\alpha)$, $G_B(\alpha)$, and $G_C(\alpha)$ change continuously within the unit interval [0,1]—that is, if each of the $G_A(\alpha)$, $G_B(\alpha)$, and $G_C(\alpha)$ sets is simply connected for any threshold value α—then simpler proofs apply, since then the suprema can be replaced with maxima realized at specific α', α'', and α''' values, and the use of the limits for $\alpha \to \alpha'$, $\alpha \to \alpha''$, and $\alpha \to \alpha'''$ can be avoided.

The scaled fuzzy Hausdorff-type metric $f(A,B)$ offers various choices for similarity measures between fuzzy sets, including,

$$s_f(A,B) = \exp(-[f(A,B)]^2) \quad (86)$$

$$t_f(A,B) = 1/(1 + [f(A,B)]^2) \quad (87)$$

and:

$$z_f(A,B) = 1/(1 + f(A,B)) \quad (88)$$

Each of these similarity measures $s_f(A,B)$, $t_f(A,B)$, and $z_f(A,B)$, takes the value of 1 for identical fuzzy sets, and the value of 0 for pairs of fuzzy sets having infinite value for the fuzzy generalizations of their Hausdorff-type distances.

In the molecular context, two fuzzy sets which are translated, rotated, or reflected versions of each other can be regarded as equivalent. For example, two fuzzy electron density clouds which can be obtained from each other by translation and rotation in the 3D space are chemically equivalent. The chemically relevant, inherent dissimilarities between two fuzzy electron densities A and B can be measured by the scaled fuzzy Hausdorff-type distance $f(A,B)$, where the relative positions of the molecules correspond to maximum superposition, minimizing their f-distance.

The notations $A_{v'}$ and $B_{v'}$ are used for translated and rotated versions of fuzzy sets A and B. A superposition-optimized variant $f_{op}(A,B)$ of the scaled fuzzy Hausdorff-type metric $f(A,B)$ is defined as:

$$f_{op}(A,B) = \inf_{A_{v'}, B_{v'}} \{f(A_{v'},B_{v'})\} \tag{89}$$

The $f_{op}(A,B)$ scaled fuzzy Hausdorff-type metric corresponds to the optimum superposition of fuzzy sets A and B if the set $\{f(A_{v'},B_{v'})\}$ contains the f-distances of all versions $A_{v'}$ and $B_{v'}$. The $f_{op}(A,B)$ function is a proper metric; a proof will be presented elsewhere.[101]

In the study of the similarity of molecules, the $f_{op}(A,B)$ distance can be used as a dissimilarity measure; by suitable transformations of $f_{op}(A,B)$, various similarity measures can be obtained.

By taking the $\mu_{X_i,L}(\mathbf{r}) = \rho_{X_i}(\mathbf{r})/\rho_L(\mathbf{r}_{\max,i})$ fuzzy membership function of Eq. 40 to describe the degree by which a point $\mathbf{r}$ belongs to molecule $A = X_i$ of a molecular family L, and using this membership function for the α-cuts involved in the definition of the scaled fuzzy Hausdorff-type metric $f_{op}(A,B)$ with respect to another molecule B, the $f_{op}(A,B)$ distance becomes a dissimilarity measure of electron densities. In turn, this measure defines various *fuzzy Hausdorff-type similarity measures* between the molecules, including,

$$s_{f_{op}}(A,B) = \exp(-[f_{op}(A,B)]^2) \tag{90}$$

$$t_{f_{op}}(A,B) = 1/(1 + [f_{op}(A,B)]^2) \tag{91}$$

and:

$$z_{f_{op}}(A,B) = 1/(1 + f_{op}(A,B)) \tag{92}$$

In some instances, only a subset of all possible versions of $A_{v'}$ and $B_{v'}$ are included in the set $\{f(A_{v'},B_{v'})\}$ when generating the supremum in Eq. 89. These cases correspond to restrictions on the possible alignments of the two molecules—for example, when comparing molecules fitting within a cavity of an enzyme, an important problem of similarity analysis in drug design. In such cases, restricted versions of similarity measures $s_{f_{op}}(A,B)$, $t_{f_{op}}(A,B)$, and $z_{f_{op}}$(A,B) are obtained.

VIII. SOME RELEVANT PROPERTIES OF MOLECULAR SHAPE ENVELOPES: T-HULLS AND INTERIOR T-AGGREGATES

The T-hull is a generalization of the convex hull of objects according to a "bias" with respect to a reference shape T. Based on local comparisons to the shape of a reference object T, the electron density T-hulls of molecules have been proposed earlier[102] for the analysis of various shape constraints in solvent–solute interactions and in biomolecular complementarity. For a given reference object T, the ordinary T-hull $\langle S \rangle_T$ of an object S is defined as the intersection of all rotated and translated versions of T which contain S. T-hulls are suggested for *relative* shape characterization of molecules, offering new tools for molecular shape and similarity analysis.[102] Several additional properties of T-hulls have also been described recently.[29,37]

Usually, in 3D chemical shape analysis, a version T_v of some reference object T is any set obtained from T by 3D translations and rotations.[29,102] Alternatively, various constrained motions, as well as additional freedoms, such as reflections, can be considered.[29,102]

In the simplest cases, the constraints (and extra freedoms) can be described by group theory. The allowed motions of T may form a group **G** of geometric transformations G, a subgroup of affine transformations (e.g., rotations) translations, reflections, collineations, and combinations thereof. If some cases, applying group theory may become cumbersome—for example, if the family **G** of allowed transformations is restricted rotations within a limited angle interval.

If a set **G** of transformations is selected, then two versions, T_v and $T_{v'}$ of reference object T are said to be **G**-equivalent if both T_v and $T_{v'}$ are derived from the reference object T by an allowed transformation. The set of **G**-equivalent versions T_v of T is denoted by:

$$V(T,G) = \{GT : G \in \mathbf{G}\} \tag{93}$$

A subset $V(T,G,S)$ of $V(T,G)$ is defined as the set that contains all those versions T_v from $V(T,G)$ which contain set S:

$$V(T,G,S) = \{T_v \in V(T,G): S \subset T_v\} \tag{94}$$

In some instances, it is advantageous to use an index set defined as:

$$I(V(T,G,S)) = \{v: T_v \in V(T,G,S)\} \tag{95}$$

Using this terminology, the **G**-constrained T-hull $\langle S \rangle_T$ of S can be written in either of the following forms,

$$\langle S \rangle_T = \bigcap_{T_v \in V(T,G,S)} T_v \tag{96}$$

or:

$$\langle S \rangle_T = \bigcap_{v \in I(V(T,G,S))} T_v \tag{97}$$

T-hulls possess important properties analogous to some of the properties of convex sets. Some related properties are also shown by a family of "plaster" sets generated as T-hulls, and their formal duals, called aggregates. The following definitions apply and some elementary properties of these sets[103] are subsequently reviewed:

Definition 1. A set B is called a T-plaster set if B is a T-hull $\langle S\rangle_T$ of some set S:

$$B = \langle S\rangle_T \tag{98}$$

The T-hull $\langle S\rangle_T$ of a set S is also called the exterior T-plaster (or, simply, the T-plaster) of S.

Definition 2. The interior T-aggregate $\rangle S\langle_T$ of a set S is the union of all $T_v \in V(T,G)$ versions of T contained in S:

$$\rangle S\langle_T = \bigcup_{T_v \in V(T,G,)\cap_S} T_v \tag{99}$$

Definition 3. The interior T-plaster $\rangle\rangle S\langle\langle_T$ of a set S is the S-relative complement of the T-aggregate $\rangle S\langle_T$ of set S:

$$\rangle\rangle S\langle\langle_T = S \setminus \rangle\rangle S\langle\langle_T = S \setminus \cup_{T_v \in V(T,G,)\cap S} T_V \tag{100}$$

A. Theorem 1

If sets A and B are T-plaster sets with respect to a reference set T, then their intersection $A \cap B$ is also a T-plaster set with respect to T. Furthermore, if $A = \langle S\rangle_T$ and $B = \langle S'\rangle_T$ then:

$$\langle S \cap S'\rangle_T \subset A \cap B \tag{101}$$

Proof:

Since A and B are T-plaster sets, there exist some sets S and S' such that $A = \langle S\rangle_T$ and $B = \langle S'\rangle_T$. Since the T-hull of the T-hull is the T-hull,[37] the relations,

$$A = \langle S\rangle_T = \langle\langle S\rangle_T\rangle_T = \langle A\rangle_T = \cap_{v \in I(V(T,G,A))} T_v \tag{102}$$

and,

$$B = \langle S'\rangle_T = \langle\langle S'\rangle_T\rangle_T = \langle B\rangle_T = \cap_{v \in I(V(T,G,B))} T_v \tag{103}$$

must hold.

(a) The relation,

$$A \cap B \subset \langle A \cap B\rangle_T \tag{104}$$

evidently holds, since every set is contained in its own T-hull.

(**b**) We show that,

$$A \cap B \supset \langle A \cap B \rangle_T \tag{105}$$

also holds. By virtue of relations 102 and 103:

$$A \cap B = \bigcap_{v \in I(V(T,G,A))} T_v \cap \bigcap_{v' \in I(V(T,G,B))} T_{v'}$$

$$= \bigcap_{v'' \in I(V(T,G,A)) \cup I(V(T,G,B))} T_{v''} = \bigcap_{v''' \in I(V(T,G,A) \cup v(T,G,B))} T_{v''} \tag{106}$$

However, since,

$$V(T,G,A) \subset V(T,G,A \cap B) \tag{107}$$

and,

$$V(T,G,B) \subset V(T,G,A \cap B) \tag{108}$$

the relations,

$$V(T,G,A) \cup V(T,G,B) \subset V(T,G,A \cap B) \tag{109}$$

and,

$$I(V(T,G,A)) \cup I(V(T,G,B)) \subset I(V(T,G,A \cap B)) \tag{110}$$

also hold, implying the reversed inclusion relation for the corresponding intersections,

$$\bigcap_{v''' \in I(V(T,G,A) \cup V(T,G,B))} T_{v'''} \supset \bigcap_{v'''' \in I(V(T,G,A \cap B))} T_{v''''} = \langle A \cap B \rangle_T \tag{111}$$

where the intersection for indices v'''' is, by definition, the T-hull of $A \cap B$. Consequently, the relation $A \cap B \supset \langle A \cap B \rangle_T$ holds.

Combining results (**a**) and (**b**) proves the first assertion of the theorem. Furthermore, if $A = \langle S \rangle_T$ and $B = \langle S' \rangle_T$, then $S \subset A$ and $S' \subset B$. Consequently,

$$S \cap S' \subset A \cap B \tag{112}$$

that implies:

$$\langle S \cap S' \rangle_T \subset \langle A \cap B \rangle_T \tag{113}$$

However, according to the first, proven assertion of the theorem, set $A \cap B$ is a T-plaster set, hence $\langle A \cap B \rangle_T = A \cap B$. Consequently, the second assertion of the theorem,

$$\langle S \cap S' \rangle_T \subset A \cap B \tag{114}$$

Q.E.D.

also holds. Q.E.D.

(*Note:* the reversed inclusion relation, $\langle S \cap S' \rangle_T \supset A \cap B$, does not necessarily hold.)

B. Theorem 2

An analogous theorem holds for interior T-aggregates, where the roles of intersections and unions are interchanged. We shall use the notations,

$$W(T,G,S) = \{T_v \in V(T,G): T_v \subset S\} \tag{115}$$

and:

$$I(W(T,G,S)) = \{v: T_v \in W(T,G,S)\} \tag{116}$$

Theorem 2: If sets A and B are interior T-aggregate sets with respect to a reference set T, then their union $A \cup B$ is also an interior T-aggregate set with respect to the same reference set T. Furthermore, if A and B are interior T-aggregates, that is, if $A = \rangle S \langle_T$ and $B = \rangle S' \langle_T$ for some sets S and S', then:

$$\rangle S \cup S' \langle_T \supset A \cup B \tag{117}$$

Proof:

Sets A and B are interior T-aggregates, hence there must exist some sets S and S' such that $A = \rangle S \langle_T$ and $B = \rangle S' \langle_T$. Since A is the union of all versions T_v which are contained in S, A itself is the union of all versions T_v which are contained in A. Consequently,

$$A = \rangle S \langle_T = \rangle A \langle_T = \bigcup_{v \in I(W(T,G,A))} T_v \tag{118}$$

and the analogous relations hold for B:

$$B = \rangle S' \langle_T = \rangle B \langle_T = \bigcup_{v \in I(W(T,G,B))} T_v \tag{119}$$

(**a**) The relation,

$$A \cup B \supset \rangle A \cup B \langle_T \tag{120}$$

evidently holds, since the interior T-aggregate of any set is contained in the set.

(**b**) We also show that

$$A \cup B \subset \rangle A \cup B \langle_T \tag{121}$$

holds.

Relations 118 and 119 imply that:

$$A \cup B = \bigcup_{v \in I(W(T,G,A))} T_v \cup \bigcup_{v' \in I(W(T,G,B))} T_{v'}$$

$$= \bigcup_{v'' \in I(W(T,G,A)) \cup I(W(T,G,B))} T_v'' = \bigcup_{v''' \in I(W(T,G,A) \cup W(T,G,B))} T_v''' \tag{122}$$

However, since

$$W(T,G,A) \subset W(T,G,A \cup B) \tag{123}$$

and,

$$W(T,G,B) \subset W(T,G,A \cup B) \tag{124}$$

the relations,

$$W(T,G,A) \cup W(T,G,B) \subset W(T,G,A \cup B) \tag{125}$$

and,

$$I(W(T,G,A)) \cup I(W(T,G,B)) \subset I(W(T,G,A \cup B)) \tag{126}$$

also hold, implying the inclusion relation:

$$\bigcup_{v''' \in I(W(T,G,A) \cup W(T,G,B))} T_{v'''} \subset \bigcup_{v'''' \in I(W(T,G,A \cup B))} T_{v''''} = \rangle A \cup B \langle_T \tag{127}$$

The union for indices v'''' is the definition of the interior T-aggregate of set $A \cup B$. Consequently, the relation $A \cup B \subset \rangle A \cup B \langle_T$ holds.

Combining results (**a**) and (**b**) proves the first assertion of the theorem.

In order to prove the second assertion, we note that if $A = \rangle S \langle_T$ and $B = \rangle S' \langle_T$ then $S \supset A$ and $S' \supset B$ also hold. Consequently,

$$S \cup S' \supset A \cup B \tag{128}$$

that implies:

$$\rangle S \cup S' \langle_T \supset \rangle A \cup B \langle_T \tag{129}$$

However, according to the first, proven assertion of the theorem, if A and B are interior T-aggregates, then the set $A \cup B$ is also an interior T-aggregate set, hence $A \cup B = \rangle A \cup B \langle_T$, implying the second assertion of the theorem:

$$\rangle S \cup S' \langle_T \supset A \cup B \tag{130}$$

Q.E.D.

(*Note:* the reversed inclusion relation, $\rangle S \cup S' \langle_T \subset A \cup B$, does not necessarily hold.)

If the reference object T is the complement of a body representing the shape properties of a solvent molecule, then the T-hull of a solute molecule S describes some of the geometrical constraints on solute–solvent interactions. Various other applications include solvation layers, and inner cavities filled with solvent molecules, such as water in proteins.

IX. SUMMARY

Similarity measures for fuzzy molecular electron densities and fuzzy electron density clouds of local molecular fragments and functional groups are discussed. Special emphasis is placed on methods designed for fuzzy objects. These techniques include additive fuzzy density fragmentation methods, macromolecular density matrix methods, similarity measures based on the Löwdin transform, a Hausdorff metric for comparing fuzzy electron densities, and T-hulls and interior T-aggregates, as tools of molecular similarity analysis.

REFERENCES

1. Carbó, R.; Leyda, L.; Arnau, M. *Int. J. Quantum Chem.* **1980**, *17*, 1185.
2. Hodgkin, E.E.; Richards, W.G. *J. Chem. Soc. Chem. Commun.* **1986**, 1342.
3. Carbó, R.; Domingo, Ll. *Int. J. Quantum Chem.* **1987**, *32*, 517.
4. Hodgkin, E.E.; Richards, W.G. *Int. J. Quantum Chem.* **1987**, *14*, 105.
5. Carbó, R.; Calabuig, B. *Comput. Phys. Commun.* **1989**, *55*, 117.
6. Carbó, R.; Calabuig, B. *Int. J. Quantum Chem.* **1992**, *42*, 1681.
7. Carbó, R.; Calabuig, B. *Int. J. Quantum Chem.* **1992**, *42*, 1695.
8. Carbó, R.; Calabuig, B.; Vera, L.; Besalu, E. In *Advances in Quantum Chemistry;* Löwdin, P.-O.; Sabin, J.R.; Zerner, M.C., Eds.; Academic Press: New York, 1994, Vol. 25.
9. Mezey, P.G. *J. Math. Chem.* **1988**, *2*, 299.
10. Leicester, S.E.; Finney, J.L.; Bywater, R.P. *J. Mol. Graph.* **1988**, *6*, 104.
11. Arteca, G.A.; Jammal, V.B.; Mezey, P.G. *J. Comput. Chem.* **1988**, *9*, 608.
12. Arteca, G.A.; Jammal, V.B.; Mezey, P.G.; Yadav, J.S.; Hermsmeier, M.A.; Gund, T.M. *J. Molec. Graphics* **1988**, *6*, 45.
13. Johnson, M.A. *J. Math. Chem.* **1989**, *3*, 117.
14. Arteca, G.A.; Mezey, P.G. *J. Phys. Chem.* **1989**, *93*, 4746.
15. Arteca, G.A.; Mezey, P.G. *IEEE Eng. in Med. & Bio. Soc. 11th Annual Int. Conf.* 1989, *11*, 1907.
16. Johnson, M.A.; Maggiora, G.M., Eds. *Concepts and Applications of Molecular Similarity;* Wiley: New York, 1990.
17. Burt, C.; Richards, W.G.; Huxley, P. *J. Comput. Chem.* **1990**, *11*, 1139.
18. Mezey, P.G. In *Concepts and Applications of Molecular Similarity;* Johnson, M.A.; Maggiora, G.M., Eds.; Wiley: New York, 1990.
19. Arteca, G.A.; Mezey, P.G. *Int. J. Quantum Chem. Symp.* **1990**, *24*, 1.
20. Mezey, P.G. In *Reviews in Computational Chemistry;* Lipkowitz, K.B.; Boyd, D.B., Eds.; VCH Publishers, New York, 1990.
21. Mezey, P.G. *J. Math. Chem.* **1991**, *7*, 39.
22. Mezey, P.G. In *Theoretical and Computational Models for Organic Chemistry*; Formosinho, S.J.; Csizmadia, I.G.; Arnaut, L.G., Eds.; Kluwer Academic Publishers, Dordrecht, 1991.
23. Good, A.; Richards, W.G. *J. Chem. Inf. Sci.* **1992**, *33*, 112.
24. Mezey, P.G. *J. Math. Chem.* **1992**, *11*, 27.
25. Mezey, P.G. *J. Chem. Inf. Comp. Sci.* **1992**, *32*, 650.

26. Dubois, J.-E.; Mezey, P.G. *Int. J. Quantum Chem.* **1992**, *43*, 647.
27. Luo, X.; Arteca, G.A.; Mezey, P.G. *Int. J. Quantum Chem.* **1992**, *42*, 459.
28. Mezey, P.G. *J. Math. Chem.* **1993**, *12*, 365.
29. Mezey, P.G. *Shape in Chemistry: An Introduction to Molecular Shape and Topology*; VCH Publishers: New York, 1993.
30. Mezey, P.G. *J. Chem. Inf. Comp. Sci.* **1994**, *34*, 244.
31. Mezey, P.G. *Int. J. Quantum Chem.* **1994**, *51*, 255.
32. Mezey, P.G. *Canad. J. Chem.* **1994**, *72*, 928. (Special issue dedicated to Prof. J. C. Polanyi.)
33. Mezey, P.G. In *Molecular Similarity and Reactivity: From Quantum Chemical to Phenomenological Approaches;* Carbó, R., Ed.; Kluwer Academic Publishers: Dordrecht, The Netherlands, 1995.
34. Mezey, P.G. In *Molecular Similarity in Drug Design*; Dean, P.M., Ed.; Chapman & Hall - Blackie Publishers: Glasgow, U.K., 1995.
35. Walker, P.D.; Mezey, P.G. *J. Comput. Chem.* **1995**, *16*, 1238.
36. Walker, P.D.; Maggiora, G.M.; Johnson, M.A.; Petke, J.D.; Mezey, P.G. *J. Chem. Inf. Comp. Sci.* **1995**, *35*, 568.
37. Mezey, P.G., Theor. Chim. Acta **1995**, *92*, 333.
38. Walker, P.D.; Mezey, P.G.; Maggiora, G.M.; Johnson, M.A.; Petke, J.D. *J. Comput. Chem.* **1995**, *16*, 1474.
39. Mezey, P.G. In *Topics in Current Chemistry;* Sen, K., Ed.; Springer-Verlag: Heidelberg, 1995, Vol. 173.
40. Mezey, P.G. *Potential Energy Hypersurfaces;* Elsevier: Amsterdam, 1987.
41. Mezey, P.G. *Int. J. Quantum Chem. Quant. Biol. Symp.* **1986**, *12*, 113.
42. Mezey, P.G. *J. Comput. Chem.* **1987**, *8*, 462.
43. Mezey, P.G. *Int. J. Quantum Chem. Quant. Biol. Symp.* **1987**, *14*, 127.
44. Mezey, P.G. *J. Math. Chem.* **1988**, *2*, 325.
45. Mezey, P.G. *Structural Chem.* **1995**, *6*, 261.
46. Walker, P.D.; Mezey, P.G. *J. Math. Chem.* **1995**, *17*, 203.
47. Mezey, P.G. In *Advances in Quantum Chemistry;* Löwdin, P.-O.; Sabin, J.R.; Zerner, M.C., Eds.; Academic Press: New York, 1996.
48. Stefanov, B.B.; Cioslowski, J. *J. Comput. Chem.* **1995**, *16*, 1394.
49. Walker, P.D.; Mezey, P.G, *Program MEDLA 93* (Mathematical Chemistry Research Unit, University of Saskatchewan, Saskatoon, Canada, 1993).
50. Walker, P.D.; Mezey, P.G. *J. Am. Chem. Soc.* **1993**, *115*, 12423.
51. Walker, P.D.; Mezey, P.G. *J. Am. Chem. Soc.* **1994**, *116*, 12022.
52. Walker, P.D.; Mezey, P.G. *Canad. J. Chem.* **1994**, *72*, 2531.
53. Mulliken, R.S. *J. Chem. Phys.* **1955**, *23*, 1833, 1841, 2338, 2343.
54. Mulliken, R.S. *J. Chem. Phys.* **1962**, *36*, 3428.
55. Mezey, P.G. *Program ADMA 95* (Mathematical Chemistry Research Unit, University of Saskatchewan, Saskatoon, Canada, 1995).
56. Mezey, P.G. *J. Math. Chem.* **1995**, *18*, 141.
57. Mezey, P.G. In *Computational Chemistry: Reviews and Current Trends;* Leszczynski, J., Ed.; World Scientific Publishers: Singapore, 1996.
58. Pilar, F.L. *Elementary Quantum Chemistry;* McGraw-Hill: New York, 1968.
59. McWeeny, R.; Sutcliffe, B.T. *Methods of Molecular Quantum Mechanics;* Academic Press: New York, 1969.
60. Löwdin, P.-O. *J. Chem. Phys.* **1950**, *18*, 365.
61. Löwdin, P.-O. *Adv. in Phys.* **1956**, *5*, 1.
62. Löwdin, P.-O. *Adv. Quantum. Chem.* **1970**, *5*, 185.
63. Massa, L.; Huang, L.; Karle, J. *Int. J. Quantum Chem.*, to be published.
64. Löwdin, P.-O. *Phys. Rev.* **1955**, *97*, 1474.
65. McWeeny, R. *Rev. Mod. Phys.* **1960**, *32*, 335.
66. Coleman, A.J. *Rev. Mod. Phys.* **1963**, *35*, 668.

67. Clinton, W.L.; Galli, A.J.; Massa, L.J. *Phys. Rev.* **1969**, *177*, 7.
68. Clinton, W.L.; Galli, A.J.; Henderson, G.A.; Lamers, G.B.; Massa, L.J.; Zarur, J. *Phys. Rev.* **1969**, *177*, 27.
69. Clinton, W.L.; Massa, L.J. *Int. J. Quantum Chem.* **1972**, *6*, 519.
70. Clinton, W.L.; Massa, L.J. *Phys. Rev. Lett.* **1972**, *29*, 1363.
71. Clinton, W.L.; Frishberg, C.; Massa, L.J.; Oldfield, P.A. *Int. J. Quantum Chem. Quantum Chem. Symp.* **1973**, *7*, 505.
72. Henderson, G.A.; Zimmermann, R.K. *J. Chem. Phys.* **1976**, *65*, 619.
73. Tsirel'son, V.G.; Zavodnik, V.E.; Fonichev, E.B.; Ozerov, R.P.; Kuznetsolirez, I.S. *Kristallogr.* **1980**, *25*, 735.
74. Frishberg, C.; Massa, L.J. *Phys. Rev. B* **1981**, *24*, 7018.
75. Frishberg, C.; Massa, L.J. *Acta Cryst. A* **1982**, *38*, 93.
76. Massa, L.J.; Goldberg, M.; Frishberg, C.; Boehme, R.F.; LaPlaca, S.J. *Phys. Rev. Lett.* **1985**, *55*, 622.
77. Frishberg, C. *Int. J. Quantum Chem.* **1986**, *30*, 1.
78. Cohn, L.; Frishberg, C.; Lee, C.; Massa, L.J. *Int. J. Quantum Chem., Quantum Chem. Symp.* **1986**, *19*, 525.
79. Massa, L.J. *Chemica Scripta* **1986**, *26*, 469.
80. Boehme, R.F.; LaPlaca, S.J. *Phys. Rev. Lett.* **1987**, *59*, 985.
81. Tanaka, K. *Acta Cryst. A* **1988**, *44*, 1002.
82. Aleksandrov, Y.Y.; Tsirel'son, V.G.; Resnik, I.M.; Ozerov, R.P. *Phys. Status Solidi, B* **1989**, *155*, 201.
83. Mezey, P.G. *Program SADMA 95* (Mathematical Chemistry Research Unit, University of Saskatchewan, Saskatoon, Canada, 1995).
84. Hellmann, H. *Einführung in die Quantenchemie;* Deuticke and Co.: Leipzig, 1937, Sec. 54.
85. Feynman, R.P. *Phys. Rev.* **1939**, *56*, 340.
86. Epstein, S.T. In *The Force Concept in Chemistry;* Deb, B.M., Ed.; Van Nostrand-Reinhold: Toronto, 1981.
87. Pulay, P. In *Applications of Electronic Structure Theory;* Schaefer, H.F., Ed.; Plenum: New York, 1977.
88. Pulay, P. In *The Force Concept in Chemistry;* Deb, B.M., Ed.; Van Nostrand-Reinhold: Toronto, 1981.
89. Zadeh, L.A. *Inform. Control* **1965**, *8*, 338.
90. Zadeh, L.A. *J. Math. Anal. Appl.* **1968**, *23*, 421.
91. Kaufmann, A., *Introduction à la Théorie des Sous-Ensembles Flous;* Masson: Paris, 1973.
92. Zadeh, L.A. In *Encyclopedia of Computer Science and Technology;* Marcel Dekker: New York, 1977.
93. Gupta, M.M.; Ragade, R.K.; Yager, R.R., Eds. *Advances in Fuzzy Set Theory and Applications;* North-Holland: Leyden, 1979.
94. Dubois, D.; Prade, H. *Fuzzy Sets and Systems: Theory and Applications;* Academic Press: New York, 1980.
95. Sanchez E.; Gupta, M.M., Eds. *Fuzzy Information, Knowledge Representation and Decision Analysis,* Pergamon Press: London, 1983.
96. Puri, M.L.; Ralescu, D.A. *J. Math. Anal. Appl.* **1986**, *114*, 409.
97. Bandemer, H.; Näther, W. *Fuzzy Data Analysis;* Kluwer: Dordrecht, 1992.
98. Wang, Z.; Klir, G.J. *Fuzzy Measure Theory;* Plenum Press: New York, 1992.
99. Klir, G.J.; Yuan, B. *Fuzzy Sets and Fuzzy Logic, Theory and Applications;* Prentice Hall PTR: Upper Saddle River, NJ, 1995.
100. F. Hausdorff, F. *Set Theory*; (Transl. by J.R. Auman), Chelsey: New York, 1957.
101. Mezey, P.G. In *Fuzzy Logic in Chemistry;* Rouvray, D.H., Ed.; Academic Press: San Diego, 1996.
102. Mezey, P.G. *J. Math. Chem.* **1991**, *8*, 91.
103. Mezey, P.G. *J. Chem. Inf. Comp. Sci.,* to be published.

ELECTRON CORRELATION IN ALLOWED AND FORBIDDEN PERICYCLIC REACTIONS FROM GEMINAL EXPANSION OF PAIR DENSITIES:

A SIMILARITY APPROACH

Robert Ponec

Advances in Molecular Similarity
Volume 1, pages 121–133

ISBN: 0-7623-0131-7

ABSTRACT

The recently proposed second-order similarity index was generalized by using the geminal expansion of pair density. This generalization, together with the incorporation of the approach into the framework of the overlap determinant method, opens the possibility of the systematic investigation of correlation effects during chemical reactions. The approach was applied to the study of selected pericyclic reactions, both forbidden and allowed. The differences in the electron and spin recoupling between the allowed and forbidden reactions are discussed.

I. INTRODUCTION

Although the basic qualitative explanation of chemical reactivity is satisfactorily described by a simple model based on the idea of independent electrons, obtaining reasonable quantitative precision necessarily requires one to complement the simple MO model by including the phenomenon of mutual coupling of electron motions, the so-called electron correlation. Such inclusion is necessary not only for the reliable description of energetic quantities as, e.g., the activation or reaction energies, but, as demonstrated by a number of examples, the inclusion of electron correlation can also considerably influence the nature and the number of critical points of the potential energy hypersurface. An example in this respect can be some cycloaddition reactions (Diels–Alder reaction, [2+2] ethene dimerization) for which the above variation in the nature of critical points (true saddle points vs. second-order saddle point) in dependence on the quality of the computational methods used was reported in a number of studies.[1–4]

Because of the richness of manifestations of correlation effects, the spectrum of studies dealing with electron correlation is extremely broad and ranges from purely computational studies (for an exhaustive review see Ref. 5) to simple qualitative investigations in which the pair density, the simplest quantity involving the effects of electron correlation, is systematically analyzed.[6–11] Among the studies attempting to apply the pair density to the analysis of chemical reactivity it is important to mention, above all, the pioneering study by Salem[12] in which the electron reorganization in allowed and forbidden pericyclic reactions was discussed in terms of pair correlation functions. The same subject was also studied by the author and co-workers using the so-called second-order similarity indices.[13–15] In addition to the expected result that electron correlation is more important in forbidden reactions than in the allowed ones, we also demonstrated that the classification introduced some time ago by Dewar,[16] in which the whole class of pericyclic processes was subdivided into the so-called one-bond and multibond ones, is indeed justified. It appears that whereas for one-bond reactions the electron correlation is important only for a forbidden reaction mechanism, in the case of multibond reactions the correlation effects become very important even for the allowed mechanism. For that reason the quantum chemical calculations of these systems are much more sensitive to the quality of the methods used. Thus, while the cyclization of butadiene to cyclobutene can be satisfactorily described at the level of the simple SCF method,[17] the analogous calculations of multibond reactions necessarily require the inclusion of electron correlation, e.g., via MCSCF or spin-coupled method.[1,3,18]

Our aim in this study is to follow up with the results of our previous study[19] based on the static description in terms of second-order similarity indices derived from geminal expansion of pair densities of the starting reactant and the final product, and to generalize it by incorporating the whole formalism into the framework of the so-called overlap determinant method.[20] The aim of this generalization is to gain more detailed insight into the nature of electron reorganization during the allowed and forbidden reactions, especially from the point of view of the differences in the extent of electron correlation during the course of concerted pericyclic processes. The main advantage of using the geminal instead of orbital expansion of pair densities consists in the specific block diagonal form of the pair density in geminal basis with individual blocks corresponding to singlet and triplet states of electron pair. This opens the possibility of complementing the previous conclusions based on the analysis of pair density[21] by the separate investigation of individual singlet and triplet states of electron pairs as a new means of the deeper insight into the process of electron and spin recoupling in the course of a chemical reaction.

II. THEORETICAL CONSIDERATIONS

The pair density $\rho(1,2)$ is generally defined as the diagonal element of second order density matrix $\rho(1,2,1',2')$ by Eq. 1, where N is the number of electrons and

$$\rho(1,2) = \frac{N(N-1)}{2} \int \Psi^2(1,2,\ldots N) d\zeta_1 d\zeta_2 \ldots d\zeta_N dr_3 dr_4 \ldots dr_N \qquad (1)$$

$d\zeta_i, dr_j$ denote the integration over spin and space coordinate of the electrons i and j, respectively. On the basis of this definition, the second-order similarity index g_{AB} of two isoelectronic molecules A and B can be defined[13] by Eq. 2 in analogy to

$$g_{AB} = \frac{\int \rho_A(1,2)\rho_B(1,2) dr_1 dr_2}{\left(\int \rho_A^2(1,2) dr_1 dr_2\right)^{1/2} \left(\int \rho_B^2(1,2) dr_1 dr_2\right)^{1/2}} \qquad (2)$$

the usual similarity index introduced some time ago by Carbo.[22] If the molecules A and B are identified with the reactant R and product P of a given reaction, then the above definition leads to the second-order similarity index g_{RP} whose exploitation for the study of pericyclic reactions was reported in previous studies.[13–15]

This static description of chemical reaction which is based only on the information about the structure of the reactant and product was subsequently generalized in the study in which the whole formalism was incorporated into the framework of the so-called overlap determinant method. Although the principles of this method are satisfactorily described in the original study,[20] we consider it useful to recapitulate briefly the basic ideas of this method to the extent necessary for the purpose of this review. Within the framework of the overlap determinant method the chemical reaction is regarded as an abstract transformation. Depending on the continuous change of a certain parameter which thus plays the role of generalized reaction coordinate, this transformation converts the structure of the reactant into the structure of the product. If now the structure of these two fundamental species is described by the approximate wave functions, Ψ_R and Ψ_P, then the above abstract transformation can be described by an arbitrary continuous function ensuring the conversion of the function Ψ_R into Ψ_P. In our study[20] we proposed for this purpose a simple trigonometric formula in which the role of the generalized reaction coordinate is played by the parameter φ varying for allowed reactions within the range $(0,\pi/2)$ and for forbidden ones within $(0,-\pi/2)$[21] (Eq. 3). On the basis of this

$$\Psi(\varphi) = \frac{1}{\sqrt{1 + S_{RP}\sin 2\varphi}} (\Psi_R \cos\varphi \pm \Psi_P \sin\varphi) \tag{3}$$

transformation relation it is then possible to introduce the pair density $\rho(1,2 \mid \varphi)$ (Eq. 4), whose values reflect the changes in the mutual coupling of electron motions

$$\rho(1,2 \mid \varphi) = \frac{N(N-1)}{2} \int \Psi^2(\varphi) d\zeta_1 d\zeta_2 \ldots, d\zeta_N dr_3 dr_4 \ldots dr_N \tag{4}$$

during the chemical reaction. The pair density (Eq. 4) can be straightforwardly expressed in the form of expansion (Eq. 5), in which the dependence on the reaction

$$\rho(1,2 \mid \varphi) = \sum_{\alpha\beta\gamma\delta} \Omega_{\alpha\beta\gamma\delta}(\varphi) \chi_\alpha(1) \chi_\beta(1) \chi_\gamma(2) \chi_\delta(2) \tag{5}$$

coordinate is concentrated into the values of the four index matrix $\Omega_{\alpha\beta\gamma\delta}(\varphi)$. However, this density is a rather complex quantity and in order to extract from it the desired information about the electron coupling it has to be subject to a subsequent analysis. One of the possibilities of such analysis is the generalization of the second order similarity index (Eq. 2) into the form (Eq. 6) in which the pair density (Eq. 5) is compared with the pair density of a certain reference standard corresponding to a hypothetical state with no electron coupling.

$$g(\varphi) = \frac{\int \rho(1,2 \mid \varphi)\rho_{ref}(1,2 \mid \varphi)dr_1dr_2}{\left(\int \rho^2(1,2 \mid \varphi)dr_1dr_2\right)^{\frac{1}{2}}\left(\int \rho_{ref}^2(1,2 \mid \varphi)dr_1dr_2\right)^{\frac{1}{2}}} \tag{6}$$

Such a standard can be in principle defined in two ways. The first arises from the proposal by McWeeny and Kutzelnigg[23] who defined the pair density of the reference standard as a product of corresponding first order density matrices (Eqs. 7 and 8),

$$\rho_{ref}(1,2 \mid \varphi) = \rho(1 \mid \varphi)\rho(2 \mid \varphi) \tag{7}$$

where

$$\rho(1 \mid \varphi) = N\int \Psi^2(\varphi)d\zeta_1 d\zeta_2 \ldots d\zeta_N dr_2 dr_3 \ldots dr_N \tag{8}$$

The second possible choice of a reference standard and the one which we use in this study is based on the proposal by Hashimoto[24] to derive the reference pair density from a one-determinantal wave function. Within this model, the pair density is given by Eq. 9 where $\rho_1(1,2 \mid \varphi)$ is the nondiagonal element of the first order

$$\rho_{ref}(1,2 \mid \varphi) = \frac{1}{2}\rho(1 \mid \varphi)\rho(2 \mid \varphi) - \frac{1}{4}\rho_1^2(1,2 \mid \varphi) \tag{9}$$

density matrix. In this study a Hashimoto type standard was used, but as also demonstrated by a direct comparison, this particular choice of standard has no qualitative effect on the resulting picture.

Having specified the reference standard, the practical applicability of the similarity index (Eq. 6) requires one to replace the general expressions for the pair densities (Eqs. 4 and 9) by the appropriate representations. One of such possibilities used in previous studies is based on the expansion in the basis of atomic orbitals (Eq. 5). Such a straightforward expansion is not, however, the only possibility for representing the pair densities. In our opinion another more convenient possibility is based on the replacement of the expansion (Eq. 5) by the alternative expansion in the basis of two-electron functions—geminals (Eq. 10). Within the framework of such an expansion the definition (Eq. 6) simplifies to Eq. 11.

$$\rho(1,2 \mid \varphi) = \sum_{\alpha\beta} \mathbf{\Gamma}_{\alpha\beta}(\varphi)\lambda_\alpha(1,2)\lambda_\beta(1,2) \tag{10}$$

$$g(\varphi) = \frac{Tr\mathbf{\Gamma}(\varphi)\mathbf{\Gamma}_{ref}(\varphi)}{\left(Tr\mathbf{\Gamma}^2(\varphi)\right)^{\frac{1}{2}}\left(Tr\mathbf{\Gamma}_{ref}^2(\varphi)\right)^{\frac{1}{2}}} \tag{11}$$

The reason for the preference of geminal expansion is that electron correlation is the phenomenon which is closely connected with the coupling of electron pairs. Also the expansion of pair density based on two-electron functions inherently describes pair behavior the most appropriately. Another important advantage of the work with the geminal expansion is that if the geminal basis is chosen so as to correspond to spin pure singlet and triplet two-electron functions, the matrices $\mathbf{\Gamma}$ have the block-diagonal form with individual blocks corresponding to singlet and triplet components (Eq. 12). From this it then follows that, in addition to global

$$\mathbf{\Gamma}(\varphi) = \mathbf{\Gamma}^s(\varphi) \oplus \mathbf{\Gamma}^t(\varphi) \tag{12}$$

similarity indices calculated from the whole pair density, it is also possible to determine "partial" similarity indices describing the similarity between the singlet and triplet components of pair densities $\rho(1,2\,|\,\varphi)$ and $\rho_{ref}(1,2\,|\,\varphi)$.

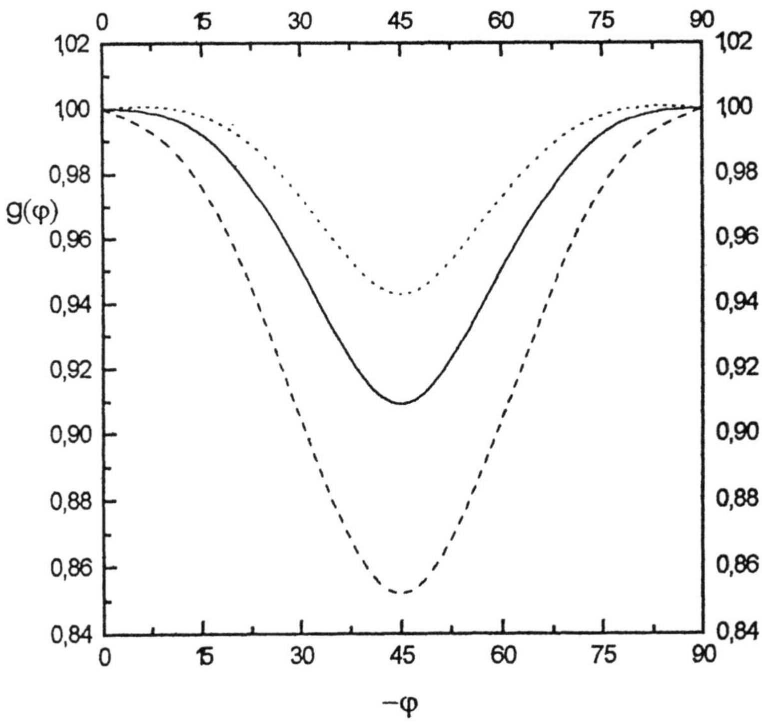

Figure 1. Calculated dependence of total (*full line*), singlet (*dashed line*), and triplet (*dotted line*) second-order similarity indices $g(\varphi)$ on the generalized reaction coordinate φ for the thermally forbidden disrotatory butadiene to cyclobutene cyclization.

Having introduced the basic philosophy of the similarity approach, we need more details about the geminal expansion of the pair density (Eq. 10). Combining Eqs. 3 and 4, the general expression for the pair density can be rewritten in the form of Eq. 13 in which $\rho_{RR}(1,2)$ and $\rho_{PP}(1,2)$ are the pair densities of the isolated reactant,

$$\rho(1,2\,|\,\varphi) = \frac{1}{(1 + S_{RP}\sin 2\varphi)} \times \{\rho_{RR}(1,2)\cos^2\varphi + \rho_{PP}(1,2)\sin^2\varphi + \rho_{RP}(1,2)\sin\varphi\cos\varphi\} \quad (13)$$

and the product $\rho_{RP}(1,2)$ is the corresponding overlap term. If we confine ourselves only to the simplest case where the reactant and the product are described by a single Slater determinant, the geminal expansions of both $\rho_{RR}(1,2)$, $\rho_{PP}(1,2)$, and $\rho_{RP}(1,2)$ can be expressed analytically. For the case of the reactant and product pair densities the corresponding formulae are given in Refs. 19 and 25, and for the remaining overlap term in the Appendix.

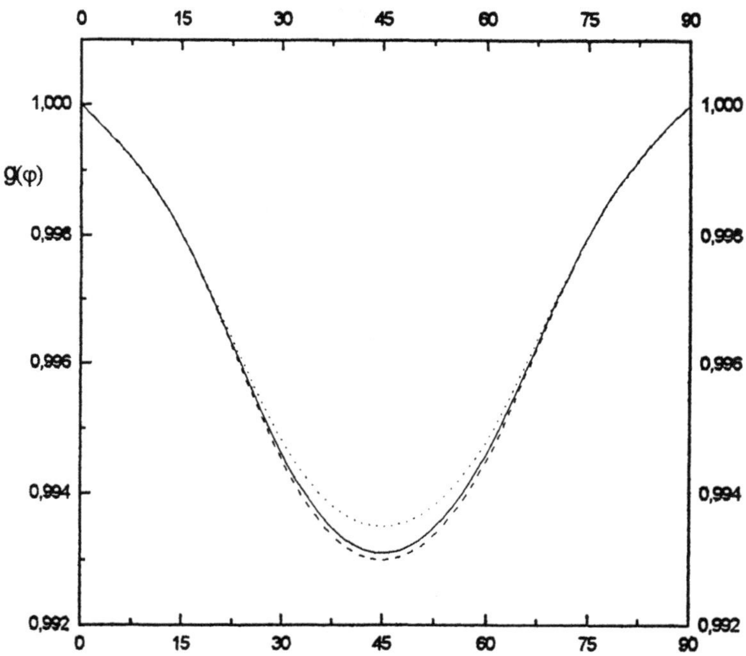

Figure 2. Calculated dependence of total (*full line*), singlet (*dashed line*) and triplet (*dotted line*) second-order similarity indices $g(\varphi)$ on the generalized reaction coordinate φ for the thermally allowed conrotatory butadiene to cyclobutene cyclization.

The above formalism was practically applied to the analysis of correlation effects in a series of selected pericyclic reactions. In order to maintain the continuity with our previous studies, the selected series was the same as in.[21] This allows us also to reduce the specification of technical details which can be found elsewhere.[13,21] Here we only specify that molecular orbitals used in the construction of the wave functions were obtained by the simple HMO method compatible with the topological nature of the overlap determinant method. The calculated dependence of similarity indices $g(\varphi)$, $g^s(\varphi)$, and $g^t(\varphi)$ on the value of the reaction coordinate φ for allowed and forbidden butadiene to cyclobutene cyclization is displayed in Figures 1 and 2.

The form of the dependence for other reactions is essentially the same except for the difference in the actual values of the indices. Because of the similarity in the form of $g(\varphi)$ vs. φ dependencies it is not necessary to display the values of the indices for all angles φ but, instead, only the values for critical points $X(\pi/4)$ and $X(-\pi/4)$ can be given. The corresponding values of similarity indices $g(\pm\,\pi/4)$, $g^s(\pm\,\pi/4)$, and $g^t(\pm\,\pi/4)$ (for allowed and forbidden reactions, respectively) are summarized in Tables 1 and 2 in Section III.

III. RESULTS AND DISCUSSION

Let us discuss the conclusions suggested by Figures 1 and 2 in Section II. First of all it is possible to see—and this conclusion holds for both forbidden and allowed reactions—that the role of the electron correlation during the reaction is not constant but varies with the position on the reaction coordinate. The greatest mutual coupling can be observed for the structures in the vicinity of the critical point $X(\pm\,\pi/4)$, which in the overlap determinant method plays the role of the transition state. This result is not surprising since it closely corresponds to the experience of practical quantum chemical calculations where the requirements on the inclusion of the electron correlation are usually higher for transition states or other structures near the top of the energy barrier than for the stable molecules near the equilibrium geometry (Table 1).

Another general conclusion that holds again for all types of reactions studied is that the qualitative parallel manifesting itself in the values of global similarity index $g(\varphi)$ for allowed and forbidden reactions is similarly reflected in general trends of "partial" similarity indices corresponding to spin pure singlet and triplet states of electron pairs. This result is parallel to what was observed in our previous study[26] dealing with the analogous study based on the use of spin-resolved similarity indices $g^{\uparrow\uparrow}(\varphi)$ and $g^{\uparrow\downarrow}(\varphi)$ and corresponding to contributions of Fermi and Coulomb correlation, respectively.

Despite all cases where the correlation in allowed and forbidden reactions acts in parallel, there are also some remarkable differences. First, it is possible to see that in an absolute sense, the mutual coupling of electron motions is generally higher in forbidden reactions than in allowed ones. Thus, if we take the value of the

Table 1. Calculated Values of Similarity Indices $g(\pm \pi/4)$, $g^s(\pm \pi/4)$, and $g^t(\pm \pi/4)$ for the Critical Structure $X(\pm \pi/4)$ in a Series of Allowed $(+\pi/4)$ and Forbidden $(-\pi/4)$ Electrocyclic Reactions

Reaction	$g^s(\pm \pi/4)$	$g^t(\pm \pi/4)$	$g(\pm \pi/4)$
butadiene → cyclobutene	0.9935	0.9930	0.9931
	0.8520	0.9428	0.9092
hexatriene → cyclohexadiene	0.9939	0.9953	0.9951
	0.9298	0.9831	0.9717
oktatetranene → cyclooktatriene	0.9951	0.9967	0.9965
	0.9602	0.9916	0.9862

Note: Upper entry corresponds to allowed and lower to forbidden reaction mechanism.

similarity index for the critical structure $X(\pm \pi/4)$ as a measure of the extent of correlation, then for all the types of the indices in the Tables 1 and 2 we find that $g(\text{allowed}) > g(\text{forbidden})$. This clearly suggests that the mutual electron coupling in allowed reactions is closer to the reference standard than for the forbidden ones. Also this conclusion is not too surprising since the greater electron coupling in forbidden reactions can be intuitively expected from the mere fact of the presence of orbital crossing taking place in this processes. The fact that this conclusion could have been expected without any calculations and only on the basis of intuitive consideration, does not detract, however, in any way from the usefulness of the proposed similarity approach. The greatest advantage of this approach is its quantitative nature which allows one to enrich the simple intuitive considerations by a certain quantitative aspect owing to which the general trends can be disclosed which would otherwise be difficult to ascertain.[13,14,21,26] Thus, the comparison of the similarity indices $g(\pm \pi/4)$ clearly suggests that for the class of allowed electrocyclic reactions the role of the electron correlation is relatively unimportant ($g(\varphi) \cong 1$ for all φ), whereas for allowed cycloadditions and sigmatropic reactions the corresponding values considerably deviate from unity and are, in fact, comparable with the values for forbidden electrocyclizations (Table 2). This result is very interesting since it provides a theoretical rationale both for the numerical observation of Houk, in which a small sensitivity of allowed electrocyclic reactions to correlation effects was reported in a study of transition state structures,[17] and also for its additional support of our earlier studies[13,14,21,26,27] confirming the legitimacy of the intuitive proposal by Dewar to include cycloadditions and sigmatropic reactions into the special class of pericyclic reactions—the so-called multibond reactions.[16]

Another interesting conclusion closely tied with the quantitative nature of the approach concerns its ability to provide an insight into the nature of electron and spin recoupling in chemical reactions. Thus, if we accept the values of the similarity indices at the critical point $X(\pm \pi/4)$ as a measure of the extent of correlation effects,

Table 2. Calculated Values of Similarity Indices $g(\pm \pi/4)$, $g^s(\pm \pi/4)$ and $g^t(\pm \pi/4)$ for the Critical Structure $X(\pm \pi/4)$ in a Series of Allowed $(+\pi/4)$ and Forbidden $(-\pi/4)$ Cycloadditions and Sigmatropic Rearrangements

Reaction	$g^s(\pm \pi/4)$	$^*g^t(\pm \pi/4)$	$g(\pm \pi/4)$
ethene dimerization	0.9703	0.9619	0.9640
2 + 2 cycloaddition	0.8520	0.9428	0.9092
Diels–Alder reaction	0.9726	0.9724	0.9724
4 + 2 cycloaddition	0.9361	0.9628	0.9572
hexatriene + ethene	0.9814	0.9836	0.9832
6 + 2 cycloaddition	0.9648	0.9796	0.9771
butadiene + butadiene	0.9755	0.9755	0.9755
4 + 4 cycloaddition	0.9637	0.9710	0.9697
Cope rearrangement	0.9568	0.9506	0.9516
3 3′ sigmatropic reaction	0.9343	0.9398	0.9384

Note: Upper entry corresponds to allowed and lower to forbidden reaction mechanism.

then it is possible to see (Tables 1 and 2) that there is a clear difference between the allowed and forbidden reactions just in the recoupling of singlet and triplet pairs. In forbidden reactions are specifically singlet pairs which are apparently more coupled, while for allowed reactions the role of electron correlation for singlet and triplet pairs is roughly the same. This result is very interesting since our conclusions seem to be supported, at least for the allowed [2+2] ethene dimerization for which the reference data are available, from the recent spin-coupled analysis.[18] The authors report that in the vicinity of transition state the spin recoupling takes place. The corresponding wave function is dominated by two modes of spin coupling, with nearly equal weights and these contributions corresponding to singlet and triplet coupling of electrons in disappearing and newly created bonds, respectively. In this connection it would be interesting to perform similar spin-coupled calculations on the thermally forbidden mechanism of the same reaction and to see whether our predicted prevalence of singlet recoupling will also be observed.

IV. SUMMARY

In summarizing the above results, it is possible to say that the presented approach represents a new, perhaps interesting attempt at the systematic study of the effects of electron and spin recoupling in chemical reactions. Even if some of the conclusions are not entirely new, we believe that the simplicity of the approach allows it to be applied to broader series of compounds and that future systematic use may contribute to better understanding of the role of electron correlation in chemical reactions.

V. APPENDIX

Let the wave functions of the reactant and the product be described by a single Slater determinant Ψ_R and Ψ_P constructed from molecular orbitals r_i, p_j (Eqs. A1, A2):

$$\Psi_R = |r_1\overline{r_1}r_2\overline{r_2}\ldots\ldots r_{N/2}\overline{r_{N/2}}| \tag{A1}$$

$$\Psi_P = |p_1\overline{p_1}p_2\overline{p_2}\ldots\ldots p_{N/2}\overline{p_{N/2}}| \tag{A2}$$

In this case the overlap term $\rho_{RP}(1,2)$ in Eq. 13 is given by Eq. A3, where A_{ij} is the

$$\rho_{RP}(1,2) = 4\sum_{i,j}^{occ} A_{ij}r_i(1)p_j(1)\sum_{i,j}^{occ} A_{ij}r_i(2)p_j(2)$$
$$-2\sum_{i,j}^{occ} A_{ij}r_i(1)p_j(2)\sum_{i,j}^{occ} A_{ij}r_i(2)p_j(1) \tag{A3}$$

minor of the matrix of overlap determinants between the molecular orbitals of R and P, and where the orbitals are expressed in harmony with the philosophy of the generalized overlap determinant method[20] in the form of usual LCAO expansion in the common basis of atomic orbitals χ (Eqs. A4, A5). Inserting these expansions

$$r_i(k) = \sum_{\mu} \rho_{\mu i}\chi_\mu(k) \tag{A4}$$

$$p_i(k) = \sum_{\mu} \pi_{\mu i}\chi_\mu(k) \tag{A5}$$

into Eq. A3 the ordinary expansion of overlap pair density in the basis of atomic orbitals can be obtained and the corresponding formulae can be found in the study.[21] However, we are not interested in such a straightforward expansion in AO basis but, instead, the alternative expansion in the basis of geminals is required. It can be shown that if the geminal basis is selected, in harmony with the study,[28] in the form of Eqs. A6–A8, the pair density $\rho_{RP}(1,2)$ can be expressed in the form of block diagonal matrix given in Table 3 where individual matrix elements $\vartheta_{\mu\nu}$ are given by Eq. A9.

$$\sigma_{\alpha\alpha}(1,2) = \chi_\alpha(1)\chi_\alpha(2) \tag{A6}$$

$$\sigma_{\alpha\beta}(1,2) = \frac{1}{\sqrt{2}}\left(\chi_\alpha(1)\chi_\beta(2) + \chi_\alpha(2)\chi_\beta(1)\right) \tag{A7}$$

Table 3. Block Diagonal Form of the Overlap Pair Density $\rho_{RP}(1,2)$ in the Basis of Singlet $(\sigma_{\alpha\alpha},\sigma_{\alpha\beta})$ and Triplet $(\tau_{\alpha\beta})$ Geminals

Basis Geminals	$\sigma_{\beta\beta}(1,2)$	$\sigma_{\beta\gamma}(1,2)$	$\tau_{\beta\gamma}(1,2)$
$\sigma_{\alpha\alpha}(1,2)$	$\vartheta^2_{\alpha\beta} + \vartheta^2_{\beta\alpha}$	$\sqrt{2}\,\vartheta_{\alpha\beta}\vartheta_{\gamma\alpha}$ $\sqrt{2}\,\vartheta_{\beta\alpha}\vartheta_{\alpha\gamma}$	0
$\sigma_{\alpha\delta}(1,2)$	$\sqrt{2}\,\vartheta_{\alpha\beta}\vartheta_{\beta\delta}$ $\sqrt{2}\,\vartheta_{\beta\alpha}\vartheta_{\delta\beta}$	$\vartheta_{\alpha\beta}\vartheta_{\gamma\delta} + \vartheta_{\alpha\gamma}\vartheta_{\beta\delta}$ $\vartheta_{\beta\alpha}\vartheta_{\delta\gamma} + \vartheta_{\gamma\alpha}\vartheta_{\delta\beta}$	0
$\tau_{\alpha\delta}(1,2)$	0	0	$3\vartheta_{\alpha\beta}\vartheta_{\gamma\delta} - 3\vartheta_{\alpha\gamma}\vartheta_{\beta\delta} +$ $3\vartheta_{\beta\alpha}\vartheta_{\delta\gamma} - 3\vartheta_{\gamma\alpha}\vartheta_{\delta\beta}$

$$\tau_{\alpha\beta}(1,2) = \frac{1}{\sqrt{2}}\left(\chi_\alpha(1)\chi_\beta(2) - \chi_\alpha(2)\chi_\beta(1)\right) \tag{A8}$$

$$\vartheta_{\mu\nu} = \sum_i^{occ} A_{ij}\rho_{\mu i}\pi_{\nu j} \tag{A9}$$

ACKNOWLEDGMENT

This work was completed within the grant project No. 203/95/0650 of the Grant Agency of the Czech Republic. The author gratefully acknowledges this support.

REFERENCES

1. Bernardi, F.; Bottoni, F.A.; Guest, M.F.; Hillier, I.H.; Robb, M.A.; Venturini, A. *J. Am. Chem. Soc.* **1988**, *110*, 3050.
2. Dewar, M.J.S.; Olivella, S.; Rzepa, H. *J. Am. Chem. Soc.* **1978**, *100*, 5650.
3. Bernardi, F.; Bottoni, F.A.; Robb, M.A.; Schlegel, H.B.; Tonachini, G. *J. Am. Chem. Soc.* **1985**, *107*, 2260.
4. Olivella, S.; Salvador, J. *J. Comput. Chem.* **1991**, *12*, 792.
5. Carsky, P.; Urban, M. *Ab initio calculations. Methods and Applications in Chemistry*, Lecture Notes in Chemistry 16, Springer Verlag, Berlin, 1980.
6. Karafiloglou, P.; Malrieu, J.P. *Chem. Phys.* **1986**, *104*, 383.
7. Smith, D.W.; Larson, E.G.; Morrison, R.C. *Int. J. Quant. Chem.* **1970**, *3*, 689.
8. Becke, D.A.; Edcombe, K.E. *J. Chem. Phys.* **1990**, *92*, 5397.
9. Lennard-Jones, J.E. *J. Chem. Phys.* **1952**, *20*, 1024.
10. Bader, R.F.W.; Stephens, M.E. *J. Am. Chem. Soc.* **1975**, *97*, 7391.
11. Hohlneicher, G.; Gutman, M. *Int. J. Quant. Chem.* **1986**, *29*, 1291.
12. Salem, L. *Nouv. J. Chem.* **1978**, *2*, 559.
13. Ponec, R.; Strnad, M. *Collect. Czech. Chem. Commun.* **1990**, *55*, 896.
14. Ponec, R.; Strnad, M. *Int. J. Quant. Chem.* **1992**, *42*, 501.
15. Ponec, R.; Strnad, M. *J. Phys. Org. Chem.* **1992**, *5*, 764.
16. Dewar, M.J.S. *J. Am. Chem. Soc.* **1984**, *106*, 209.
17. Houk, K.N.; Yi, Li; Evanseck, J.D. *Angew. Chem. Int. Ed.* **1992**, *31*, 682.

18. Karadakov, P.; Gerratt, J.; Cooper, D.L.; Raimondi, M. *J. Chem. Soc. Faraday Trans.* **1994**, *90*, 1643.
19. Strnad, M.; Ponec, R. *Int. J. Quant. Chem.* **1994**, *49*, 35.
20. Ponec, R. *Collect. Czech. Chem. Commun.* **1985**, *50*, 1121.
21. Ponec, R.; Strnad, M. *Collect. Czech. Chem. Commun.* **1993**, *58*, 1751.
22. Carbo, R.; Leyda, L.; Arnau, M. *Int. J. Quant. Chem.* **1980**, *17*, 1185.
23. McWeeny, R.; Kutzelnigg, W. *Int. J. Quant. Chem.* **1968**, *2*, 187.
24. Hashimoto, K. *Int. J. Quant. Chem.* **1982**, *21*, 861.
25. Ponec, R.; Strnad, M. *Int. J. Quant. Chem.* **1994**, *50*, 43.
26. Ponec, R.; Strnad, M. *Chem. Papers* **1994**, *48*, 72.
27. Ponec, R.; Strnad, M. *Collect. Czech. Chem. Commun.* **1990**, *55*, 2363.
28. Smith, D.W.; Fogel, S.J. *J. Chem. Phys.* **1965**, *43*, S91.

CONFORMATIONAL ANALYSIS FROM THE VIEWPOINT OF MOLECULAR SIMILARITY

Josep M. Oliva, Ramon Carbó-Dorca, and Jordi Mestres

Advances in Molecular Similarity
Volume 1, pages 135–165

ISBN: 0-7623-0131-7

ABSTRACT

Different approaches to exact overlap quantum molecular self-similarity measures (QMSMs) are used to analyze the charge density redistribution due to torsional rotations. For this purpose, four different approximations have been employed: (1) fitting the electron density using gaussian *s* functions, (2) constructing the electron density using atom-centered single-gaussian functions, (3) using a fitted function from quantum atomic self-similarity measures, and (4) calculating a sum of quantum atomic self-similarity measures.

The *n*-alkanes family has been chosen to test the behavior of the different approximations to QMSM as compared to energy profiles when torsional angles are rotated. The results presented in this contribution reveal that: (1) the use of exact QMSMs appears to be a useful methodology to accurately quantify the charge density redistribution of torsional profiles under a given level of theory; and (2) the use of several approximations to the exact QMSM can serve to tackle the well-known difficult task of performing a detailed analysis of the torsional hypersurface, emerging as a promising tool for a fast and wide survey in the search for those regions where local minima (and in particular, the global minimum) are located. In this sense, differences between conformational and rotational profiles have been clarified. For this series of *n*-alkanes, it is shown that electronic energy and overlap quantum molecular self-similarity measure profiles are analogous when the rotational approximation is used, while they become opposite if a conformational approximation is employed.

I. INTRODUCTION

It is widely established that the three-dimensional structure of molecules cannot only be described by a single frozen geometry, but by the ensemble of conformations they can adopt. In fact, the properties of molecules strongly depend on their conformational flexibility which becomes an essential fact in any approach to computer-aided drug design.[1] However, when dealing with large molecules, a wide exploration of the conformational space may represent a difficult task because of the presence of a huge number of local minima along the potential energy hypersurface. When the number of torsional angles increases, it is practically impossible to perform an exhaustive systematical search to locate the global minimum, due to computational time requirements. Moreover, once a theoretical level has been chosen, even finding the global minimum at this level does not ensure that the structure found at other theoretical levels will be the same. A final additional difficulty in conformational problems is that the representation of the conformational energy profile in the gas phase may be far away from the one perturbed by a solvent or under the effects of the proteinic environment when bounded to a receptor.

Due to the above mentioned inherent difficulties in dealing with this problem, the main objective of any conformational search will be the efficient scanning of the full conformational space in order to identify all thermally accessible conformations and locate the region containing a potential well around the global minimum. Sometimes the goal is focused into reducing the number of low-energy regions under consideration to a computationally manageable number. For this purpose, a variety of methods have been described to identify minimum energy conformations.[2,3] Alternatively, stochastic strategies have been recently adapted to deal with this multiple minima problem. Among them, simulated annealing[4,5] and genetic algorithms[6] appear to be useful approaches.

The study of the changes undergone by a molecule under torsional rotations are usually evaluated by the obtention of its energy, used as a molecular descriptor. In this way, the size of the conformational problem often restricts calculations to the evaluation of some empirical force fields. For large biological molecules, application of quantum mechanical semiempirical methods is limited and ab initio methods become prohibitive. Recently, the variation of molecular hardness and chemical potential has been also used to analyze those changes produced under torsional rotations.[7]

This contribution presents a new technique to approach the conformational problem. It is based on the fact that a torsional rotation always produces a change in the relative structural parameters (distances and angles) between atoms in the molecule, inducing a charge density redistribution. It seems obvious that the analysis of this phenomenon will give an idea of the evolution of changes suffered by the molecule, from an electronic density point of view.

At this point, it is necessary to stress the difference between rotational and conformational analyses. In the former, when rotating any of the active torsional angles of the molecule, no nuclear relaxation is allowed; that is, there is no geometry reoptimization of the molecule at each point of the torsional hypersurface. Notwithstanding, in the conformational approach, a molecular relaxation is allowed in such a way that in the torsional hypersurface every point will correspond to a constrained energy minimum. In other words, in the rotational analysis, all internal coordinates of the molecule are kept fixed except the active torsional angles, whereas in the conformational analysis all internal coordinates are altered during the geometry optimization process, except the same active torsional angles which define the independent variables of the conformational surface. The differences between the use of these two approximations from the viewpoint of the electron density redistribution will be clarified.

In a more exact quantitative level, it has been recently shown that exact overlap quantum molecular self-similarity measures (QMSMs) can be employed as molecular descriptors to quantify the degree of concentration of any given charge density distribution[8] and, in particular, to its use in the differentiation of several conformational, configurational, and constitutional isomeric systems.[9] The main drawback of this approach consists in the evaluation of exact QMSMs, which are computationally very demanding. In the present chapter several approximations will be proposed in order to speed up the QMSM calculation applied to the conformational analysis of different test cases. A discussion of the algorithm performances and viability will be also given.

The aim of this contribution is twofold: (1) the use of exact QMSMs is revealed as an excellent methodology to quantitatively study the charge density redistribution undergone by torsional rotations and, (2) the use of approximations to the exact QMSMs can serve to perform fast conformational analyses from the molecular similarity viewpoint.

II. APPROXIMATIONS TO EXACT QUANTUM MOLECULAR SIMILARITY MEASURES

The exact QMSM was originally defined by Carbó et al.[10] as,

$$S_{IJ}(\mathbf{r}_I, \mathbf{r}_J; \theta) = \int\int \rho_I(\mathbf{r}_1)\theta(\mathbf{r}_1,\mathbf{r}_2)\rho_J(\mathbf{r}_2)d\mathbf{r}_1 d\mathbf{r}_2 \tag{1}$$

where ρ_I and ρ_J are, respectively, the electron density distributions of two molecules I and J; $\theta(\mathbf{r}_1,\mathbf{r}_2)$ is a positive definite operator depending on two-electron coordinates; and $\mathbf{r}_I$ and $\mathbf{r}_J$ represent the coordinates of molecules I and J. When $\theta(\mathbf{r}_1,\mathbf{r}_2) = \delta(\mathbf{r}_1 - \mathbf{r}_2)$, Eq. 1 becomes an overlap integral between two electron density distributions which quantifies the shared concentration of electron density distributions of molecules I and J.[11] In the particular case that $I = J$, S_{II} becomes a measure of the concentration of the electron density distribution of molecule I and, thus, it can be taken as a molecular descriptor.[8] In order to simplify our notation and due to the fact that only overlap quantum self-similarity measures (S_{II}) will be computed, throughout this work we will use the general notation QMSM to denote these particular overlap quantum self-similarity measures.

From the computational ab initio calculations point of view, exact QMSM (hereafter EQMSM) present a serious problem: the computational cost of the integrals involved in Eq. 1 depends on N_b^4, N_b being the number of basis functions.[12] This is the reason why, in order to lower the computational time due to expensive integrals appearing in EQMSM, different approximations will be surveyed. The behavior of these approximations will then be tested when performing an exhaustive analysis of the conformational hypersurface in a given molecule.

A. QMSM from Fitted Densities

In order to circumvent the above exposed N_b^4 problem for the computation of EQMSM, the electronic first-order density can be approximated by a linear expression using a set of gaussian s-type functions $\{g_k(\mathbf{r})\}$:[12]

$$\rho_I(\mathbf{r}) \approx \sum_{k \in I} a_k g_k(\mathbf{r}) \tag{2}$$

Substitution of Eq. 2 into Eq. 1 yields an approximation to the EQMSM of chemical system I:

$$S_{I,I} \approx \sum_{k \in I} \sum_{l \in I} a_k a_l \int g_k(\mathbf{r}) g_1(\mathbf{r})\, d\mathbf{r} \tag{3}$$

If N_f is the number of gaussian functions used in the fitting of the density (Eq. 2), once the electron density has been fitted, evaluation of QMSM becomes a N_f^2-dependent process in comparison with the N_b^4-dependent process in ab initio EQMSM calculations. Thus, the computational time used in QMSM calculations is considerably lowered when using fitted densities.[12,13] An improved algorithm for performing a density fitting restricted to have positive a_k coefficients has been recently adapted.[14]

Hereafter, QMSM using fitted densities will be denoted as FQMSM. Under the conformational approach, a density fitting will be performed at each point of the torsional profile, and the appropriate FQMSM computed. However, when the rotational approximation is employed, only one density fitting is performed and all FQMSM of the different rotamers are computed within the same density fitting, rotating the $\{g_k(\mathbf{r})\}$ functions centered at each atom of the molecule. The consequences of this approximation will be discussed in Section III.

B. The Atom-Centered Single-Gaussian Approximation

The molecular electron density can be also approximated by summing up the contributions of the constituent atomic electron densities of each molecule as:

$$\rho_I(\mathbf{r}) = \sum_{i \in I} \rho_i(\mathbf{r}) \tag{4}$$

These atomic electron densities (ρ_i) are represented by an atom-centered single-gaussian function,

$$\rho_i(\mathbf{r}) = \alpha_i \exp\left(-\beta_i \left| \mathbf{r} - \mathbf{R}_i \right|^2\right) \tag{5}$$

where $\mathbf{R}_i$ is the nuclear coordinate position of atom i and the coefficients α_i (which depend on the effective charge of atom i) and β_i for any distinct atom are obtained using a procedure previously described[15] which ensures that integration of each ρ_i over all space returns the atomic number of electrons. This atom-centered single-gaussian approximation will be referred to as ACSGA.

C. Fitted Function from Quantum Atomic Similarity Measures

In this section a new approximation to QMSM is introduced. In order to distinguish between molecular and atomic self-similarities, capital and small letters will be used throughout, i.e., $\{S_{I,I}\}$ and $\{s_{i,i}\}$ will denote overlap quantum self-similarities of molecule I and atom i, respectively.

Atomic self-consistent field (SCF) energies from hydrogen to xenon, can be fitted to a potential function depending only on the atomic number, as shown in Figure 1a:

$$-E_i \approx 0.5246(Z_i)^{2.3901} \tag{6}$$

In the same way, computations of overlap quantum atomic self-similarity measures (QASMs) from exact first-order atomic density functions[8] lead to the obtention of a potential fitted function depicted in Figure 1b,

$$s_{i,i} \approx 0.0676(Z_i)^{3.3921} \tag{7}$$

where Z_i is the atomic number of atom i. Arranging Eqs. 6 and 7, an approximate connection between atomic energies and QASMs can be obtained:

$$-E_i \approx 3.5131\,(s_{i,i})^{0.7041} \tag{8}$$

Atomic SCF densities and energies were obtained by means of the ATOMIC program[16] at the ROHF level of calculation[17] with a double-ζ basis set over Slater-type orbitals (STO).[18] Exact overlap quantum atomic self-similarity measures were computed using the program SEMAT.[19]

Equation 7 provides a good approximation to QASM values, but in order to evaluate QMSM it is necessary to involve crossed terms between different atoms of a given molecule, i.e., the QASM between two atoms at a given distance R. Taking into account Eq. 7, a new formula for approximate QASM is put forward:

$$s_{i,j} \approx 0.0676(Z_iZ_j)^{1.6961}\,f(R) \tag{9}$$

Thus, QASM of two atoms at a given distance can be approximated by a function that depends on both atomic numbers, Z_i and Z_j, and the distance between atoms R. The function $f(R)$ behaves approximately as a negative exponential, having an exact solution only for the ground state of the hydrogen atom (Figure 2):

$$s_{H,H}(R) = \frac{e^{-2R}}{24\pi}(4R^2 + 6R + 3) \tag{10}$$

The long-range behavior of $\rho(\mathbf{r})$ for both atoms and molecules has been discussed by a number of authors.[20] The results of these studies show that the charge density, at a sufficiently large distance from all nuclei, decays exponentially according to $\rho(\mathbf{r}) \approx \exp[-(2\varepsilon)^{1/2}\mathbf{r}]$, where ε is the first ionization potential of the system. Thus, as a first approximation, $f(R)$ was chosen to be $\exp(-R)$ in all calculations. Studies on the dependence of $f(R)$ depending on each particular pair of atoms (as the one presented in Figure 2 between a pair of Hs) are being done in our laboratory.

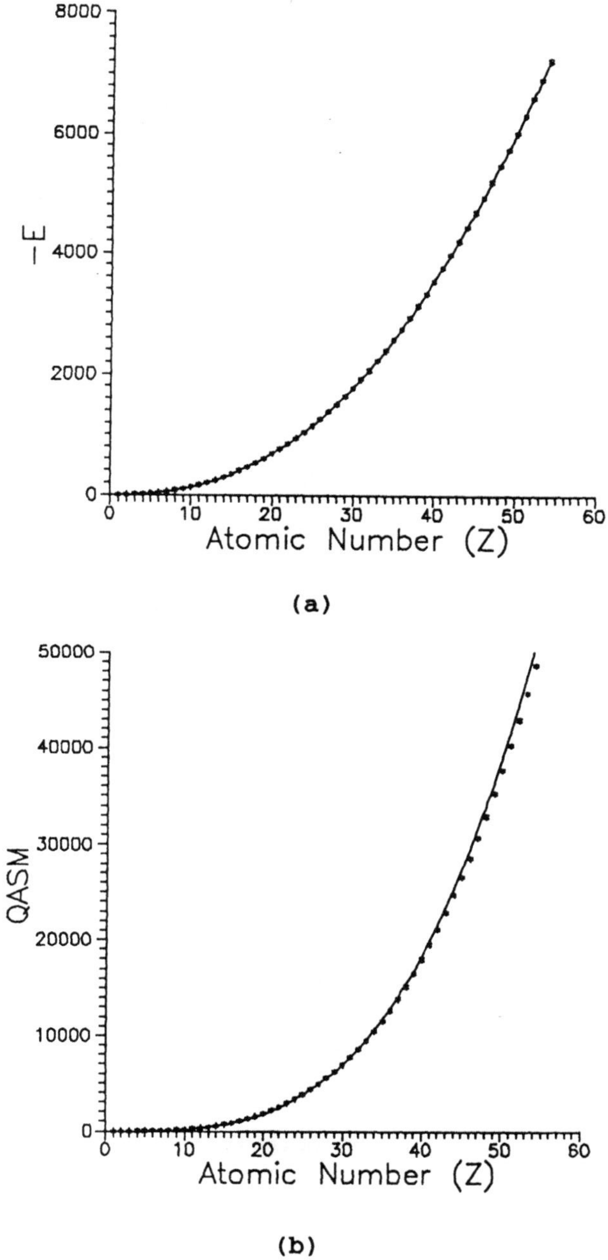

Figure 1. Relationships between (**a**) atomic number and atomic energy (in hartrees) and (**b**) atomic number and quantum atomic self-similarity measure (in au).

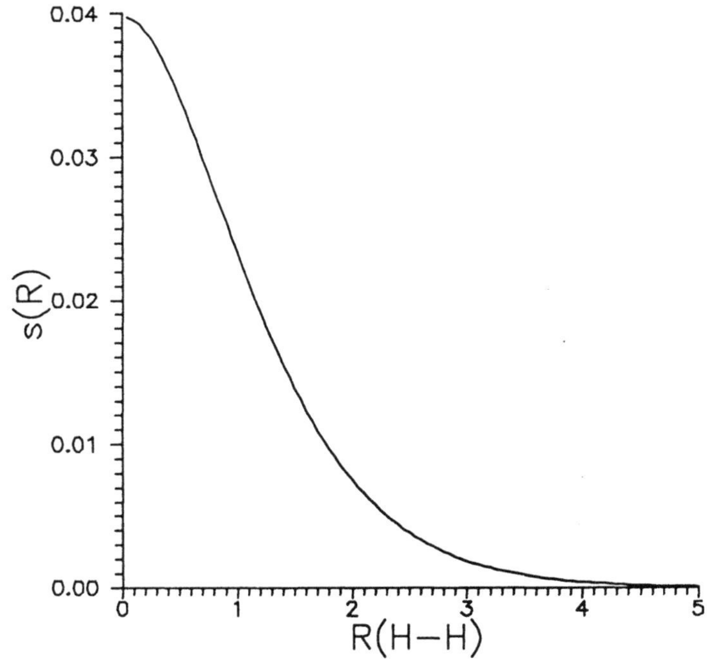

Figure 2. Electron density overlapping between two hydrogen atoms depending on their interatomic distance (in au).

Therefore, an approximate QMSM can be defined as,

$$S_{I,J} \approx \sum_{i=1}^{N_I} \sum_{j=1}^{N_J} s_{i,j} \tag{11}$$

where $s_{i,j}$ are approximate QASM between atoms i of molecule I and atom j of molecule J (as defined in Eq. 9), and N_I and N_J are the number of atoms of molecules I and J, respectively. As can be seen, Eq. 11 becomes an approximate analogous expression to the definition of exact QMSM given in Eq. 1. Hereafter, the QMSM approach as calculated from Eq. 11 will be denoted as FQASM.

D. Sum of QASM

An exact evaluation of QASMs can be obtained by calculating the integral,

$$s_{i,j} = \int \rho_i(\mathbf{r} - \mathbf{R}_i)\, \rho_j(\mathbf{r} - \mathbf{R}_j)\, d\mathbf{r} \tag{12}$$

where $\rho_i(\mathbf{r} - \mathbf{R}_i)$ and $\rho_j(\mathbf{r} - \mathbf{R}_j)$ are the atomic electron densities of atom i of molecule I centered at $\mathbf{R}_i$ and atom j of molecule J centered at $\mathbf{R}_j$, respectively.[8] Obtention of QMSMs (as defined in Eq. 11) from $s_{i,j}$ computed from Eq. 12 will be denoted as SQASM. Note that the above presented Eq. 9 is an approximation to the integral given by Eq. 12. In fact, sums of $s_{i,i}$ QASMs were already used as first-order molecular descriptors.[9] These $s_{i,i}$ QASM values were recently reported in a table[8] to be used as an incredibly fast approach to exact QMSMs. This approach may be useful for families of molecules with different stoichiometry, but the singular differences between QMSMs of a set of conformational, configurational, or constitutional isomers are due to the $s_{i,j}$ QASM terms.[9] Thus, although s_{ij} QASMs have much smaller values than $s_{i,i}$ QASMs, they play a fundamental role for discerning small changes in atomic density distributions at a given interatomic distance. In order to speed up calculation of SQASM, an atomic single-ζ basis set[18] was used throughout this work when referred to this particular approximation.

In the next section, the ensemble of EQMSM and the different approximations proposed to EQMSM will be applied to test cases of molecules up to four dihedral angles.

III. CONFORMATIONAL ANALYSIS OF *n*-ALKANES

In order to understand the charge density redistributions undergone during torsional rotations we must focus, first, our attention to those variations in the relative structural parameters (distances, angles, and torsional angles) which take place between the constituent atoms of a molecule. Torsional distortions from a given local minimum structure modify the molecular electron density due to the fact that atomic interactions change: while some of them become weaker, other interactions become stronger, and even new interactions may emerge giving rise to a somehow different electron density overlapping and, consequently, a different charge density distribution.

For this purpose, ethane and propane were taken as prototype molecules to gain an insight into the relationship between structural changes and density redistributions. Geometry optimizations were performed by means of the GAUSSIAN 92 program,[21] at the semiempirical AM1 and ab initio HF/3-21G levels of theory. For the sake of clarity, the structures of these alkanes are depicted in Figure 3, where torsional angles are indicated by arrows.

Table 1 gathers the most important structural parameters (computed at the HF/3-21G level of theory) involved in the staggered and eclipsed conformers of ethane and propane, which are taken as simple examples of one and two torsional angle rotation problems, respectively. In both cases, the main structural differences that deserve to be noted are: (1) the eclipsed conformers have longer C–C bonds, and (2) the eclipsed conformers have larger bond angles among those atoms where repulsive steric interactions become evident. Similar results have been observed for larger *n*-alkanes. For instance, in *n*-butane the C_2–C_2 distances (see Figure 3c) for the two energy minima (*trans* and *gauche*) and the two saddle-point conformations

Table 1. Structural Parameters[a] for the HF/3-21G Optimized Minimum (Staggered) and Maximum (Eclipsed) Energy Points of Ethane and Propane[b]

n-Alkane	*Parameter*	*Staggered*	*Eclipsed*
Ethane	C–H	1.084	1.083
	C–C	1.542	1.556
	C–C–H	110.80	111.22
Propane	C_1–H	1.085	1.083
	C_1–C_2	1.541	1.559
	C_1–C_2–H	111.19	111.27
	C_1–C_2–C_1	111.60	114.28

Notes: [a] Distance in Å and angles in degrees.
[b] See Figures 3a and 3b for atom labels.

that separate them (*A* and *syn*) are found to be 1.5404, 1.5432, 1.5573, and 1.5675 Å, respectively.

The consequences of these structural changes on the energy and QMSM torsional profiles can be envisaged in the ensemble of results collected in Table 2. EQMSM, fitted densities, and FQMSM were computed with the program MESSEM[22] from

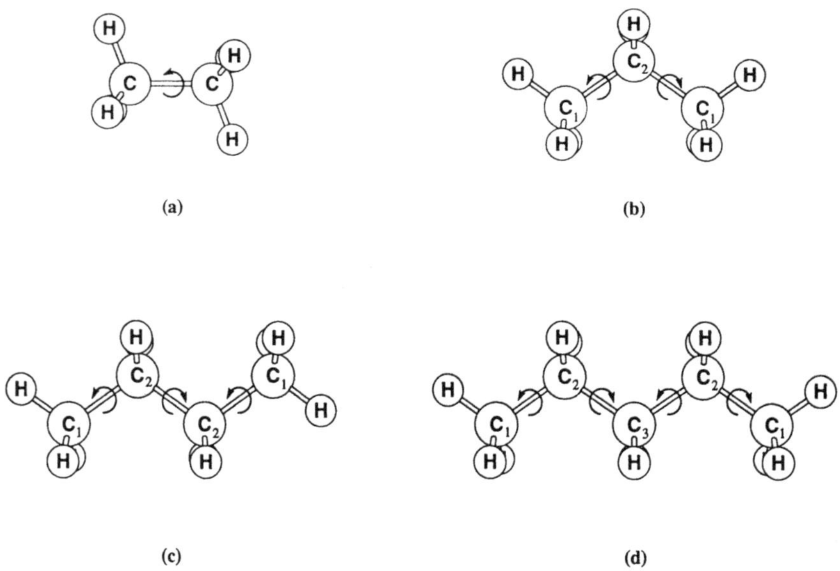

Figure 3. Structures of the energy minimum conformer for (**a**) ethane, (**b**) propane, (**c**) *n*-butane, and (**d**) *n*-pentane. Active dihedral angles are marked with arrows.

Table 2. Energies[a] and EQMSM[b] at the HF/3-21G Level of Theory for Various Structures of Ethane and Propane[c]

n-Alkane	*Torsional Angles*	*Energy*	*EQMSM*	*FQMSM*	*FQASM*	*SQASM*
Ethane	180	−78.79395	62.80458	62.71350	62.23708	64.86965
	0 (conformer)	−78.78957	62.79672	62.70288	62.19428	64.86030
	0 (rotamer)	−78.78935	62.80753	62.71489	62.23728	64.86989
Propane	180,180	−117.61330	94.14846	94.01397	94.20291	97.29091
	0,0 (conformer)	−117.60214	94.13018	93.99133	94.06142	97.26508
	0,0 (rotamer)	−117.60097	94.16409	94.02673	94.20621	97.29273

Notes: [a]In hartrees.
[b]In au.
[c] QMSM values obtained using the FQMSM, FQASM, and SQASM approximations to EQMSM are also Included for comparison

ab initio HF/3-21G densities. Conformational and rotational FQMSM, ACSGA, FQASM and SQASM approximations, were calculated using the program CONFORM.[23]

Values of Table 2 show that, from an energetic point of view, the formation of repulsive steric interactions when going from the staggered to the eclipsed conformers is translated in an energy destabilization. On the other hand, the evolution of QMSM profiles depends on the torsional approach under consideration (vide supra). If a conformational approach is used, allowance of the nuclear relaxation implies that the electron density distribution is globally depleted when going from staggered to eclipsed conformers as a consequence of the longer C–C bonds and larger C–C–C bond angles in the eclipsed conformers, thus, smaller QMSM values are obtained. However, when a rotational approach is employed a reverse trend is found due to the fact that all structural parameters are kept frozen throughout the torsional profile and steric contacts become more evident, producing a larger electron density overlapping; consequently, the total electron density distribution is globally concentrated, which implies obtention of larger QMSM values. Thus, at first sight it seems that (at least for this class of "nonpolar" torsional rotations) energy and QMSM torsional profiles are opposite if a conformational approach is used, while they are analogous if a rotational approach is used. As will be shown below, this interesting result could serve to perform fast conformational analyses from an electron density redistribution viewpoint by qualitatively locating those regions where energetic local minima are found.

The crossed $s_{i,j}$ terms in the generation of approximate EQMSM from SQASM (Eqs. 11 and 12) reflect the electron density overlapping between atoms in a molecule. In ethane, the sum of the overlap atomic self-similarity measures ($s_{i,i}$)

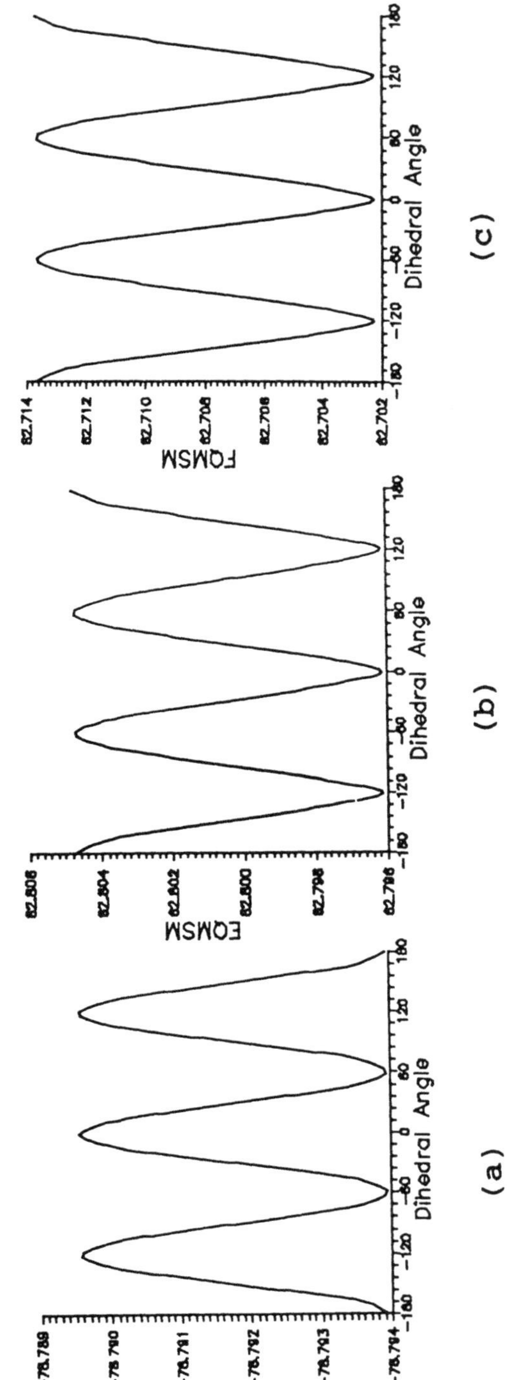
Energy
−78.789
−78.790
−78.791
−78.792
−78.793
−78.794
−180
−120
−60
0
60
120
180
Dihedral Angle
(a)
EQMSM
82.806
82.804
82.802
82.800
82.798
82.796
(b)
FQMSM
82.714
82.712
82.710
82.708
82.706
82.704
82.702
(c)

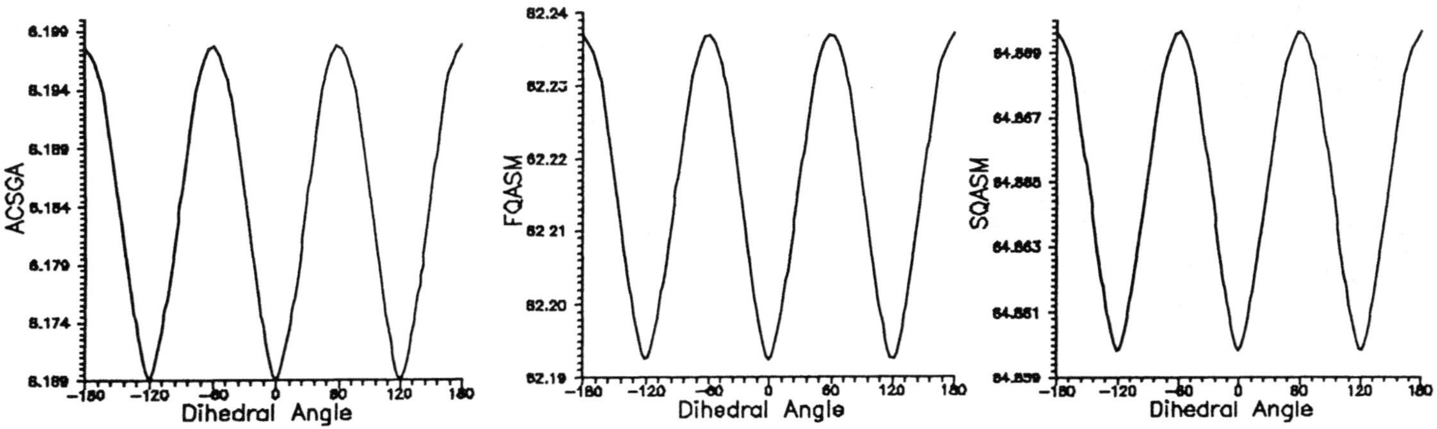

(d) (e) (f)

Figure 4. Ethane torsional profiles using the conformational approach. Dihedral angle (in degrees) is plotted against (**a**) HF/3-21G energy, (**b**) EQMSM, (**c**) FQMSM, (**d**) ACSGA, (**e**) FQASM, and (**f**) SQASM.

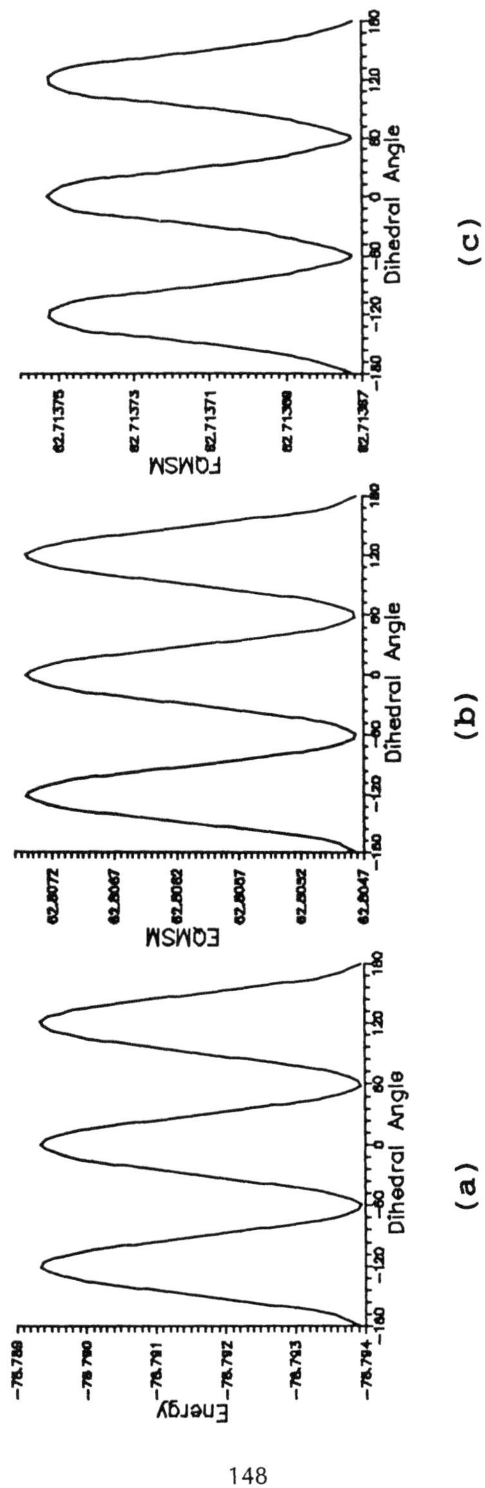
Energy
Dihedral Angle
(a)
EQMSM
Dihedral Angle
(b)
FQMSM
Dihedral Angle
(c)

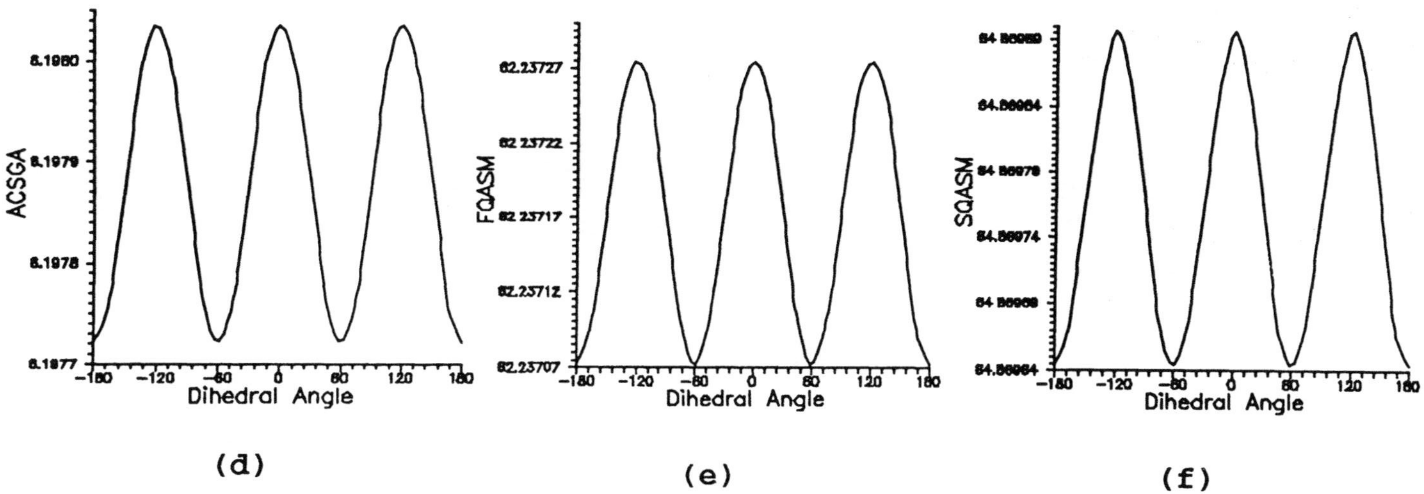

Figure 5. Ethane torsional profiles using the rotational approach. Dihedral angle (in degrees) is plotted against (**a**) HF/3-21G energy, (**b**) EQMSM, (**c**) FQMSM, (**d**) ACSGA, (**e**) FQASM, and (**f**) SQASM.

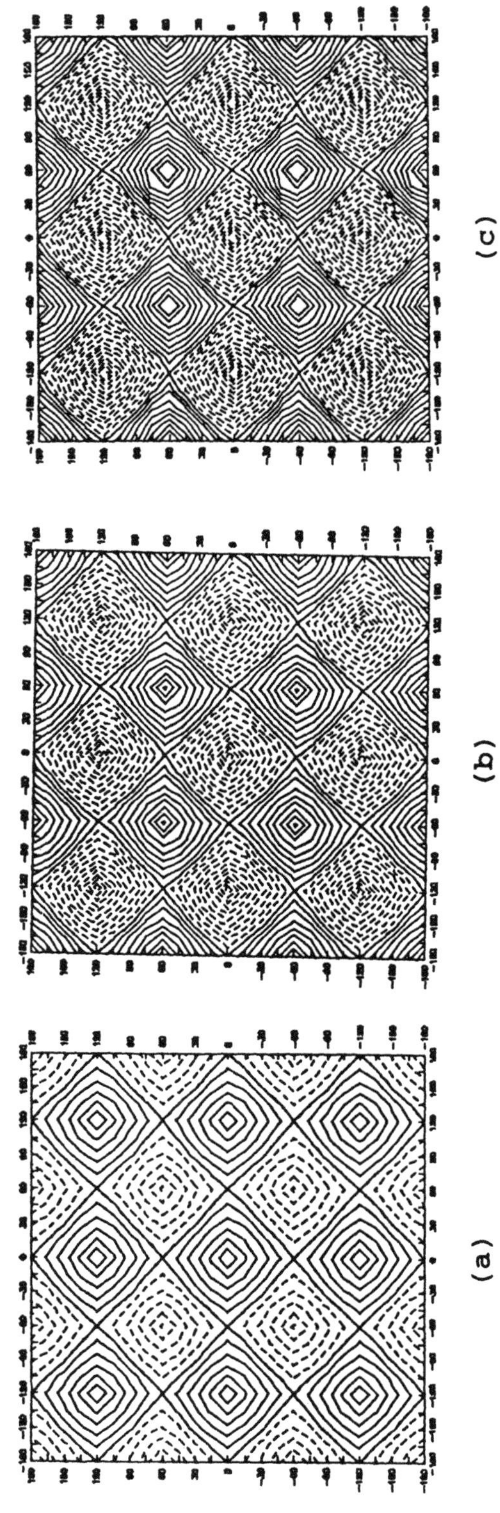

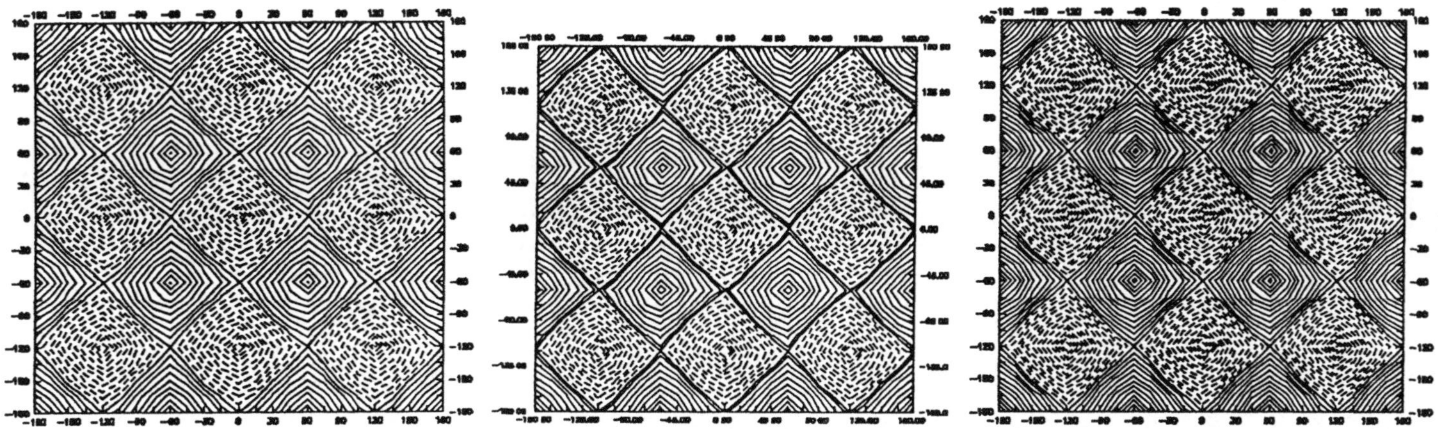

(d) (e) (f)

Figure 6. Propane torsional topological surfaces using the conformational approach. Dihedral angles (in degrees) are plotted against (**a**) HF/3-21G energy, (**b**) EQMSM, (**c**) FQMSM, (**d**) ACSGA, (**e**) FQASM, and (**f**) SQASM.

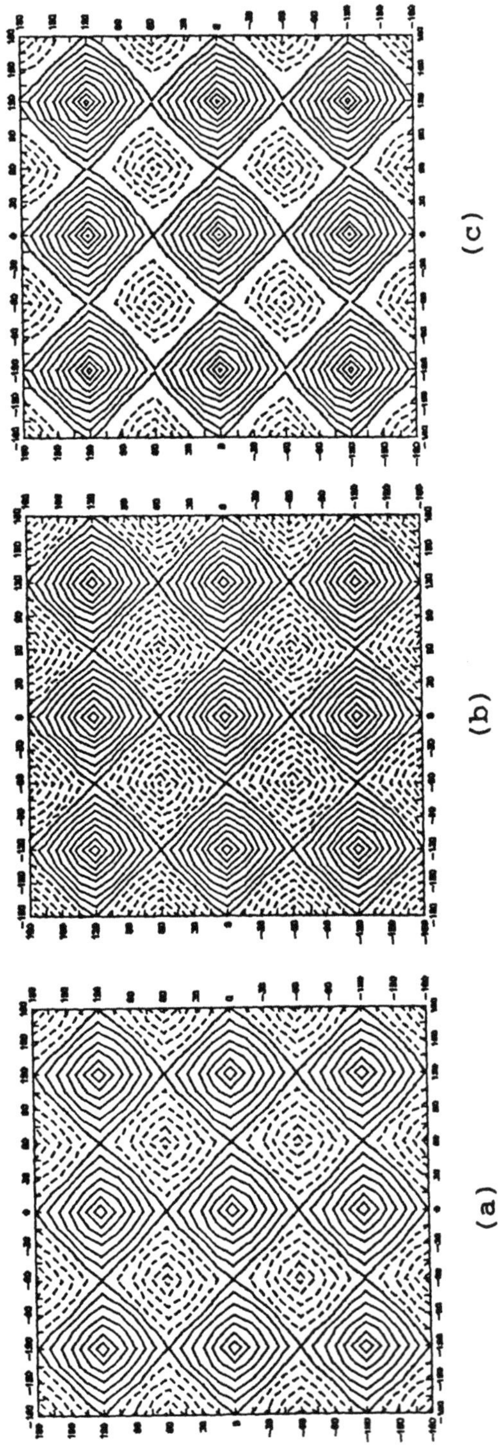
(a)
(b)
(c)

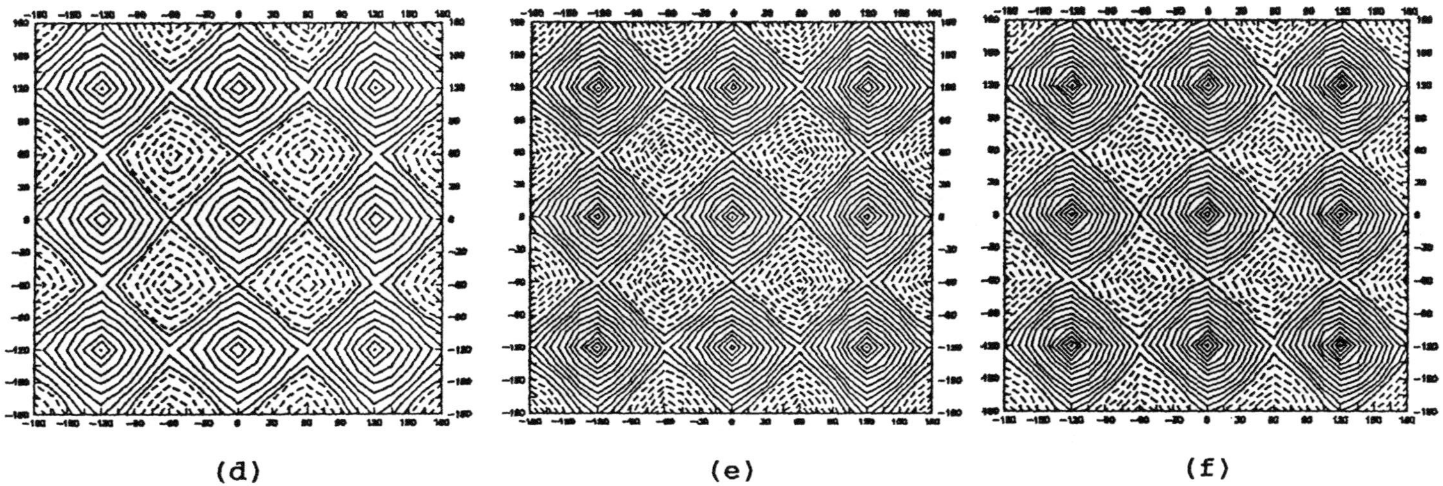

Figure 7. Propane torsional topological surfaces using the rotational approach. Dihedral angles (in degrees) are plotted against (**a**) HF/3-21G energy, (**b**) EQMSM, (**c**) FQMSM, (**d**) ACSGA, (**e**) FQASM, and (**f**) SQASM.

leads to a value of 63.9252. In this case the $s_{i,j}$ terms contribute to the final SQASM value with 0.9445 and 0.9351 for the staggered (energy minimum) and eclipsed (energy maximum) conformers, respectively. These results clearly agree with the above commented trend in structural parameters when going from the staggered to the eclipsed conformers (see Table 1): longer C–C distance and larger C–C–H bonds are translated in small electron density overlappings. It is made even more evident in propane where $s_{i,i}$ terms sum up to 95.8480, which means that $s_{i,j}$ terms contribute with 1.4429 and 1.4171 for the energy minimum and energy maximum conformers, respectively.

To obtain a more visual information, Figures 4 and 5 depict the different energy and QMSM torsional profiles for ethane under the conformational and rotational approaches, respectively, and Figures 6 and 7 present the same information by plotting the two-dihedral angle topological surfaces of propane. Comparing first the energy and EQMSM profiles, we recover the above-stated relationship depending on the use of conformational or rotational approach. Furthermore, taking the EQMSM torsional profile as a reference for the four approximations to EQMSM proposed in this work, it can be seen in all figures that a good qualitative agreement is found among the five QMSM profiles. In all cases the QMSM stationary point regions correspond with the same stationary point regions found in the energy profile. In this sense, to make Figures 6 and 7 more clear to distinguish between energy minima regions from energy maxima regions, the value corresponding to an approximate saddle point (e.g., the point with dihedral angles of 0° and 60°) has been subtracted to all values of the topological surface. Thus, in these two figures encircled regions having solid lines denote energy maxima regions while regions having dashed lines locate energy minima regions. The confirmation of the good correspondence of all the different approximations to EQMSM with the energy profile encourages their use in performing extremely fast conformational analyses.

Once the behavior of the conformational and rotational approaches to torsional profiles from an electron density redistribution point of view has been clarified and the confidence of the different approximations to EQMSM ensured, we can go one step further and apply this methodology to larger *n*-alkanes. The next *n*-alkane of the family is *n*-butane. Butane has usually been taken as a key to understanding torsional interactions about carbon–carbon single bonds because these interactions seem to be central in all methods in molecular modeling.[24] As a result, the barrier to rotation about the C_2–C_2 bond (see Figure 3c) has been extensively studied theoretically.[25]

First, different potential energy surface points were optimized at the ab initio HF/3-21G level of theory. The optimizations of the two energy minima (*trans* and *gauche* with respect to the two terminal methyl groups) were carried out without constraints, and those for points connecting the two energy minima (the so-called *A* and *syn*) were constrained only with regard to the dihedral angle, which maintained the eclipsed conformation (methyl–hydrogen for *A* and methyl–methyl for *syn*). Table 3 collects the optimized energies and EQMSM values for the different

Table 3. C_2–C_2 Bond Distances[a], Energies[b], and EQMSM[c] at the HF/3-21G Level of Theory for Different Potential Energy Surface Points of *n*-Butane

Point	C_2–C_2	*Energy*	*EQMSM (conformer)*	*EQMSM (rotamer)*
trans	1.5404	−156.43247	125.49208	125.49208
A	1.5573	−156.42673	125.48473	125.50184
gauche	1.5432	−156.43124	125.48673	125.49174
syn	1.5675	−156.42285	125.47338	125.51929

Notes: [a] In Å.
[b] In hartrees.
[c] In au.

points. For the sake of comparison, also included are the EQMSM values obtained under the rotational approach (taking the *trans* structure as the initial structure). As rationalized earlier, the EQMSM conformational values describe a torsional profile opposite to the energy profile. This is in perfect agreement with the C_2–C_2 bond rearrangements suffered under the torsional rotation (which appear to be the main structural distortions): the longer the C_2–C_2 bond, the more depleted is the electron density distribution and the smaller the EQMSM value obtained. On the other hand, the EQMSM rotational profile recovers the original energy profile. In this case, due to the fact that a nuclear relaxation is not allowed at each point of the torsional profile (the C_2–C_2 bond is kept fixed at 1.5404 Å), steric contacts are stronger, the atomic electron density overlapping is larger, and, consequently, larger values of EQMSM are found.

In a more qualitative sense, we are going to focus our attention on the study of the charge density redistribution due to the C_2–C_2 torsional rotation and for this purpose the two C_1–C_2 torsional angles will be constrained to 180° (see Figure 3c). The results of this study are depicted in Figure 8. As stated above, under the rotational approach, energy and EQMSM torsional profiles (Figures 8a and 8b) look very similar. The use of the SQASM approximation (Figure 8c) begins to present some problems in reproducing the shoulder of the energy profile due to the *A* rotamer (dihedral angle at −120° and 120°). However, the FQASM approximation (as introduced in Eq. 9 using $f(R) = \exp(-R)$) is not capable of describing the steric contact present in the *A* rotamer structure and, consequently, it becomes inefficient for locating the *A* and *gauche* rotamer regions (Figure 8d).

To solve this problem, an alternative strategy has to be devised. The success of the fast approximations to EQMSM is based on the ability to recognize steric contacts which, from an electron density viewpoint, are located by computing the atomic electron density overlapping. If this overlapping is poorly described, location of energy minima and maxima can be unsuccessful. In the *n*-butane rotational study, it seems that it is the case for the FQASM approximation, basically caused

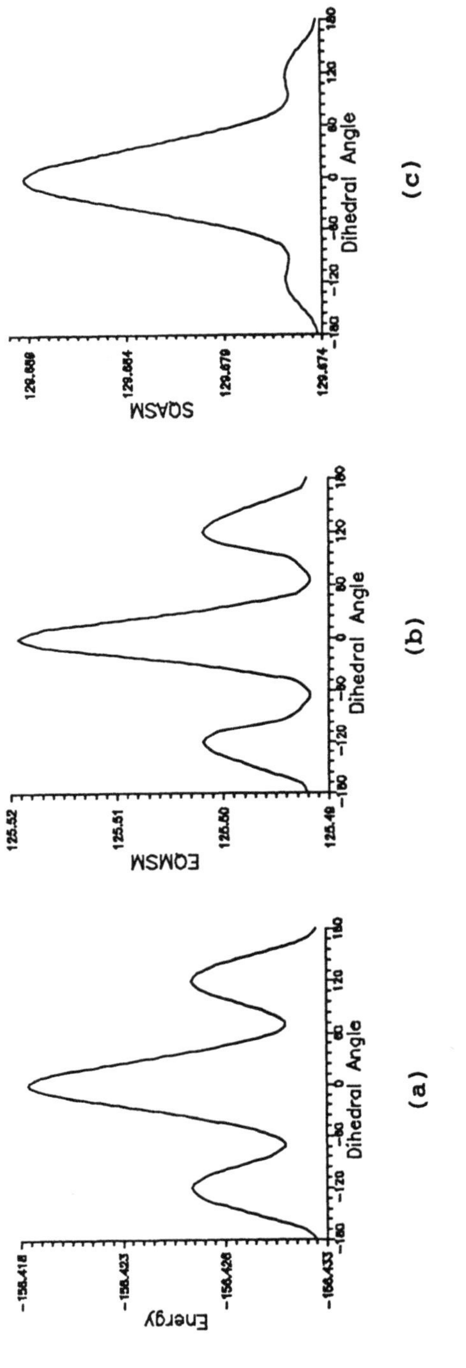
Energy
-156.418
-156.423
-156.428
-156.433
-180
-120
-60
0
60
120
180
Dihedral Angle
(a)
EQMSM
125.52
125.51
125.50
125.49
-180
-120
-60
0
60
120
180
Dihedral Angle
(b)
SQASM
129.689
129.684
129.679
129.674
-180
-120
-60
0
60
120
180
Dihedral Angle
(c)

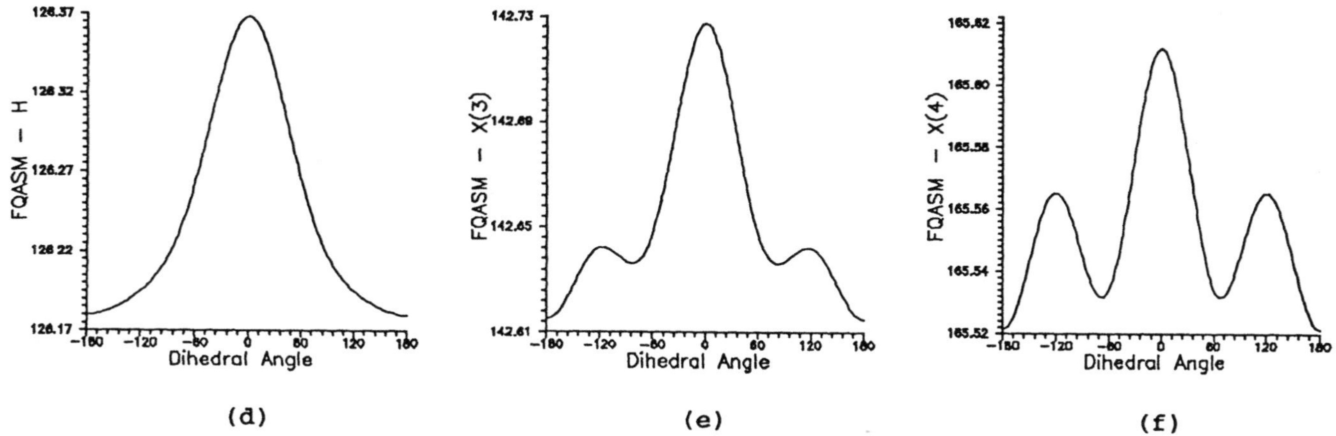

Figure 8. *n*-Butane C_2–C_2 torsional profiles using the rotational approach. Dihedral angle (in degrees) is plotted against (**a**) HF/3-21G energy, (**b**) EQMSM, (**c**) SQASM, (**d**) FQASM, (**e**) FQASM where Hs attached to C_2 were substituted for dummy 3-electron atoms (X(3)), and (**f**) FQASM where Hs attached to C_2 were substituted for dummy 4-electron atoms (X(4)).

Table 4. Formal Dihedral Angles[a] Together with Energies[b] and EQMSM[c] at the HF/3-21G Level of Theory for the Four Conformers of *n*-Pentane

Conformer	C_1–C_2–C_3–C_2	*Energy*	*EQMSM*
trans	180/180	−195.25156	156.83716
gauche	± 60/180	−195.25033	156.83101
G^+G^+	± 60/±60	−195.24916	156.82294
G^+G^-	± 60/∓60	−195.24569	156.82321

Notes: [a] In degrees.
[b] In hartrees.
[c] In au.

by the fact that the overlapping is mainly due to hydrogen–hydrogen contacts. Because the main objective is to locate repulsive steric contacts, in order to exaggerate electron density overlaps it is suggested that we substitute hydrogens with other atoms of higher atomic number. Figures 8e and 8f illustrate the FQASM torsional profiles when hydrogens attached to C_2 were substituted with 3- and 4-electron atoms, respectively. As can be seen, steric contact regions are now clearly encountered and FQASM profiles become similar to the original energy profile (Figure 8a).

The four established conformers of *n*-pentane[26] were also optimized at the HF/3-21G level of theory and an exact quantitative electron density redistribution analysis was performed. Results are gathered in Table 4 and show that the more packed structures have the more depleted electron density distributions (smaller EQMSM values). Of mention is the fact that although the G^+G^- conformer is energetically far more destabilized than the other three, it presents a concentration of the total electron density distribution similar to that found in the G^+G^+ conformer. Some simple arguments based on the conformational analysis of hydrocarbons (mainly butane and pentane) have been recently reported[27] in an attempt to explain the backbone-dependent and backbone-independent rotamer preferences of protein side chains. Thus, the methodology used in this work could be of great help in quantifying the magnitude of the steric repulsions and their role in the structural packing.

The *n*-pentane was also taken to perform a rotational analysis considering only the two C_1–C_2–C_3–C_2 dihedral angles (see Figure 3d). An energy torsional three-dimensional surface obtained at the HF/3-21G level of theory is presented in Figure 9 (top). In this figure, the four regions of minimum steric contacts can be clearly identified. These are those regions from which geometry optimizations would lead to the four conformers mentioned above. For comparison, the corresponding FQASM torsional surface has been included (Figure 9, bottom). Under this approximation, the strategy presented above for *n*-butane was also used in order to exaggerate steric contacts. The qualitative agreement of both surfaces is evident

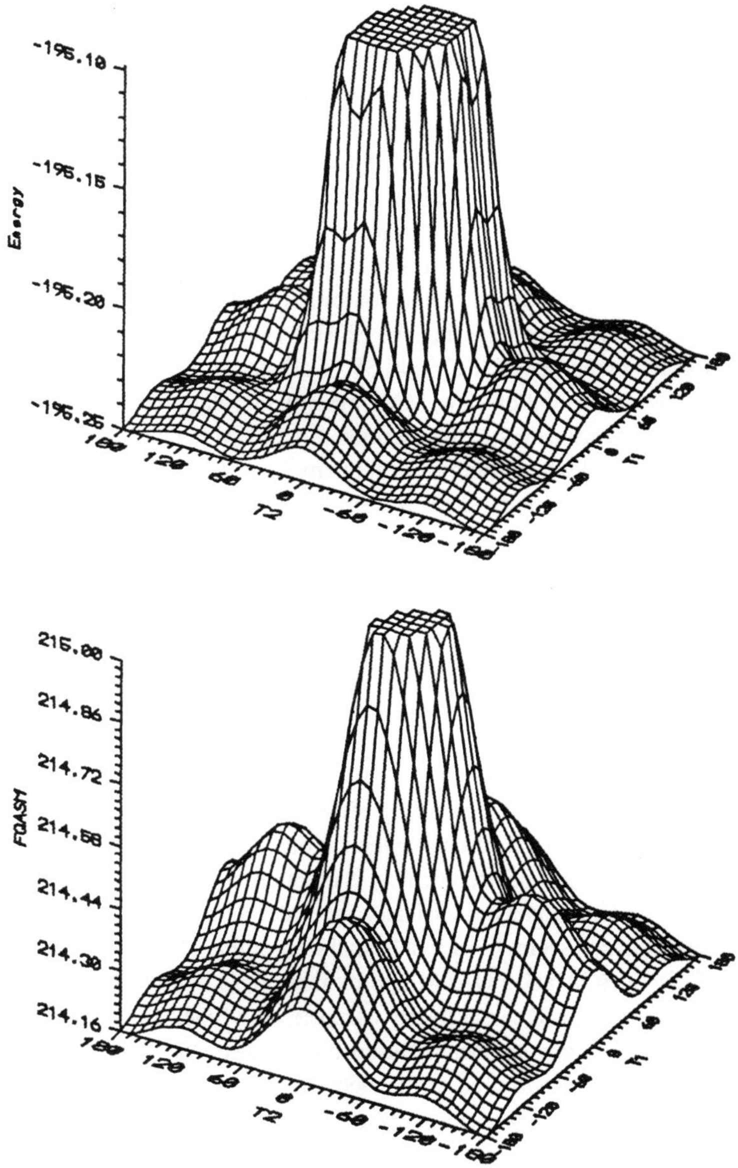

Figure 9. *n*-Pentane C_2–C_3 torsional 3D surfaces using the rotational approach. Dihedral angles (in degrees) are plotted against HF/3-21G energy (**top**) and FQASM where Hs attached to C_2 were substituted for dummy 3-electron atoms and Hs attached to C_3 for dummy 5-electron atoms (*bottom*).

and (what is really important) the four different regions are found at the same places but in an extremely fast way.

At this stage it becomes necessary to present a comparative computational cost test to show the advantage of performing conformation analyses from the molecular similarity viewpoint and by using some of the approximations employed in this work. Table 5 collects the required computational time employed to perform a systematic rotational analysis of the four *n*-alkanes. A dihedral step of 10° was taken for all calculations. Molecular mechanics computations by means of the MM3[28] force field are also included. This was seen to be necessary due to the general use of these types of force fields in current conformational analyses. The MM3 results presented were performed by using the SPARTAN[29] program. All computations were performed on an IBM RISC/6000-355 workstation.

From an energetic point of view, MM3 systematic rotational analyses appear to be an order of magnitude faster than semiempirical AM1 calculations. Even more dramatic is the effect when going from semiempirical to ab initio calculations. As an example, for *n*-pentane the difference in these computational times is about 2 orders of magnitude. It must be stressed that rotational analyses require only single point energy calculations. If conformational analyses are needed, the time required for all the structure optimization gradient cycles should be added.

From a QMSM point of view, the results in Table 5 show that the use of different approximations to EQMSM without a qualitative loss of accuracy is widely justified due to computational time requirements. The use of the FQMSM approximation is a compromise between the goal of significantly accurate QMSM values and the computational cost. However, its use is only correct in torsional analyses using the conformational approach (where a fitting of the electron density has to be performed at each point of the torsional surface); when a rotational approach is employed, the fact that the fitting of the electron density is uniquely done at the original structure makes this approximation symmetrically incorrect under a torsional rotation. In another perspective, it seems clear that from the ensemble of computational timings the use of the ACSGA and FQASM approximations is highly recommended and their computational speed can perfectly compete with MM3 calculations.

It is of interest to study the linear relationships between the *n*-alkane constitution and its energy and concentration of the electron density distribution based on the fact that the *n*-alkane family is constructed by systematically substituting a H by a CH_3 fragment. For this purpose, the energy and EQMSM of *n*-alkanes up to 10 carbons were calculated at the HF/3-21G level of theory. The results are depicted in Figure 10 and show the perfect correspondence between energy and EQMSM values. Linear least-square fittings of the values obtained gave rise to the following equations,

$$S = -0.807426 \cdot E - 0.811933 \tag{13}$$

Table 5. Energy and QMSM Rotational Computation[a,b]

		Energy			*QMSM*				
n-Alkane	*No. Rotamers*	*MM3*	*AM1*	*HF/3-21G*	*EQMSM*	*FQMSM*	*ACSGA*	*FQASM*	*SQASM*
Ethane	36	6.5	61.2	302	1241	0.61	0.09	0.08	6.34
Propane	1296	246	2332	15422	197821[c]	18.72	1.84	1.54	609
Butane	46656	8865	93312	905126	1.95×10^{7}[c]	1387	181	156	51062
Pentane	1679616	335923[c]	3.5×10^{6}[c]	1.18×10^{8}[c]	1.63×10^{9}[c]	62052	6998	6532	2.73×10^{6}[c]

Notes: [a] In CPU seconds.
[b] In all cases the dihedral step was set to 10 degrees.
[c] Extrapolated values.

$$E = -38.819024 \cdot n - 1.156620 \tag{14}$$

$$S = 31.343499 \cdot n + 0.122849 \tag{15}$$

where n, E, and S are the number of carbon atoms, the energy, and the EQMSM, respectively. Each one of these equations presents a regression coefficient of, at least, 0.999999 which can be considered as a guarantee for extrapolation validity.

On the other hand, by taking only into account results from methane, ethane. and propane it is possible to obtain a very accurate value of a given property (P) by simply summing up the perturbation induced from substituting a H by a CH_3 fragment:

$$P_n = \sum_{k=0}^{n-1} (n-k)\, P^{(k)} \tag{16}$$

In Eq. 16 $P^{(0)}$ is the entire contribution of methane to the property; $P^{(1)}$ is the perturbation induced by the formation of a C–C bond; $P^{(2)}$ is the perturbation induced by the formation of a second C–C bond, and so on. For the energy and EQMSM it has been found that the series converges very quickly and that contri-

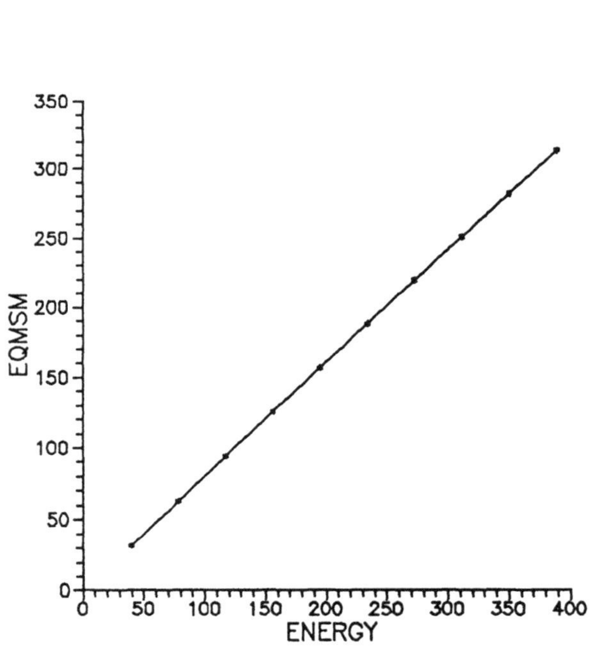

Figure 10. Linear relationship for *n*-alkanes between electronic energy and overlap quantum molecular self-similarity measure.

butions of orders larger than 2 are negligible. In these two cases, these contributions are found to be:

$$E^{(0)} = -39.97688 \quad S^{(0)} = 31.48084$$

$$E^{(1)} = 1.15981 \quad S^{(1)} = -0.15710$$

$$E^{(2)} = -0.00228 \quad S^{(2)} = 0.02014$$

As can be seen, E and S terms have opposite signs for all contributions and this provides us with additional interesting information. Taking methane as reference, the chemical significance of these terms can be rationalized. Formation of the ethane molecule is destabilized with respect to two isolated methane molecules ($E^{(1)} > 0$) and, at the same time, the formation of a C–C bond is translated in a global depletion of the electron density distribution with respect to the concentration of the electron density distributions of two isolated methane molecules ($S^{(1)} < 0$). An opposite extrapolated argument can be found for the second-order terms (in this $E^{(2)} < 0$ and $S^{(2)} > 0$).

IV. CONCLUSIONS

The study of the charge density redistribution due to torsional rotations represents another example of the application of methodological aspects of quantum molecular similarity. This methodology is emerging as a very useful tool in performing quantitative studies, at a given theoretical level, of any kind of charge density redistribution problem and it is being shown in the series of latest works developed in our laboratory.[30]

The set of results obtained in this contribution can be summarized in the following points: (1) calculation of EQMSMs appears to be a very good methodology for quantifying the evolution of the concentration of the molecular electron density distribution under torsional rotations; (2) several fast approximations to EQMSM have been proposed and their accuracy with respect to EQMSM analyzed; (3) the use of these approximations to EQMSM as an extremely fast alternative strategy for identifying steric contacts has been successfully applied when performing conformational analyses; and (4) several general equations reflecting the linear relationships between the n-alkane constitution, the electronic energy, and the EQMSM have been reported.

However, these results hold only for the particular electronic nature of the torsional rotations in n-alkanes. The behavior of the charge density redistributions in "polar" torsional rotations is expected to evolve in a different way as the one found here for "nonpolar" torsional rotations due to the formation of hydrogen bridges and long-range polar interactions. This will be the subject of future investigations.

ACKNOWLEDGMENTS

Many helpful comments from Dr. Miquel Solà are gratefully acknowledged. One of us (J.M.O.) benefits from a grant provided by the Generalitat de Catalunya under project no. BQF92/11.

REFERENCES

1. Leach, A.R. In *Molecular Similarity in Drug Design*, Dean, P.M., Ed.; Blackie Academic: London, 1995, pp. 57–88.
2. Howard, A.E.; Kollman, P.A. *J. Med. Chem.* **1988**, *31*, 1669.
3. Leach, A.R. In *Reviews in Computational Chemistry*, Lipkowitz, K.B.; Boyd, D.B., Eds.; VCH Publishers: New York, 1991, Vol. II, pp. 1–55.
4. Wilson, S.R.; Cui, W.; Moskowitz, J.W.; Schmidt, K.E. *Tetrah. Lett.* **1988**, *29*, 4373.
5. Wilson, S.R.; Cui, W. *Biopolymers* **1990**, *29*, 225.
6. Judson, R.S. In *Reviews in Computational Chemistry*, in press (and references therein).
7. (a) Chattaraj, P.K.; Nath, S.; Sannigrahi, A.B. *J. Phys. Chem.* **1994**, *98*, 9143. (b) Cárdenas-Jirón, G.I.; Lahsen, J.; Toro-Labbé, A. *J. Phys. Chem.* **1995**, *99*, 5325. (c) Cárdenas-Jirón, G.I.; Toro-Labbé, A. *J. Phys. Chem.* **1995**, *99*, 12730.
8. Solà, M.; Mestres, J.; Oliva, J.M.; Duran, M.; Carbó, R. *Int. J. Quantum. Chem.*, in press.
9. Mestres, J.; Solà, M.; Carbó, R. *Sci. Gerund.*, in press.
10. Carbó, R.; Leyda, L.; Arnau, M. *Int. J. Quantum Chem.* **1980**, *17*, 1185.
11. Besalú, E.; Carbó, C.; Mestres, J.; Solà, M. *Top. Curr. Chem.* **1995**, *173*, 31.
12. Mestres, J.; Solà, M.; Duran, M.; Carbó, R. *J. Comp. Chem.* **1994**, *15*, 1113.
13. Mestres, J.; Solà, M.; Besalú, E.; Duran, M.; Carbó, R. In *Molecular Similarity and Reactivity: From Quantum Chemical to Phenomenological Approaches*; Carbó, R., Ed.; Kluwer Academic: 1995, pp. 75–85.
14. Constans, P.; Carbó, R. *J. Chem. Inf. Comput. Sci.*, in press.
15. (a) Rohrer, D.C. In *Molecular Similarity and Reactivity: From Quantum Chemical to Phenomenological Approaches*, Carbó, R., Ed.; Kluwer Academic: 1995, pp. 141–161. (b) Mestres, J.; Rohrer, D.C., submitted for publication.
16. Roos, B.; Salez, C.; Veillard, A.; Clementi, E. *A General Program for Calculation of Atomic SCF Orbitals by the Expansion Method.* Technical Report RJ-518, IBM Research (1968). ATOMIC is a completely new updated version by R. Carbó.
17. Roothaan, C.C.J.; Bagus, P.S. *Methods in Computational Physics*, Academic Press: New York, 1963, Vol. 2, pp. 17–95.
18. Clementi, E.; Roetti, C. *At. Data Nucl. Data Tables*, **1974**, *14*, 177.
19. SEMAT: a Program for Calculating Exact Quantum Atomic Similarity Measures, Oliva, J.M.; Carbó, R., IQC-UdG, Girona, CAT, 1993.
20. (a) Ahlrichs, R. *Chem. Phys. Lett.* **1972**, *15*, 609. (b) Hoffmann-Ostenhof, M.; Hoffmann-Ostenhof, T. *Phys. Rev. A* **1977**, *16*, 1782. (c) Tal, Y., *Phys. Rev. A*, **1978**, *18*, 1781. (d) Katriel, J.; Davidson, E.R. *Proc. Natl. Acad. Sci. USA* **1980**, *77*, 4403. (e) Bader, R.F.W. In *Atom in Molecules: A Quantum Theory*; Oxford University Press: Oxford, 1990, pp. 45–47.
21. GAUSSIAN 92. Revision G.1, Frisch, M.J.; Trucks, G.W.; Head-Gordon, M.; Gill, P.M.W.; Wong, M.W.; Foresman, J.B.; Johnson, B.G.; Schlegel, H.B.; Robb, M.A.; Replogle, R.S.; Gomperts, R.; Andrés, J.L.; Raghavachari, K.; Binkley, J.S.; Gonzales, C.; Martin, R.L.; Fox, D.J.; Defrees, D.J.; Baker, J.; Stewart, J.J.P.; Pople, J.A., Gaussian Inc., Pittsburgh, PA, 1992.
22. MESSEM: a Density-based Molecular Similarity Program. Mestres, J.; Solà, M.; Besalú, E.; Duran, M.; Carbó, R., IQC-UdG, Girona, CAT, 1994.

23. CONFORM: a QMSM Rotational Analysis Program, Mestres, J.; Oliva, J.M., IQC-UdG, Girona, CAT, 1995.
24. Burkert, U.; Allinger, N.L. *Molecular Mechanics*; ACS Monograph 177; American Chemical Society: Washington, DC, 1981.
25. (a) Radom, L.; Lathan, W.A.; Hehre, W.J.; Pople, J.A. *J. Chem. Soc.* **1973**, *95*, 693. (b) Peterson, M.R.; Csizmadia, I.G. *J. Am. Chem. Soc.* **1978**, *100*, 6911. (c) Allinger, N.L.; Profecta. S. *J. Comput. Chem.* **1980**, *1*, 181. (d) Darsey, J.A.; Rao, B.K. *Macromolecules* **1981**, *14*, 1575. (e) van Catledge, F.A.; Allinger, N.L. *J. Am. Chem. Soc.* **1982**, *104*, 6272. (f) Raghavachari, K. *J. Chem. Phys.* **1984**, *81*, 1383. (g) Steele, D. *J. Chem. Soc., Faraday Trans. 2* **1985**, *81*, 1077. (h) Wiberg, K.B.; Murcko, M.A. *J. Am. Chem. Soc.* **1988**, *110*, 8029.
26. (a) Pitzer, K.S. *Chem. Rev.* **1940**, *27*, 39. (b) Abe, A.; Jernigan, R.L.; Flory, P.J. *J. Am. Chem. Soc.* **1966**, *88*, 631. (c) Pitzer, R.M. *Acc. Chem. Res.* **1983**, *16*, 201. (d) Mencarelli, P. *J. Chem. Ed.* **1995**, *72*, 511.
27. Dunbrack, R.L., Jr.; Karplus, M. *Nature Struct. Biol.* **1994**, *1*, 334.
28. (a) Allinger, N.L.; Yuh, Y.H.; Lii, J.-H. *J. Am. Chem. Soc.* **1989**, *111*, 8551. (b) Allinger, N.L.; Li, F.; Yan, L.; Tai, J.C. *J. Comput. Chem. Soc.* **1990**, *11*, 868.
29. Spartan 4.0, Wavefunction, Inc., 1995.
30. (a) Solà, M.; Mestres, J.; Carbó, R.; Duran, M. *J. Am. Chem. Soc.* **1994**, *116*, 5909. (b) Solà, M.; Mestres, J.; Duran, M.; Carbó, R. *J. Chem. Inf. Comput. Sci.* **1994**, *34*, 1047. (c) Solà, M.; Mestres, J.; Carbó, R.; Duran, M. In *QSAR and Molecular Modelling: Concepts, Computational Tools, and Biological Applications*; Prous Publishers, in press. (d) Solà, M.; Mestres, J.; Carbó, R.; Duran, M. *J. Chem. Phys.*, in press. (e) Mestres, J.; Solà, M.; Carbó, R.; Luque, F.J.; Orozco, M. *J. Phys. Chem.*, in press. (f) Torrent, M.; Duran, M.; Solà, M. *Adv. Mol. Sim.* (in this volume).

HOW SIMILAR ARE HF, MP2, AND DFT CHARGE DISTRIBUTIONS IN THE $Cr(CO)_6$ COMPLEX?

Maricel Torrent, Miquel Duran, and Miquel Solà

Advances in Molecular Similarity
Volume 1, pages 167–186

ISBN: 0-7623-0131-7

ABSTRACT

A procedure based on quantum molecular similarity measures (QMSM) has been used to compare electron densities obtained from conventional ab initio and a wide variety of density functional methodologies (including both pure and hybrid models) at their respective optimized geometries. This method has been applied to chromium hexacarbonyl, a transition metal system with a considerable bulk of experimental and theoretical data. Results show that Hartree–Fock density is transcended by correlated densities because of the well-known problems of the Hartree–Fock level of theory to describe correctly the metal–CO bonds in carbonyl complexes in which the metal has the oxidation state 0. Among density functional methods, a careful comparison has allowed us to classify the set of functionals under study into subsets.

I. INTRODUCTION

The one-electron density distribution, $\rho(\mathbf{r})$, of an electronic state is a function of the three spatial variables that gives the number of electrons per unit volume present in this state. Its formula in terms of the wavefunction Ψ is given by:[1]

$$\rho(\mathbf{r}) = N\int \ldots \int |\psi(\mathbf{x}_1, \mathbf{x}_2, \ldots, \mathbf{x}_N)|^2 \, ds_1 d\mathbf{x}_2 \ldots d\mathbf{x}_N \tag{1}$$

The fundamental properties of the electron density have been recognized since the initial stages of quantum chemistry. This function is a physical observable upon which other molecular properties, directly or indirectly, depend. For instance, the density functional formalism[2] derived from the landmark work of Thomas and Fermi[3] is based on the Hohenberg–Kohn theorem[4] which is the basis of modern density functional theory (DFT), and states that all ground-state molecular properties, and in particular the energy, can be expressed as functionals of the electron density. Likewise, relevant chemical information can be gathered from the electron density maps and from the gradient and Laplacian of the electron density as shown by Bader.[5] Furthermore, the total electronic density and its gradient can be used to construct an electron localization function (ELF)[6] which also provides a reliable visualization of atomic shell structure and core, binding, and lone electron pairs in molecular systems. Moreover, given that the electron density is an observable, any theoretical method in the exact limit should reproduce the same electron density, and therefore the same molecular properties. For this reason, a reasonable comparison between different methodologies has been carried out by making a systematic study of the electron density difference maps obtained from the methods being compared.[7]

From the applications given above, it is clear that there has been much attention paid to electron density over the years. Another quite widespread use of electron density functions can be found in the calculations of the quantum similarity between molecules.[8] In particular, one of the most widely used definitions of quantum molecular similarity measures (QMSM) between two chemical systems $\{I,J\}$ having electron densities $\rho_I(\mathbf{r})$ and $\rho_J(\mathbf{r})$ is given by the integral,[9]

$$Z_{IJ}(\Theta) = \int\int \rho_I(\mathbf{r}_1)\, \Theta(\mathbf{r}_1,\mathbf{r}_2)\, \rho_J(\mathbf{r}_2)\, d\mathbf{r}_1\, d\mathbf{r}_2 \tag{2}$$

where $\Theta(\mathbf{r}_1,\mathbf{r}_2)$ is a positive definite operator depending on two-electron coordinates. In the particular case that $\Theta(\mathbf{r}_1,\mathbf{r}_2)$ is the Dirac δ function $\delta(\mathbf{r}_1 - \mathbf{r}_2)$, substitution in Eq. 2 yields the formula to calculate the overlap-like similarity:

$$Z_{IJ} = \int \rho_I(\mathbf{r})\, \rho_J(\mathbf{r})\, d\mathbf{r} \tag{3}$$

Likewise, the repulsion-like similarity is given by:

$$Z_{IJ}\,(r_{12}^{-1}) = \int\int \rho_I(\mathbf{r}_1)\, \frac{1}{\mathbf{r}_{12}}\, \rho_J(\mathbf{r}_2)\, d\mathbf{r}_1\, d\mathbf{r}_2 \tag{4}$$

Other operators can be used depending on the information being requested. Once the QMSM has been calculated it is possible to define an Euclidean distance between the molecular electronic distributions $\rho_I(\mathbf{r})$ and $\rho_J(\mathbf{r})$ as:[9]

$$d_{IJ} = [Z_{II} + Z_{JJ} - 2Z_{IJ}]^{1/2} \tag{5}$$

Since the value of the distance given by Eq. 5 depends on the relative spatial orientation of molecular electron distributions $\rho_I(\mathbf{r})$ and $\rho_J(\mathbf{r})$, their mutual orientation is optimized in order to maximize Z_{IJ}, which is equivalent to minimize the d_{IJ} value. A final d_{IJ} value of zero means that charge density distributions $\rho_I(\mathbf{r})$ and $\rho_J(\mathbf{r})$ are equivalent, while larger d_{IJ} values correspond to a smaller similarity.

So far, comparisons between charge density distributions have been performed by analyzing charge density difference contours only at a fixed geometry for all levels of theory,[7a,b] and then reflecting only those changes explicitly due to electronic relaxation. The main interest in using QMSM instead of depicting electron density differences between charge density distributions $\rho_I(\mathbf{r})$ and $\rho_J(\mathbf{r})$, is the fact that with this methodology the analysis can be performed at any desired geometry, and in particular at the optimized geometry corresponding to each methodology employed, thus accounting for both nuclear and electronic relaxation. Therefore, the procedure used here, which was already employed in a recent work on small organic molecules,[10] is deemed to be a proper extension to the standard analysis of the electron density difference maps.

Transition metal carbonyl complexes have been of interest to experimental and theoretical chemists for a long time.[11,12] The interest stems partly from the fact that CO may act as both a σ-base through the 5σ-carbon lone-pair orbital, and as a π-acid through the $2\pi^*$-orbital. It has been established that a proper description of the metal–CO bond in carbonyl complexes with the metal bearing a zero-oxidation state requires an extensive treatment of electron correlation.[13,14] The available ab initio approaches for describing electron correlation at post-HF level range from Møller–Plesset second-order perturbation theory (MP2) to coupled cluster theory with single and double excitations and a perturbative treatment of triple excitations (CCSD(T)).[15,16] With the most accurate CCSD(T) method, good results for transition metal systems have been obtained,[17,18] but the computational costs are very high and limit the size of the systems that can be studied.

Very recently, Jonas et al.[19] have computed harmonic force fields of nine transition metal carbonyls, namely those involving chromium, iron, and nickel using Hartree–Fock (HF), MP2, and gradient-corrected density functionals (BP86 and BLYP). They concluded that DFT results are in very good agreement with available experimental data, whereas HF results are inadequate and MP2 results are satisfactory only for 5*d* and (partly) for 4*d* transition metal complexes, but not for 3*d* transition metal complexes, which is the actual case of chromium hexacarbonyl. In particular, it was pointed out that the DFT-BP86 approach is superior to the HF and MP2 methodologies because it provides more reliable results at computational costs that are intermediate between those of HF and MP2 methods.

Metal carbonyls of the chromium group have been theoretically studied earlier[14,17,18,20] with most emphasis on molecular structures and binding energies. Today, there is a fair understanding of the problems associated with HF calculations on transition metal complexes. It has been recognized[21] that HF calculations on such complexes yield an energetic separation between $d^{n+1}s^1$ configuration and $d^{n+2}s^0$ and d^ns^2 configurations which is strongly overestimated. A related problem is the poor representation of the transition metal–ligand bond lengths in SCF calculations. These bonds tend to be far too long for carbonyl complexes, as has been extensively documented.[22] One way to understand this aspect is the incorrect preference of the HF model for 4*s*-occupation instead of 3*d* occupation, leading to important Pauli repulsions even at large Cr–CO bond lengths. The π-bond develops optimal strength at distances which are too short for the bulky 5σ (the C lone pair), which at such distance already has considerable repulsion with the valence electrons of Cr.[21]

This review extends these earlier studies mainly in two directions. First, a detailed and systematic comparison of the electronic densities corresponding to the optimized geometry for each methodology is given in order to elucidate the correlation effects on the metal–CO bond. The correlation effects diminish this repulsion and allow shortening of the metal–CO bond. Although most of the previous studies have focused on an analysis of the molecular orbitals, they have not investigated it in terms of the electronic density. Second, we carefully revise the behavior of DFT for $Cr(CO)_6$, which is reported to be adequate for the case of a large variety of functionals ranging from local to nonlocal approaches and including both pure and hybrid methods. Not all functionals lead to the same conclusions since some of them can be as inadequate as HF. QMSM are very helpful because they allow one to classify these functionals in subsets according to their ability to properly describe electronic distributions.

The main goal pursued when comparing electron densities from different DFT methodologies is to discover the disadvantages and benefits of the different available density fuctionals, and thus assist researchers in building more accurate functionals. Moreover, these studies can also help us understand the successes and failures of DFT in some metal–ligand chemical interactions, and also to understand how nonlocal corrections influence the calculated electron density. In the analyses performed here, 10 methodologies, namely Hartree–Fock, 1 correlated ab initio method, and 8 density functional formalisms have been investigated.

II. COMPUTATIONAL DETAILS

Standard HF, frozen-core MP2, and DFT calculations have been performed by means of the Gaussian 92 program.[23] A basis set of a triple-ζ quality and (6,2,1,1,1,1,1,1/3,3,1,2/3,1,1) contraction scheme for the metal[24] and double-ζ with a polarization function (6-31G*) for ligands[25] has been used throughout.

QMSM have been obtained from the Gaussian 92 electron densities using the MESSEM program.[26] For MP2, generalized densities [27] have been used. Likewise, DFT electron densities have been calculated from SCF-converged Kohn–Sham orbitals. All QMSM are overlap-like and have been obtained through use of Eq. 3. In a previous study,[28] it was shown that overlap measures are more scattered over a large range of values than repulsion similarities, and consequently they are more suitable to quantify small changes in electron density distributions. However, the process of maximizing the similarity was carried out using repulsion-like similarity measures as defined by Eq. 4. The reason is due to the fact that the presence of the Coulomb operator smoothes the electron density surface and reduces the cusps of electron density at the nuclei, making the process of optimization easier since gradient components are smaller.[8g]

An approximate density instead of the exact density has been used in order to eliminate the need of evaluating costly four-index integrals as found in Eqs. 3 and 4. Details of this methodology have been given elsewhere.[8g,h] The set of fitting functions has been chosen to be the same as the squared molecular *s*-type renormalized basis functions. The validity of such approximation can be assessed from the values obtained when total overlap-like self-similarity at the Hartree–Fock optimized geometry and total overlap-like similarity between HF and MP2 at their respective optimized geometries are computed using exact and fitted densities. It has been found that small differences (0.1 and 0.02%, respectively) appear when the exact density is substituted by a fitted density, thus supporting the accuracy of this procedure. Bader topological analyses[5] have been performed through use of the ELECTRA program.[29] All calculations have been run on IBM RISC/6000 350 workstations.

A brief description of all functionals used is given as follows. DFT methods can be divided into pure and hybrid, the latter making use of the exact Hartree–Fock exchange. They are named by concatenating two keywords: on the left, a local exchange functional (S[30]), with or without a nonlocal correction (B[31]), combined on the right with a correlation correction to the local functional (LYP,[32] P86,[33] or VWN[34]). HFS and HFB are keywords for exchange functionals used without a nonlocal correlation correction. As far as hybrid methods are concerned, different mixtures of the exact Hartree–Fock exchange with DFT exchange–correlation are available via keywords BHH,[35] BHHLYP,[32,35] B3P86,[33,36] and B3LYP.[32,36]

III. RESULTS AND DISCUSSION

We shall begin our discussion by considering the geometrical parameters for the $Cr(CO)_6$ and CO molecules corresponding to the 10 methodologies investigated. A brief comparison on dipole moments of CO will conclude this first section. After that, a proper comparison of these methodologies in terms of QMSM is carried out: first, we discuss the effects of electronic relaxation on Euclidean distances and depict contours of electron density differences for $Cr(CO)_6$; and second, both nuclear and electronic relaxation effects on the Euclidean distance matrix are carefully examined and Bader analyses of the electron density of this molecule at the different levels of theory considered are presented.

A. Electronic Structure

Geometries

Table 1 gathers the experimental[37] and computed structural parameters for this highly symmetric chromium complex. An interesting consequence of the octahedral symmetry of this molecule is that there is a clear grouping pattern of the σ- and π-bonds of Cr–C and C–O. There are essentially two types of metal–ligand bonds. First, the $d(t_{2g})$ + CO π-bond is formed largely as a result of $2\pi^*$-backdonation (π-metal–ligand bond). The second bonding type is the well-known 5σ-donation to e_g and a_{1g} of Cr (σ-metal–ligand bond). It has been shown that, apart from these two main bonds, a secondary interaction exists involving hybridization of σ- and π-orbitals,[38] the latter bond being less important than the former (caused by a smaller overlap and a larger difference between orbital energies). Such a third bond is made up of three t_{1u} orbitals, and is formed by mixing σ-orbitals of one set of carbonyls (longitudinals) with π-orbitals of another set (transverse) through *p*-orbitals of the metal. The longitudinal (parallel to *p*) metal–ligand bond has σ-character, while the transverse bond bears π-character. Hence, this p + CO (σ + π)-bond can be denoted as σ + π.

As seen from Table 1, HF leads to an inaccurate description in two directions: the C–O bond length is predicted to be too short, whereas the Cr–C distance is overestimated by 0.078 Å. These discrepancies with respect to experimental data are both due to the problems associated with the insufficient backdonation from metal $d(t_{2g})$ to CO($2\pi^*$) at this level of theory. Noteworthy, results from the other methodologies indicate that this deficiency is corrected when correlation is introduced. Thus, at the MP2 level not only is backdonation taken into account, but it fails in emphasizing this effect by excess, which is not an unusual behavior of the MP2 method.[7b,10] The local functionals SVWN and HFS come to the same error. It is not until gradient corrections are included that the accuracy of such parameters increases. For instance, the average error of Cr–C and C–O distances for the five functionals with a Becke's nonlocal correction is about 0.015 and 0.014 Å, respectively, whereas for CCSD(T) it is twice as much (0.021 and 0.037 Å, respectively).

Another interesting point which provides information about the efficiency of a given method (in order to properly describe the backdonation) concerns the comparison of the C–O distance between free CO and CO belonging to a transition metal system as a ligand bonded to the chromium atom. One expects that the C–O distance increases from the free molecule to the fragment as an obvious result of the bond order reduction. Experimentally, the C–O distance[37,39] increases by 0.013 Å. From values of Table 1, all methods correctly take into account this increase, the only exception being HF which yields a C–O bond length for the ligand just 0.005 Å longer. It is clear that correlation effects are crucial when studying the nature of the metal–ligand bond in carbonyl complexes. This notwithstanding, the CCSD(T) approach is overcome by DFT methods; the former produces an increase of 0.044 Å, while the latter methods stay within a reasonable 0.010–0.014 Å range. MP2 yields an increase of 0.017 Å.

Dipole Moments

The dipole moment of CO has been a long-time favorite for evaluating the performance of various theoretical methods, and a large number of calculations have appeared over the years.[40–42] This molecule has a very special charge density distribution with a remarkable charge transfer from C to O and a large opposing polarization of the positive charge on C. These two effects counteract leading to dipole moments close to zero and a complicated charge density distribution. Therefore, the correct sign for the dipole moment of CO is difficult to reproduce. The HF result, for instance, predicts the wrong sign.[43] While this discrepancy is partly due to the small absolute value of the experimental dipole moment[44] and the usual overestimation at the SCF level, the dipole moment of CO has proven to be sensitive to the amount of correlation included in the wave function.[42] It has been shown previously that DFT is successful in computing the proper dipole direction of this molecule.[45–47] From values of Table 1 it is found that HF yields the erroneous direction for the dipole moment; BHH and BHHLYP also fail to provide the correct sign to the dipole moment, although the error is quantitatively smaller than in HF. Conversely, MP2 gives the correct direction but slightly exaggerates the dipole moment. With the exception of the local functionals SVWN and HFS, the other

Table 1. Bond Distances[a] and Dipole Moments[b] for $Cr(CO)_6$ and CO

	$Cr(CO)_6$		*CO*	
	R(Cr–C)	*R(C–O)*	*R(C–O)*	μ
HF	1.992	1.119	1.114	–0.104
MP2	1.877	1.168	1.151	0.074
CCSD(T)	1.939[c]	1.178[c]	1.134[d]	0.059[e]
HFS	1.888	1.167	1.153	0.084
SVWN	1.857	1.155	1.141	0.073
HFB	1.973	1.175	1.161	0.058
BP86	1.901	1.164	1.150	0.057
BLYP	1.925	1.164	1.150	0.057
B3LYP	1.915	1.150	1.138	0.024
BHH	1.876	1.131	1.121	–0.010
BHHLYP	1.922	1.134	1.124	–0.025
Expt.	1.918[f]	1.141[f]	1.128[g]	0.048[h]

Notes: [a] In Å.
[b] In au.
[c] Ref. 17.
[d] Ref. 19.
[e] Ref. 42.
[f] Ref. 37.
[g] Ref. 39.
[h] Ref. 44.

DFT approaches yield results close to the experimental value (0.048 au). In particular, BP86 and BLYP are shown to provide a reliable charge density distribution for this molecule. Interestingly, gradient-corrected DFT methods produce dipole moments which are better than the MP2 one, and in some cases they are even as good as that yielded by the CCSD(T) procedure.[42]

B. Analysis in Terms of QMSM

Analysis at a Fixed Geometry

As commented in the Introduction, the difference between two results arising from two methodologies in a molecule is directly related to the dissimilarity between the respective electronic distributions of this molecule, computed with the two methodologies being compared: the larger the distance, the larger the difference in these two electron densities. Therefore, the values of the distance yield a quantitative measure of how similar two methodologies are in the molecule under study. In this way it is possible to compare different methodologies, which is the main purpose of this work.

Table 2. Euclidean Distance Matrices[a] for the $Cr(CO)_6$ Molecule Computed at a Fixed Geometry[b] for the Different Methodologies Analyzed

Level	*HF*	*MP2*	*HFS*	*SVWN*	*HFB*	*BP86*	*BLYP*	*B3LYP*	*BHH*	*BHHLYP*
HF	0.0000									
MP2	0.0975	0.0000								
HFS	0.2482	0.2238	0.0000							
SVWN	0.2392	0.2138	0.0374	0.0000						
HFB	0.1664	0.1296	0.1237	0.1034	0.0000					
BP86	0.1600	0.1269	0.1273	0.1077	0.0100	0.0000				
BLYP	0.1712	0.1356	0.1292	0.1054	0.0141	0.0224	0.0000			
B3LYP	0.1396	0.1131	0.1404	0.1204	0.0316	0.0265	0.0346	0.0000		
BHH	0.1217	0.1179	0.1288	0.1183	0.0742	0.0693	0.0806	0.0574	0.0000	
BHHLYP	0.0854	0.0837	0.1819	0.1664	0.0837	0.0781	0.0872	0.0566	0.0608	0.0000

Notes: [a] In au.
[b] Experimental.

It is found that in most systems where correlation energy is of utmost importance, Hartree–Fock density is defective. As seen in the previous section, in the $Cr(CO)_6$ molecule correlation energy becomes essential. In order to assess the degree of viability of the 10 methods under study toward the obtention of correct electron distributions, in Table 2 we have gathered the distance between them when only electronic relaxation is taken into account. From these results, DFT approaches can be grouped into: (1) local functionals (SVWN and HFS), and (2) nonlocal functionals. The latter group can be divided into two different subsets: (2.1) one with the hybrids BHH and BHHLYP, and (2.2) another with HFB, BP86, BLYP, and B3LYP.

What local functionals SVWN and HFS (group 1) have in common is that their functionals depend only on the very $\rho(\mathbf{r})$, so it is not at all surprising that they yield very similar density distributions, both being the furthest ones from HF and MP2. The problem is mainly due to the poor description around nuclei as a consequence of ignoring the effects of the gradient of the density, ∇_{P}. These effects are basic in this region of large gradient. In particular, the electron density at the Cr nucleus is clearly underestimated by both SVWN (2104.069 au) and HFS (2103.879 au) methodologies, whereas all other procedures provide a higher density (Table 3). The small value of the density at the nuclei has a very important effect on the similarity integral, which results in large distances between SVWN and HFS as compared to other methodologies. When nonlocal corrections (i.e. derivatives of the density) are included, representations of nuclear cusps are improved. As revealed by values of Table 3, the nonlocal correction of Becke to the exchange functional is essential to solve this problem (HFS vs. HFB), whereas corrections to the correlation functional are not so decisive (HFS vs. SVWN or HFB vs. BLYP).

Table 3. Values of the Electron Density[a] in the Nucleus of Chromium (ρ_{Cr}), Carbon (ρ_c), and Oxygen (ρ_o) for the Different Methodologies at the Experimentally Reported Geometry[b]

Method	ρ_{Cr}	ρ_c	ρ_o
HF	2113.590	119.232	291.015
MP2	2113.612	119.022	291.170
HFS	2103.879	117.711	289.209
SVWN	2104.069	117.817	289.328
HFB	2113.457	118.510	290.460
BP86	2114.168	118.534	290.507
BLYP	2113.421	118.505	290.469
B3LYP	2112.734	118.614	290.507
BHH	2108.692	118.488	290.149
BHHLYP	2113.480	118.880	290.753
HFB	2113.457	118.510	290.460

Notes: [a] In au.
[b] Ref. 37.

The hybrid HH functionals (BHH and BHHLYP, subset 2.1) not only are quite similar to each other, but they are also the closest ones to HF and MP2. As previously seen from Table 1, among all DFT methods BHH and BHHLYP are precisely those yielding the worst description of electronic distributions (wrong sign of the dipole moment for the CO molecule).

Although the B3LYP functional makes use of the exact HF exchange so it is a hybrid functional, it behaves like most pure gradient-corrected functionals selected here (HFB, BP86, and BLYP, subset 2.2). Thus, according to our analysis it has to be considered for systems like $Cr(CO)_6$ as a member of this subgroup, instead of the hybrid 2.1 subset.

Since the Euclidean distance matrix collected in Table 2 has been computed at the experimental geometry of $Cr(CO)_6$ for all levels of theory considered (i.e., the geometry has been kept fixed), it is possible to perform an additional analysis by means of density difference maps (Figure 1). These maps show the difference between densities obtained using a given method [namely, SVWN (a), BHH (b), BP86 (c) and MP2 (d)] and the density yielded by the HF methodology. The effect of correlation is very similar for all cases, and can be summarized in mainly three points:

1. An increase of the electron density in the $3d(e_g)$ orbitals which possess the symmetry suitable for interacting with the 5σ of CO (and which are located at the cross-shaped region around the chromium atom, depicted by the solid

line), together with a reduction of density in the 5σ orbital, due to the donation from 5σ to $3d(e_g)$ (see dashed-line region near C nucleus along the Cr–C bond).

2. An increase of the electron density of the $2\pi^*$-orbitals of CO, together with a reduction of the electron density of the $3d(t_{2g})$ orbitals, which exhibit π-symmetry (and which are depicted by the four dashed lobes alternating with the arms of the central cross). Noticeably, the $2\pi^*$-orbital has a greater coefficient for C than for O; hence, the change in π-backdonation due to inclusion of correlation effects is more remarkable for the former atom (see the two lobes depicted in solid lines around the C nucleus).
3. A reduction of the electron density at nuclei, except for the case of the MP2 method.

In conclusion, the main effect of introducing correlation in this molecule is to withdraw density out of the σ region and $d(t_{2g})$ orbitals and to accumulate it on the π^* of CO and $d(e_g)$ orbitals favoring the donation from the ligand to the metal atom, which in turn causes a correct feed–backdonation (synergetic effect).

On the other hand, the features mentioned in the preceding paragraphs about the different groups and subsets of functionals are reflected in these maps. First, the SVWN-HF plot (Figure 1a) reveals that local functionals underestimate the electron

(a)

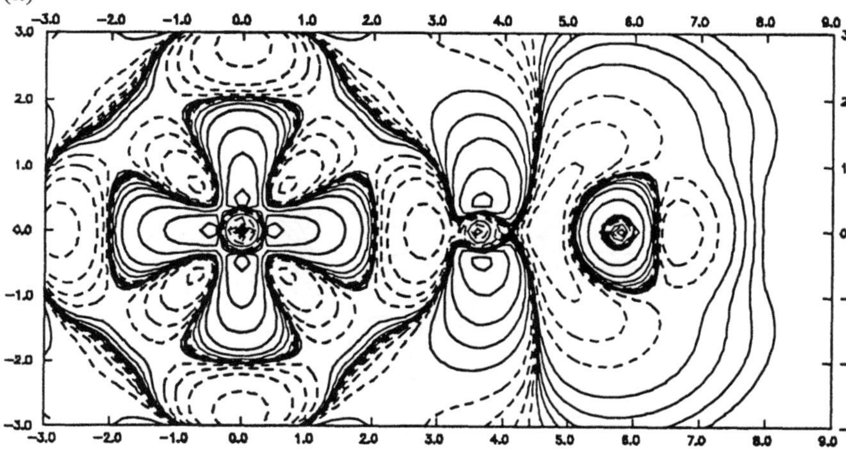

(continued)

Figure 1. Plots of electron density differences comparing densities obtained from the Hartree–Fock methodology with those computed at SVWN (**a**), BHH (**b**), BP86 (**c**), and MP2 (**d**) levels, for the Cr(CO)6 molecule at its experimental geometry. In these maps the chromium atom is on the left, the carbon atom in the middle, and the oxygen on the right. The minimum contour is 1 × 10^{-4} au and they increase to 2, 4, 8, 20, 40, 80, . . . × 10^{-4} au. Dashed lines correspond to negative values, that is, points where Hartree–Fock density is larger.

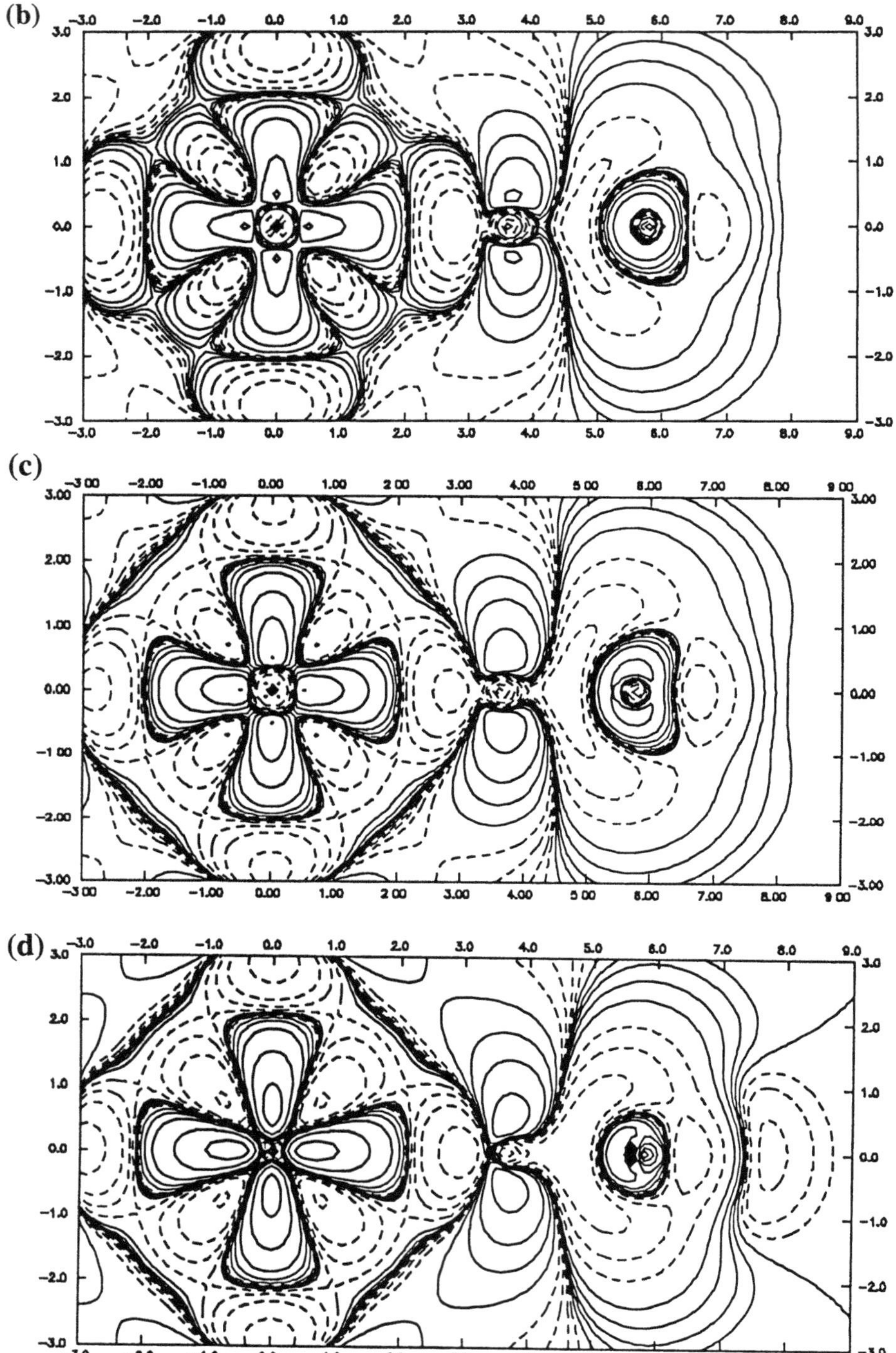

Figure 1. (Continued)

density at the Cr nucleus (small negative region in the center of the metal atom). Second, the density difference map between HF and BHH hybrid functionals (Figure 1b) is quite smooth, showing that the BHH density does not differ too much from that arising from HF. In particular, it is the number of concentric lines and their spacing which allows one to reach such a conclusion.

In our discussion about geometrical parameters we have pointed out that an alternative way of measuring the backdonation effect in a given method lies in evaluating the lengthening of the CO distance when changing from the free CO molecule to the ligand CO bonded to metal. We can visualize this effect using a technique which considers the $(CO)_6$ cage resulting from withdrawing the central Cr atom. Let us suppose an O_h symmetry for the cage and the same C–O distance as in the experimentally reported structure for the $Cr(CO)_6$ complex ($d_{CO} = 1.141$ Å). If we depict, for a given methodology, the electron density difference between the density of the whole $Cr(CO)_6$ molecule and the density of such a cage, maps as those shown in Figure 2 are obtained. It is worth noting that the BP86 and MP2

(a)

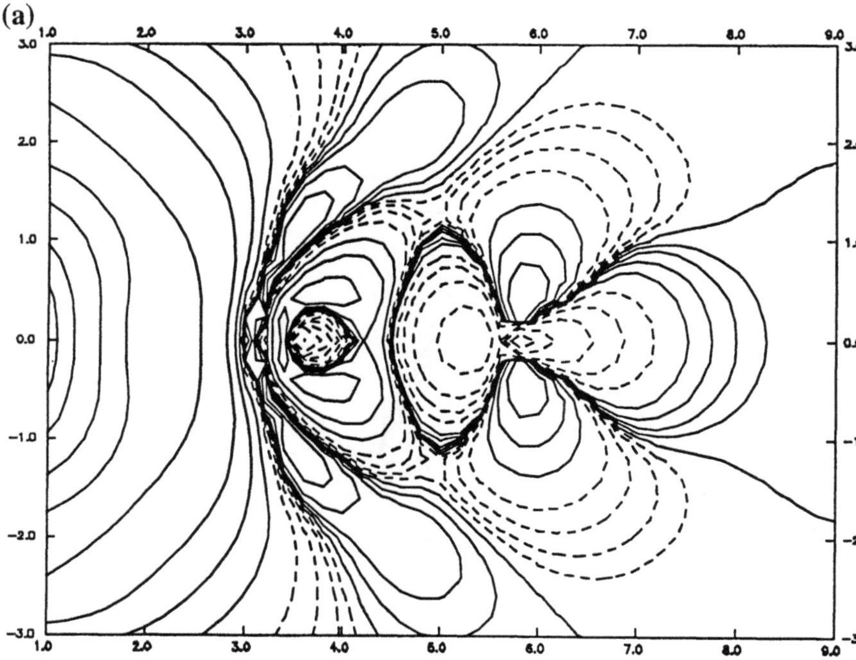

(continued)

Figure 2. Plots of electron density differences comparing densities for the $Cr(CO)_6$ molecule and the $(CO)_6$ cage at the experimental geometry of the former system, obtained from Hartree–Fock (**a**), BP86 (**b**), and MP2 (**c**) methodologies. The minimum contour is 1×10^{-4} au and they increase to 2, 4, 8, 20, 40, 80, . . . $\times 10^{-4}$ au. Solid lines correspond to positive values, that is, points where the density for $Cr(CO)_6$ is larger than for $(CO)_6$.

(b)

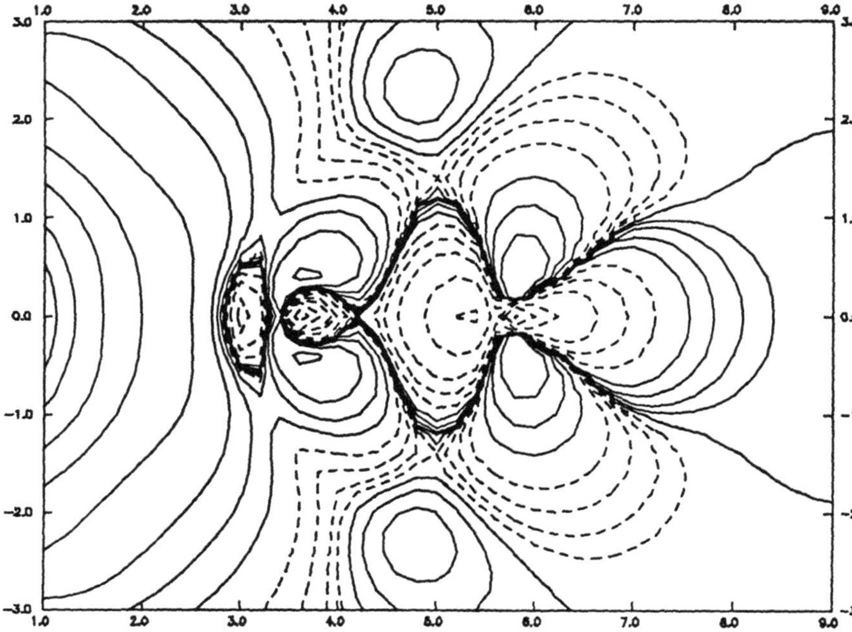

(c)

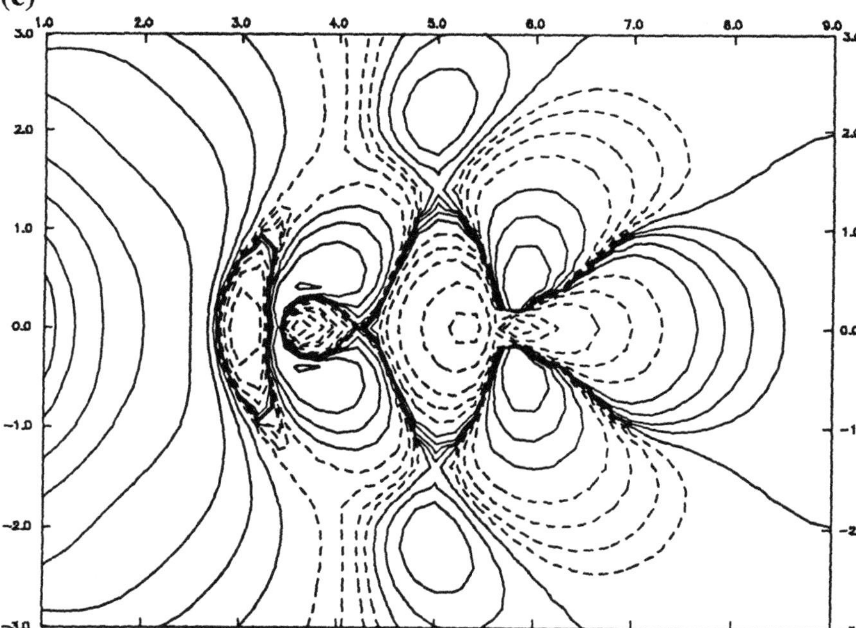

Figure 2. (Continued)

maps (Figures 2b and 2c) show a pattern similar to that obtained when comparing HF and correlated densities in the whole $Cr(CO)_6$ complex (Figures 1c and 1d). The HF map (Figure 2a) shows the same effect, but clearly diminished. The main effect observed when rearranging the electron density from $(CO)_6$ to $Cr(CO)_6$ is that HF overemphasizes the density of the CO σ-orbital and underestimates the density located at CO π^*-orbitals. Thus, this method is once again defective. On the basis of the similarities between DFT and MP2 plots of Figure 2, it seems clear to us that correlation effects in DFT methods are included to some extent.

Analysis at Optimized Geometries

To gain more insight into the nature of differences in charge density distributions obtained from the different methodologies analyzed, we have performed an analysis of $Cr(CO)_6$ electron densities at the optimized geometries for each method. The analysis presented here includes both electronic and nuclear relaxation, whereas the study carried out in the last section, accounted only for the electronic relaxation (fixed geometry).

As expected, if both types of relaxation are allowed (Table 4), distances d_{IJ} increase although it is also certain that they grow in a different proportion. Interestingly, the order and classification of methodologies according to Table 4 is no longer the same as that provided by Table 2. Thus, the largest differences in electron densities corresponds now to the HFB, BHH pair (30.0885 au) and the HFB, SVWN pair (30.4586 au), while at fixed geometry such distances were small or intermediate (0.0742 and 0.1034 au, respectively). It must be pointed out that HF gives a large distance to any method analyzed; HF always appears at $d_{IJ} > 12$ au and can be considered as a method quite separate from the others.

Table 4. Euclidean Distance Matrices[a] for the $Cr(CO)_6$ Molecule Computed at the Optimized Geometry Corresponding to Each Methodology Employed Accounting for Both Nuclear and Electronic Relaxation

Level	*HF*	*MP2*	*HFS*	*SVWN*	*HFB*	*BP86*	*BLYP*	*B3LYP*	*BHH*	*BHHLYP*
HF	0.0000									
MP2	23.9031	0.0000								
HFS	21.5814	4.3303	0.0000							
SVWN	28.4673	11.7062	14.8840	0.0000						
HFB	12.5899	27.5128	25.9173	30.4586	0.0000					
BP86	19.2233	8.1518	4.1141	17.7695	24.2395	0.0000				
BLYP	12.5673	16.1338	12.8801	23.2822	19.1289	9.4162	0.0000			
B3LYP	18.2474	9.5874	5.8688	18.7151	23.4888	2.4844	8.4739	0.0000		
BHH	28.3073	12.3676	15.1761	3.6175	30.0885	17.7611	23.0101	18.5216	0.0000	
BHHLYP	19.6500	8.3543	5.6464	17.2662	24.3764	4.7759	11.1483	3.5441	16.9793	0.0000

Notes: [a] In au.

Another interesting feature is that HFB also yields large distances compared to the other tested methodologies ($d_{IJ} > 19$ au), indicating that this functional is not reliable enough for studying chromium hexacarbonyl and related transition metal systems. Although HFB yields good densities at fixed geometry, when densities at optimized geometries are computed it behaves inaccurately. In fact, from a structural point of view (see Table 1), HFB has already been shown to be the worst of the 8 DFT approaches here selected. On the other hand, despite the SVWN density being initially the nearest to HFS (local group 1), when nuclear relaxation is allowed, it becomes very different from the HFS and similar to the hybrid BHH result. Not only are the SVWN and BHH results very similar to each other, but they are also different from results of any other method. As seen in a previous work,[10] conclusions from charge density analyses at fixed geometry cannot be extrapolated to optimized systems. Thus, while the largest difference between HF and DFT methods corresponds to HFS if only electronic relaxation is considered, when both nuclear and electronic relaxation are allowed, then HFS behaves similarly to the subset 2.2, the largest deviation from the HF result shown for SVWN. One can say that large density differences at a fixed geometry do not always imply large structural and charge density differences in the optimized molecules. For this reason, an analysis of density differences at a fixed geometry may provide different conclusions to those arising from analyses performed at optimized geometries.

With respect to the analysis of nonoptimized $Cr(CO)_6$, only the subgroup 2.2 of nonlocal DFT functionals partially keeps up its integrity. Thus, BP86, B3LYP, and BLYP can still be considered in the same subset, but it is found that HFB no longer belongs to this group when both types of relaxation are taken into account. Moreover, now this subset grows due to the incorporation of two new related functionals: BHHLYP and, surprisingly, HFS. In its turn, the latter becomes very close to MP2 and yields better results than SVWN.

We can conclude that these five functionals (BP86, B3LYP, BLYP, BHHLYP, and HFS) would be those most reliable for studying systems like $Cr(CO)_6$, since they present large distances to HF and are quite close to MP2, especially BP86, B3LYP, and BLYP which show an adequate behavior both at fixed and optimized geometries. Among them, HFS becomes very recommendable because, in addition, it is computationally inexpensive due to its local character.

Finally, in Table 5 we offer an analysis of the charge density distributions obtained from the different methodologies studied from Bader's theory point of view.[5] The tendency followed by the HF C–O bond length, which is shorter than the correlated bond lengths, is reproduced by distances from C to the C–O bond critical point. It is found that when correlation is included such distances are larger ($d_{C\text{-}BCP}(\text{corr}) > 0.371$ Å) in all cases. Furthermore, due to the fact that the HF C–O bond length is shorter, the density at the bond critical point is larger for the HF method as compared to correlated methodologies: $\rho^{HF} < \rho^{corr}$. An additional consequence of the shorter HF bond length is that charge depletion ($\nabla^2\rho > 0$) becomes clearly exaggerated at this level: 1.466 au in front of a DFT average ranging between

Table 5. Bader Analysis for the $Cr(CO)_6$ Molecule at the Optimized Geometry Corresponding to Each Level Studied[a]

	Cr–C Bond			*C–O Bond*		
Method	$d_{BCP\text{-}C}$	ρ_{BCP}	$\nabla^2\rho_{BCP}$	$d_{C\text{-}BCP}$	ρ_{BCP}	$\nabla^2\rho_{BCP}$
HF	1.070	0.077	0.582	0.371	0.497	1.466
MP2	0.958	0.118	0.413	0.385	0.435	0.983
HFS	0.959	0.114	0.414	0.388	0.441	0.817
SVWN	0.938	0.122	0.440	0.385	0.454	0.938
HFB	1.026	0.090	0.403	0.388	0.432	0.809
BP86	0.973	0.109	0.433	0.385	0.443	0.933
BLYP	0.988	0.103	0.406	0.386	0.444	0.895
B3LYP	0.993	0.102	0.481	0.381	0.460	1.053
BHH	0.972	0.111	0.555	0.376	0.483	1.234
BHHLYP	1.009	0.097	0.553	0.376	0.479	1.247

Notes: [a] Distances in Å, densities in au, and Laplacian in au.

0.8–1.0 au (without considering the hybrid HH functionals). On the other hand, HF overestimation of Cr–C bond distance is also reflected, first in a larger value of the $d_{BCP\text{-}C}$ distance ($d^{HF} > d^{corr}$), second in the computed density at the bond critical point, which becomes smaller than for correlated methodologies: $\rho^{corr} > 0.077$ au, and third in the value of the Laplacian: $\nabla^2\rho^{HF} > \nabla^2\rho^{corr}$.

IV. CONCLUSIONS

It has been shown that distances obtained from quantum molecular similarity measures can be a useful tool in analyzing charge density distribution differences within a series of methodologies, allowing the analysis to be performed at the optimized geometry corresponding to each methodology. Although we had come to a similar conclusion in a previous study on small organic molecules including atoms up to fluorine,[10] it is interesting to point out that the validity of such a procedure can also be extended to transition metal systems.

The use of electron density difference contours is undeniably practical to illustrate differences at a fixed geometry (in this case, at the experimentally reported geometry), but can lead to conclusions that are no longer valid at the optimized geometries. For instance, if only electronic relaxation is taken into account, the largest difference between HF and DFT methods corresponds to HFS, whereas when both nuclear and electronic relaxation are allowed, HFS behaves similarly to the subset of nonlocal functionals including BP86, B3LYP, and BHHLYP (subset 2.2), the largest distance to HF being now for SVWN.

Among the DFT formalisms studied here, the local SVWN shows a qualitatively poor description, whereas the nonlocal functionals of the aforementioned subset offer more accurate densities, correctly accounting for the π-backbonding in the metal–CO coordination. Furthermore, the latter methods correct the overestimated ionicity present in Hartree–Fock electron densities, and are as adequate as MP2, if not better, for describing charge density distributions in the $Cr(CO)_6$ complex.

The main conclusion of this work is that, although DFT surpasses HF, only a particular kind of functional is shown to be very accurate for describing transition metal–hexacarbonyl systems. Indeed, BP86, B3LYP, and BLYP seem to be quite suitable, according to our analysis performed at both fixed and optimized geometries. If the second analysis is taken into account, then BHHLYP and HFS functionals must also be included among the recommended methods. In particular, the latter functional offers the additional advantage of being inexpensive from a computational point of view and, therefore, probably the most recommended for such studies.

The analysis presented in this work will be applied to other cases of interest, which will be reported in the near future. More research on these points is underway in our laboratory.

ACKNOWLEDGMENTS

This work was financially supported by the Spanish DGICYT through Project No. PB92-0333. Valuable discussions with Dr. J. Mestres are most appreciated.

REFERENCES

1. (a) Löwdin, P.O. *Phys. Rev.* **1955**, *97*, 1474. (b) McWeeny, R. *Proc. Roy. Soc. A* **1955**, *232*, 114. (c) McWeeny, R. *Proc. Roy. Soc. A* **1956**, *235*, 496. (d) McWeeny, R. *Proc. Roy. Soc. A* **1959**, *253*, 242.
2. (a) Parr, R.G.; Yang, W. *Density-Functional Theory of Atoms and Molecules*; Oxford University: New York, 1989. (b) Ziegler, T. *Chem. Rev.* **1991**, *91*, 651.
3. (a) Fermi, E.Z. *Z. Phys.* **1928**, *48*, 73. (b) Thomas, L.H. *Proc. Camb. Philos. Soc.* **1927**, *23*, 542.
4. Hohenberg, P.; Kohn, W. *Phys. Rev. B* **1964**, *136*, 864.
5. (a) Bader, R.F.W. *Acc. Chem. Res.* **1985**, *18*, 9. (b) Bader, R.F.W. *Atoms in Molecules: A Quantum Theory*; Clarendon: Oxford, 1990. (c) Bader, R.F.W.; Gillespie, R.J.; MacDougall, P. J. *J. Am. Chem. Soc.* **1988**, *110*, 7329.
6. Becke, A.D.; Edgecombe, K.E. *J. Chem. Phys.* **1990**, *92*, 5397.
7. (a) Wang, J.; Eriksson, L.A.; Boyd, R.J.; Shi, Z.; Johnson, B.G. *J. Phys. Chem.* **1994**, *98*, 1844. (b) Wang, J.; Shi, Z.; Boyd, R.J.; Gonzalez, C.A. *J. Phys. Chem.* **1994**, *98*, 6988. (c) Solà, M.; Mestres, J.; Carbó, R.; Duran, M. *QSAR and Molecular Modeling: Concepts, Computational Tools and Biological Applications*; Prous: Barcelona, 1995, pp. 403–406.
8. (a) Cioslowski, J.; Fleischmann, E.D. *J. Am. Chem. Soc.* **1991**, *113*, 64. (b) Ciolowski, J.; Challacombe, M. *Int. J. Quantum Chem., Quantum Chem. Symp.* **1991**, *25*, 81. (c) Ciolowski, J. *J. Am. Chem. Soc.* **1991**, *113*, 6756. (d) Ortiz, J.V.; Ciolowski, J. *Chem. Phys. Lett.* **1991**, *185*, 270. (e) Ciolowski, J. *Theor. Chim. Acta* **1992**, *81*, 319. (f) Solà, M.; Mestres, J.; Duran, M.; Carbó, R. *J. Chem. Inf. Comput. Sci.* **1994**, *34*, 1047. (g) Mestres, J.; Solà, M.; Duran, M.; Carbó, R.

Molecular Similarity and Reactivity: From Quantum Chemistry to Phenomenological Approaches; Kluwer: Dordrecht, 1995, pp. 89–111. (h) Mestres, J.; Solà, M.; Duran, M.; Carbó, R. *J. Comp. Chem.* **1994**, *15*, 1113. (i) Solà, M.; Mestres, J.; Duran, M.; Carbó, R. *J. Am. Chem. Soc.* **1994**, *116*, 5909.

9. (a) Carbó, R.; Arnau, M.; Leyda, L. *Int. J. Quantum Chem.* **1980**, *17*, 1681. (b) Carbó, R.; Calabuig, B. *Int. J. Quantum Chem.* **1992**, *42*, 1681.
10. Solà, M.; Mestres, J.; Carbó, R.; Duran, M. *J. Chem. Phys.*, in press.
11. Werner, H. *Angew. Chem.* **1990**, *102*, 1109; *Angew. Chem. Int. Ed. Engl.* **1990**, *29*, 1077.
12. Davidson, E.R.; Kunze, K.L.; Machado, F.B.C.; Chakravorty, S.J. *Acc. Chem. Res.* **1993**, *26*, 628.
13. Faegri, K.; Almlöf, J. *Chem. Phys. Lett.* **1984**, *107*, 121.
14. Persson, B.J.; Roos, B.O.; Pierloot, K. *J. Chem. Phys.* **1994**, *101*, 6810.
15. Barlett, R.J. *Annu. Rev. Phys. Chem.* **1981**, *32*, 359.
16. Raghavachari, K.; Trucks, G.W.; Pople, J.A.; Head-Gordon, M. *Chem. Phys. Lett.* **1989**, *157*, 479.
17. (a) Barnes, L.A.; Rosi, M.; Bauschlicher, C.W. *J. Chem. Phys.* **1991**, *94*, 2031. (b) Barnes, L.A.; Liu, B.; Lindh, R. *J. Chem. Phys.* **1993**, *98*, 3978.
18. (a) Ehlers, A.W.; Frenking, G. *J. Am. Chem. Soc.* **1994**, *116*, 1514. (b) Ehlers, A.W.; Frenking, G. *Organometallics* **1995**, *14*, 423.
19. Jonas, V.; Thiel, W. *J. Chem. Phys.* **1995**, *102*, 8474.
20. Blomberg, M.R.A.; Brandemark, U.B.; Siegbahn, P.E.M.; Wennerberg, J.; Bauschlicher, C.W. *J. Chem. Phys.* **1991**, *94*, 2031.
21. Baerends, E.J.; Rozendaal, A. *Quantum Chemistry: The Challenge of Transition Metals and Coordination Chemistry*; Veillard. A., Ed.; Kluwer: Dordretch, 1986, pp. 159–177.
22. Demuynck, J.; Strich, A.; Veillard, A. *Nouv. J. Chim.* **1977**, *1*, 217.
23. Frisch, M.J.; Trucks, G.W.; Head-Gordon, M.; Gill, P.M.W.; Wong, M.W.; Foresman, J.B.; Johnson, B.G.; Schlegel, H.B.; Robb, M.A.; Replogel, R.S.; Gomperts, E.S.; Andres, J.L.; Raghavachari, K.; Binkley, J.S.; Gonzalez, C.; Martin, R.L.; Fox, D.J.; Defrees, D.J.; Baker, J.; Stewart, J.J.P.; Pople, J.A.; GAUSSIAN 92-DFT, Revision G.1, Gaussian, Pittsburg, PA, 1992.
24. Wachters, A.J.H. *J. Chem. Phys.* **1985**, *82*, 299.
25. (a) Hehre, W.J.; Ditchfield, R.; Pople, J.A. *J. Chem. Phys.* **1972**, *56*, 2257. (b) Hariharan, P.C.; Pople, J.A. *Theor. Chim. Acta* **1973**, *28*, 213. (c) Gordon, M.S. *Chem. Phys. Lett.* **1980**, *76*, 163.
26. Mestres, J.; Solà, M.; Besalú, E.; Duran, M.; Carbó, R. MESSEM, Girona, CAT, 1993.
27. (a) Handy, N.C.; Schaefer III, H.F. *J. Chem. Phys.* **1984**, *81*, 5031. (b) Wiberg, K.B.; Hadad, C.M.; LePage, T.J.; Breneman, C.M.; Frisch, M.J. *J. Phys. Chem.* **1992**, *96*, 671.
28. Solà, M.; Mestres, J.; Oliva, J.M.; Duran, M.; Carbó, R. *Int. J. Quantum Chem.* **1996**, *58*, 361.
29. J. Mestres, ELECTRA, Girona, CAT, 1994.
30. Slater, J.C. *Phys. Rev.* **1951**, *81*, 385.
31. Becke, A.D. *Phys. Rev. A* **1988**, *38*, 3098.
32. Lee, C.; Yang, W.; Parr, R.G. *Phys. Rev. B* **1988**, *37*, 786.
33. Perdew, J.P. *Phys. Rev. B* **1986**, *33*, 8822. Erratum, ibid. **1986**, *34*, 7406.
34. Vosko, S.H.; Wilk, L.; Nusair, M. *Can. J. Phys.* **1980**, *58*, 1200.
35. Becke, A.D. *J. Chem. Phys.* **1993**, *98*, 1372.
36. Becke, A.D. *J. Chem. Phys.* **1988**, *88*, 2547.
37. Jost, A.; Rees, B. *Acta Cryst.* **1975**, *B31*, 2649.
38. Arratia-Perez, R.; Yang, C.Y. *J. Chem. Phys.* **1985**, *83*, 4005.
39. Huber, K.P.; Herzberg, G.P. *Constants of Diatomic Molecules*; Van Nostrand Reinhold: New York, 1979.
40. Feller, D.; Boyle, C.M.; Davidson, E.R. *J. Chem. Phys.* **1987**, *86*, 3224.
41. Frisch, M.J.; Del Bene, J.E. *Int. J. Quantum Chem.* **1989**, *23*, 363.
42. Scuseria, G.E.; Miller, M.D.; Jensen, F.; Geertsen, J. *J. Chem. Phys.* **1991**, *94*, 6660.
43. Laaksonen, L.; Pyykkö, P.; Sundholm, D. *Comp. Phys. Rep.* **1986**, *4*, 313.
44. Muenter, J.S. *J. Mol. Spectrosc.* **1975**, *55*, 490.

45. Wang, J.; Shi, Z.; Boyd, R.J.; Gonzalez, C.A. *J. Phys. Chem.* **1994**, *98*, 6988.
46. (a) Johnson, B.G.; Gill, P.M.W.; Pople, J.A. *J. Chem. Phys.* **1993**, *98*, 5612. (b) Murray, C.W.; Laming, G.J.; Handy, N.C.; Amos, R.D. *Chem. Phys. Lett.* **1992**, *199*, 551.
47. (a) Jones, R.O.; Gunnarsson, O. *Rev. Mod. Phys.* **1989**, *61*, 689. (b) Baerends, E.J.; Vernooijs, P.; Rozendaal, A.; Boerrigter, P.M.; Krijn, M.; Feil, D.; Sundholm, D. *J. Mol. Struct. (Theochem)* **1985**, *133*, 147.

QUANTUM MOLECULAR SIMILARITY MEASURES (QMSM) AND THE ATOMIC SHELL APPROXIMATION (ASA)

Pere Constans, Lluís Amat,
Xavier Fradera, and Ramon Carbó-Dorca

Advances in Molecular Similarity
Volume 1, pages 187–211

ISBN: 0-7623-0131-7

ABSTRACT

First-order electron density similarity measures for large molecules are straightforward and can be efficiently computed if the *atomic shell approximation* (ASA) is used. Within this approximation the molecular electron distributions are represented by simple superpositions of spherical atomic contributions. A new algorithm to optimally select shells fitting known electron distributions and an empirical scheme to construct molecular densities by summing atomic fragments are presented. The accuracy of both ASA procedures is analyzed comparing approximated and ab initio QMSM.

I. INTRODUCTION

Molecules, as quantum objects, are completely described by the set of reduced density matrices arising from successive integration of their attached spin–space N electron wave functions, $\Psi(\mathbf{x}_1, \ldots, \mathbf{x}_N)$, being the s order reduced density matrix given by:

$$\rho_s(\mathbf{x}_1, \ldots, \mathbf{x}_s;\mathbf{x}'_1, \ldots, \mathbf{x}'_s) = \binom{N}{S} \int \Psi^*(\mathbf{x}_1, \ldots, \mathbf{x}_N)\Psi(\mathbf{x}'_1, \ldots, \mathbf{x}'_N) d\mathbf{x}_{s+1} \ldots d\mathbf{x}_N \tag{1}$$

Sets of functions belonging to different molecules could be compared and similarity measures among them mathematically established. Similarities are cognitive relations for ordering and classifying object qualities, and their measure can reveal aspects of accessible human knowledge. The classical understanding of chemical systems as physical, three-dimensional entities can be recovered by means of the diagonal part of the spin independent first-order density matrices, or briefly, the electron densities of probability, which are expressed, removing superfluous indices, as:[1]

$$\rho(\mathbf{r}) = N \int \Psi^*(\mathbf{x}_1,\mathbf{x}_2, \ldots, \mathbf{x}_N)\Psi(\mathbf{x}_1,\mathbf{x}_2, \ldots, \mathbf{x}_N) ds_1 d\mathbf{x}_2 \ldots d\mathbf{x}_N \tag{2}$$

The spatial electron density function $\rho(\mathbf{r})$ and its derivatives provide the means for a definition of atoms in molecules,[2] the identification of chemical bonds, and rigorous quantification of chemical concepts as covalent bond order, steric crowding, electronegativity, or bond hardness.[3]

A quantum molecular similarity measure (QMSM) based on these real space electron densities is generally defined as,[4]

$$z_{AB} = \int\int \rho_A(\mathbf{r}_A)\Theta(\mathbf{r}_A,\mathbf{r}_B)\rho_B(\mathbf{r}_B) d\mathbf{r}_A d\mathbf{r}_B \tag{3}$$

where ρ_A and ρ_B are the electron densities of two arbitrary molecules A and B, and Θ is a positive definite operator. Since the set of functions (Eq. 1) and consequently function (Eq. 2) parametrically depends, in the Born–Oppenheimer approximation, on the nuclei coordinates, the measure z_{AB} for any considered molecular geometry is assumed to be taken at the mutual positioning of both molecules which maximizes the integral (Eq. 3). This conceptually simple similarity measure is impractical for drug design purposes because of its computational difficulty. Within the LCAO approach, first-order electron densities are given as double sums over pairs of basis functions in the form,

$$\rho(\mathbf{r}) = \sum_{ij}^{n} D_{ij}\varphi_i^*(\mathbf{r})\varphi_j(\mathbf{r}) \tag{4}$$

where D_{ij} are the density matrix coefficients, $\varphi_i(\mathbf{r})$ and $\varphi_j(\mathbf{r})$ are the atomic orbitals, and n is the number of these basis functions. Every evaluation of z_{AB} in the maximization procedure requires $n_A^2 n_B^2$ computations of many center integrals, together with a cumbersome transformation of the elements D_{ij} under molecular rotation. CNDO-like approximations—computations based on a discrete representation of electron densities, computationally more attainable definitions of similarity,[5] or fittings of electron density to simpler spherical functions[6]—have been proposed with the aim to extend similarity measures based on quantum mechanics to pharmacological design.

Since the First Girona Seminar, where several works were presented exploring this last strategy,[7] important advances have been done in our laboratory in the representation of electron densities as superposition of spherical atomic shells, eliminating deficiencies, both theoretical and computational, that the simple least-squares fitting (LSF) presents. The theoretical restriction imposed on the set of variational coefficients, i.e. to be non-negative, has led to the development of a fitting scheme for approximating electron densities, the *atomic shell approximation* (ASA), where shells are optimally selected from a nearly complete functional space.[8] Solving this theoretical constraint in the ASA procedure fixes the computational drawbacks: exponent optimization; nearly linear dependencies; the need for several basis sets to optimally reproduce different calculated densities; and arbitrary assignments of shells in an atom, which could distort the resulting charge distribution within a molecule. Moreover, the ASA opens an avenue for modeling promolecules, i.e., molecular electron representations built on atomic contributions. Therefore, sharp electronic distributions may be diffused by atomic vibrations, or conformational movements may be allowed during the similarity maximization, giving a more realistic vision of molecules. In this latter case, atoms and their attached electrons can be displaced from the original position to construct different conformations. This is, strictly speaking, an extrapolation since the density is initially computed at a single conformational arrangement; thus densities for the rest of the conformations are obtained starting from this initial density. In such a case, it is likely that the nonphysically reliable density obtained by simple LSF could fail.

Now, at the time of concluding the Second Girona Seminar, one can regard ASA as more than a computational device to approximate first-order QMSM integrals. ASA is an accurate physical model useful to extend QMSM to real problems in pharmacological drug research. The present work is concerned with the ASA and its ability to accurately calculate overlap QMSM based on first-order density functions. The complete ASA fitting scheme will be presented, empirical ASA approaches made by summing atomic fragments of density analyzed, and deviations of approximated QMSM from ab initio values quantified.

II. ATOMIC SHELL APPROXIMATION

Electron distributions of atoms in field-free space are spherically symmetric[9] and expressible in terms of integral transforms over the radial coordinate, such as:

$$\rho_a(\mathbf{r}) = \int_0^\infty f_i(\zeta) e^{l} - \zeta \,|\, \mathbf{R}_a - \mathbf{r} \,|^{l} d\zeta \tag{5}$$

In the case of a Gaussian kernel, the approximation of the integral (Eq. 5) by a finite sum leads to electron densities expressed by a superposition of spherical shells in the form,

$$\rho_a(\mathbf{r}) \approx \sum_i n_i S_i(\mathbf{R}_a - \mathbf{r}) \tag{6}$$

where shells $S_i(\mathbf{R}_a - \mathbf{r})$ are defined as,

$$S_i(\mathbf{R}_a - \mathbf{r}) \equiv \left(\frac{\zeta_i}{\pi}\right)^{3/2} e^{-\zeta_i(\mathbf{R}_a - \mathbf{r})^2} \tag{7}$$

in order to identify coefficients n_i with shell populations. Approximation (Eq. 6) together with the idealization of molecular densities built on spherical atomic shells constitutes the ASA, whose molecular electron distributions appear as:

$$\rho_{ASA}(\mathbf{r}) = \sum_a \sum_{i \in a} n_i S_i(\mathbf{R}_a - \mathbf{r}). \tag{8}$$

This portable representation of electron densities has been widely used when simple functional forms were required, such as the treatment of X-ray crystallographic data,[10] or in molecular shape characterization.[11] Equation 8 can also be used to compute molecular wave functions from n S-like orbitals.[12] When these representations are applied to QMSM computations, a great simplification is reached with both the number of involved basis functions and integral complexity being greatly reduced. The following sections show how to obtain the shells S_i and the respective occupations n_i for any molecule, while quantifying at the same time the errors of such approximation by comparison with ab initio QMSM. In Section II.A we present a new algorithm which optimally selects shells from a nearly complete functional space and approximate known molecular electron densities, $\rho(\mathbf{r})$. Section II.B analyzes the construction of $\rho_{ASA}(\mathbf{r})$ based on the approximate additivity and invariance of atomic densities in molecular environments. This rough representation of molecular densities is still useful to compute QMSM with acceptable accuracy when densities are not available, as in the case of large systems, or when they are not worthwhile to compute, as in a first selection of similar compounds in a structural database search.

A. Density Fitted Atomic Shells

Having a discrete or functional representation of the electron density of a system, $\rho(\mathbf{r})$—the best approach in a least-squares sense—$\rho_{ASA}(\mathbf{r})$, in terms of a complete set of functions $S_i(\mathbf{R}_a - \mathbf{r})$, requires only the lineal minimization of the quadratic error integral function:

$$\varepsilon^2(\mathbf{n}) = \int(\rho(\mathbf{r}) - \sum_{a}\sum_{i\in a} n_i S_i(\mathbf{R}_a - \mathbf{r}))^2 d\mathbf{r} \tag{9}$$

Nearly complete spaces of Gaussian functions can be generated selecting exponents in a geometric sequence,[13]

$$\zeta_i = \alpha\beta^i \tag{10}$$

together with an implicit dependence of the generators α and β with respect to the basis size n, postulated by Ruedenberg et al.[14] as,

$$\ln \ln \beta = b \ln n + b' \tag{11}$$

and,

$$\ln \alpha = a \ln(\beta - 1) + a' \tag{12}$$

to ensure a successful approach to completeness when n is increased. These *even-tempered* sequences, which are a simple and elegant way to construct truncated basis sets, avoid cumbersome nonlinear optimizations and take control over possible linear dependencies.[14] A simple two-dimensional search over generators α and β gives no significant improvement with respect to a fully variational solution optimizing all the exponent series.[14] The parameters α and β are optimized for different sizes of the basis sets and the constants in Eqs. 11 and 12 are obtained by a linear regression.[8] The values given by these equations, called *regularized even-tempered* parameters, differ very little from the optimized ones, having the interesting advantage, besides the theoretical correctness, of allowing different basis sizes and a quality fitting exploration in the implementation of the ASA.

Coefficients **n** are subject to the physical constraints derived from the fact that $\rho_{ASA}(\mathbf{r})$ is a density of probability function. These constraints are the normalization condition,

$$\sum_i n_i = N \tag{13}$$

and the set of inequalities,

$$n_i \geq 0 \ \forall i, \tag{14}$$

assuring a positive valued $\rho_{ASA}(\mathbf{r})$ in its whole domain. Restriction (Eq. 13) can be introduced using a Lagrange multiplier formalism. Then the restricted minimum $\mathbf{n}_0'$, denoted by primes, of the quadratic error integral function $\varepsilon^2(\mathbf{n})$ accomplishes the linear equation,

$$\mathbf{S}\mathbf{n}_0' = \mathbf{t}' \tag{15}$$

where the elements of the overlap matrix **S** are,

$$s_{ij} = \int S_i(\mathbf{r}) S_j(\mathbf{r}) d\mathbf{r} \tag{16}$$

and vector **t′** is the sum:

$$\mathbf{t}' = \mathbf{t} + \lambda\mathbf{m} \tag{17}$$

The elements of vector **t** are the overlap integral of the $\rho(\mathbf{r})$ to be fitted by the basis functions in the new representation, $S_i(\mathbf{r})$, being:

$$t_i = \int \rho(\mathbf{r}) S_i(\mathbf{r}) d\mathbf{r} \tag{18}$$

And finally, the elements of **m**, taking into account the normalization condition, are given by,

$$m_i = \int S_i(\mathbf{r}) d\mathbf{r} = 1 \tag{19}$$

and the Lagrange multiplier λ is given by the products:

$$\lambda = (N - \mathbf{m}^T\mathbf{S}^{-1}\mathbf{t})(\mathbf{m}^T\mathbf{S}^{-1}\mathbf{m})^{-1}. \tag{20}$$

Coefficients solving Eq. 15 can be expressed, in terms of the Cramer's rule, by,

$$n_{0,i}' = (S_{1i}t_1' + S_{2i}t_2' + \ldots + S_{ni}t_n')\det|\mathbf{S}|^{-1} \tag{21}$$

where S_{ij} is the cofactor of the element s_{ij} in the metric **S**. Since **S** is a positive definite matrix, and consequently det|**S**| positive valued, non-negative coefficient values constraints (Eq. 14) are equivalent to:

$$S_{1i}t'_1 + S_{2i}t'_2 + \ldots + S_{ni}t'_n \geq 0 \ \forall i \quad (22)$$

This set of inequalities establishes intricate relationships which, once a system and its attached density function ρ(**r**) are given, indicates that physically acceptable ASA fitted densities will lie in some subspaces from the nearly complete function space. The ASA algorithm, presented in the following section, is an original way to optimally localize such subspaces, or, in other words, to minimize $\varepsilon^2(\mathbf{n})$ constrained to the set of conditions in Eqs. 13 and 14. The subsequent two sections that follow examine the results of this methodology when applied to atomic and molecular systems, respectively.

Algorithm Scheme

Since the error quadratic integral function $\varepsilon^2(\mathbf{n})$ is a quadratic form, its minimum $\mathbf{n}'_0$ can be expressed in terms of an arbitrary vector **n** by the equation,

$$\mathbf{n}'_0 = \mathbf{n} - \mathbf{S}^{-1}\nabla\varepsilon^2(\mathbf{n}) \quad (23)$$

where the gradient at **n** is given by:

$$\nabla\varepsilon^2(\mathbf{n}) = 2(\mathbf{nS} - \mathbf{t}') \quad (24)$$

Choosing the arbitrary point **n** with all the components positive, and taking the direction **p**,

$$\mathbf{p} = \mathbf{S}^{-1}\nabla\varepsilon^2(\mathbf{n}) \quad (25)$$

the shortest *approaching path* from the point **n** to the minimum $\mathbf{n}'_0$, it is possible to define a new point $\mathbf{n}_1'$ in **p** given by:

$$\mathbf{n}_1' = \mathbf{n} - \xi\mathbf{p}. \quad (26)$$

The parameter $\xi \in [0,1]$ is the largest step through the descending path that keeps the coefficients positive. Analyzing every component at the intersection,

$$0 = n_i - \xi_i p_i, \ \forall i \quad (27)$$

it can be defined as a subset of ξ_i values,

$$\xi_k^{(+)} = n_k p_k^{-1} \wedge p_k > 0 \ \forall k \quad (28)$$

for the positive components of the approaching path **p** only, giving the maximum step for the considered component. Obviously, no restriction exists if a component

p_j is negative because the corresponding coefficient $n'_{j,1}$ always will be positive. Then, taking ξ as,

$$\xi \equiv \underset{k}{\text{Min}}\ (1, \xi_k^{(+)}) \tag{29}$$

forces the new point $\mathbf{n}'_1$ to have positive or zero components. Since the path $\mathbf{p}$ directly conduces to the minimum, the new set of coefficients will decrease the function $\varepsilon^2(\mathbf{n})$.

At this step of the iterative process the functions with null coefficients and positive slope at $\mathbf{n}'_1$ are discarded. This is so because they would have negative coefficients in a differential steepest-descent displacement from $\mathbf{n}'_1$. Afterwards, a new approaching path is computed:

$$\mathbf{p}_r = \mathbf{S}_r^{-1} \nabla_r \varepsilon^2(\mathbf{n}'_{1,r}) \tag{30}$$

The dimension of the problem has been reduced as indicated in expression (Eq. 30) by the subindices r. In the way previously shown, a new step ξ and a new point $\mathbf{n}'_{2,r}$ are computed. Then after expanding $\mathbf{n}'_{2,r}$ to a whole dimensioned vector $\mathbf{n}'_2$, maintaining the original zero values for the discarded functions, a computation of the gradient at this improved $\mathbf{n}'_2$ is performed, closing the second iteration. The process stops when ξ equals one—the minimum reached in a possible subspace—and when all the slopes of shells with zero occupancies are positive, the conditions of a restricted minimum. In this manner, as shown in Table 1, not only a minimum is found in a problem subspace, accomplishing,

$$\mathbf{n}'_{0,r} = \mathbf{S}_r^{-1} \mathbf{t}'_r \tag{31}$$

Table 1. Schematic Description of the ASA Algorithm[a]

- Compute Integrals $\mathbf{t}$ and $\mathbf{S}$
- Compute λ and $\mathbf{t}'$
- Initialize $\mathbf{n}$ and $\nabla\varepsilon^2(\mathbf{n})$
- DO
    - For i (*if* $n_i = 0$ and $\nabla_i \varepsilon^2(\mathbf{n}) > 0$ discard shell i)
    - Establish Reduced Dimension $\mathbf{t}_r$, $\mathbf{S}_r$, and $\mathbf{S}_r^{-1}$
    - Compute λ_r
    - Compute $\mathbf{n}'^{1}_r$
    - Expand $\mathbf{n}'^{1}_r$ to $\mathbf{n}'^{1}$
    - If $\xi < 1$ DO Continue
    - If (for i ($n'^{1}_i = 0$ and $\nabla_i \varepsilon^2(\mathbf{n}'^{1}) > 0$)) DO exit
- End DO
- Minimum $\mathbf{n}'^{0} = \mathbf{n}'^{1}$

Note: [a] Nomenclature explained in text.

but also the best subspace, i.e., the best fitting function from all possible combinations of basis set functions, is obtained.

Referring to the computational efficiency of this algorithm, two considerations must be taken into account. First, it is worthwhile to realize that an important computational simplification can be introduced removing constraint (Eq. 13), i.e. using **t** instead of the more expensive **t**′, during the localization of compatible subspaces. Since the original density function strictly obeys the electron normalization, any flexible enough fitting expansion will freely reproduce this constraint and, consequently, this imposition does not influence the final selection of functions. Constraint (Eq. 13) can be introduced once this first selection is done, allowing further iterations if necessary. The second consideration refers to Eq. 23, which might yield numerical inaccuracies, reflected in abnormally large values for the gradient components. In such a case the solution could be refined since the compatible subspace is already determined, solving directly the linear system (Eq. 31). Even when the number of matrix inversions to be performed during the iterative procedure is large, the computational cost for this restricted fitting is only slightly greater than the simple LSF. This is because symmetric matrix inversion is a fast process compared to integral evaluation.

Fitting Spherical Systems: The Argon Atom

The closed-shell argon atom has a completely spherical electron distribution, and therefore is a suitable example for testing the flexibility of the restricted ASA function. The density to be fitted was computed at the MP2/6-311G* level of theory. Spanning a nearly complete space with 50 functions generated from even-tempered parameters,[8] the computed ASA density, composed of 22 shells or selected functions, has an associated quadratic error integral value $\varepsilon^2(\mathbf{n})$ of 6.94×10^{-6} with density scaled to one. Such scaling improves the convergence of the algorithm, especially if the initial fitting space is large. The maximum of the function at the nucleus has a value of 46824.18 au, which is 0.9 units over the ab initio 46823.28, and thus being the greatest local difference. The radial distribution presented in Figure 1 is defined as,

$$D(r) \equiv r^2 \int_0^{2\pi} \int_0^{\pi} \rho(r,\varphi,\phi) \sin\varphi \, d\varphi \, d\phi, \tag{32}$$

for the ab initio and the ASA functions. A complete agreement for the first two shells is found, shells now in the sense of Parr et al.,[15] while some slight differences appear in the outer region of argon. Since $\rho(\mathbf{r})$ decreases rapidly in the neighborhood of the nucleus, one finds that at the distance of 1 au from it the value is only 5.2 au, and values close to zero are found at greater distances. For this reason, this region of large distances has an unnoticeable effect in an unweighted $\varepsilon^2(\mathbf{n})$. This is the reason for the differences at greater distances and not the existence of high quantum number electrons, which prevent neither the spherical symmetry of electron distri-

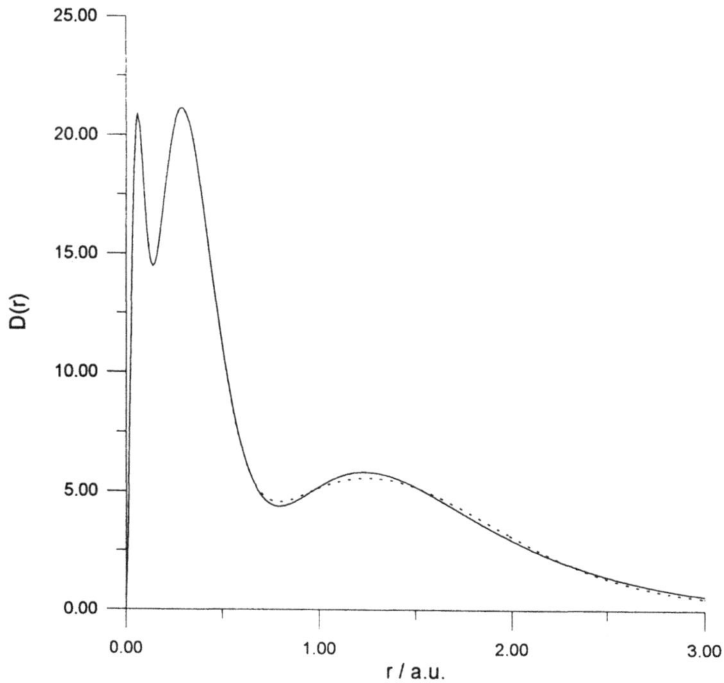

Figure 1. Radial electron distribution $D(r)$ for argon. The MP2/6-311G* distribution is solid line and the ASA is dashed line.

bution in atoms nor their representation by $1S$ functions. This is also in agreement with the well-established practice of using only $1S$ Slater or Gaussian functions for spherical orbitals.[16]

Fitting Molecular Systems: The Boron Trichloride Molecule

Atoms in molecules no longer have spherical electron distributions. Nevertheless, superposition of spherical atomic shells is still accurate, especially for QMSM purposes, as can be seen in the next example. The electron density for the boron

Table 2. Number of Functions for the Different Densities of the Boron Trichloride Molecule[a]

Number of Functions	*STO-3G*	*3-21G*	*6-21G*	*6-31G**	*6-311G***
Basis functions	32	48	48	72	100
Primitives	96	96	144	184	179
Fitting functions	140	140	140	140	140

Note: [a] The number of initial functions for the ASA fitting is also showed, corresponding to 35 functions per atom.

Table 3. HF Densities for the Boron Trichloride Molecule[a]

	STO-3G	*3-21G*	*6-21G*	*6-31G**	*6-311G***
Shells	42	49	61	61	66
Shells on B	9	10	13	13	15
Shells on Cl	11	13	16	16	17
ε^2	3.0105E-5	1.7625E-5	6.3197E-6	8.0264E-6	8.0198E-6
Error in S(A,A)	0.0211%	0.0022%	0.0009%	0.0008%	0.0008%

Note: [a] Shells, quadratic errors integrals, and errors in self-similarity.

trichloride molecule, with partial boron–chlorine double bonds, has been computed at different levels of theory at its D_{3h} optimized geometry. The ASA algorithm is independent of these levels of theory since shells are optimally and automatically selected to describe a particular density from a nearly complete space. Table 2 gathers the number of primitives for every basis set whose square is the number of terms in the ab initio density, and the considered basis set size to span a nearly complete space for the ASA fitting, corresponding to 35 functions, generated from parameters in Ref. 8, per atom. Table 3 and Table 4 collect the results of the fitting computations, namely, the number of shells or selected functions and the quadratic integral error $\varepsilon^2(\mathbf{n})$, and the error in the self-similarity for an evaluation of the quality of the ASA function. The immediate conclusion from these tables is that when using the ASA there is an important reduction in the number of functions used to express the density function which, together with the fact that these functions are 1S Gaussians, immediately gives an idea of the important reduction in the time needed to compute QMSM. Such simplification does not prevent the generation of QMSM with an acceptable accuracy, as can be seen observing the different errors. As in the previous example, $\varepsilon^2(\mathbf{n})$ is computed with density scaled to one and is nearly constant for the different orbital basis sets. The increase in the number of shells when improving the wave function quality is another remarkable aspect of the ASA procedure, showing that it is a *systematic* and *universal* method. Slightly better

Table 4. MP2 Densities for the Boron Trichloride Molecule[a]

	STO-3G	*3-21G*	*6-21G*	*6-31G**	*6-311G***
Shells	43	52	61	62	66
Shells on B	10	10	13	14	15
Shells on C1	11	14	16	16	17
ε^2	2.9225E-5	1.6823E-5	5.8198E-6	6.4506E-6	6.3815E-6
Error in S(A,A)	0.0206%	0.0021%	0.0008%	0.0008%	0.0007%

Note: [a] Shells, quadratic errors integrals, and errors in self-similarity.

values for the more precise densities is just a consequence of the optimization of the even-tempered parameters, which were obtained from atomic 6-311G* densities. This selection of shells also gives atomic populations, unambiguously defined in ASA, in agreement with chemical intuition. For the boron atom in the MP2/6-311G** fitting, the atomic population is –0.003 au, in agreement with the expected value. Four acceptable resonant structures can be written down for the boron trichloride molecule, three of them involving double bonds with positive chlorines and the other with partial ionic single bonds with negative chlorines, making the total charge transfer negligible.[17] Exemplifying the importance of a good selection of shells, one can regard the LSF density, computed using the whole 140 function basis set and without positive valued constraint to coefficients (i.e., a lower value in $\varepsilon^2(\mathbf{n})$, which gives a boron charge of –1.10 au) quite far away of what it is expected.

B. Empirical Atomic Shells

A really fast computation of QMSM which could be applied to pattern recognition in 3D structural databases should be extremely simplified and should avoid the need of density computations. Empirical molecular densities can be modeled as simple sums of atomic contributions, having for the so-called *promolecular* electron density:

$$\rho_{EASA}(\mathbf{r}) = \sum_a \rho^a_{EASA}(\mathbf{R}_a - \mathbf{r}), \tag{33}$$

Several functional forms for the shell structure of atoms, $\rho^a_{EASA}(\mathbf{R}_a - \mathbf{r})$, will be analyzed in the present work. The first strategy, based in CNDO-like densities, uses a simple nS STO function per atom, being,

$$\rho^a_{EASA}(\mathbf{R}_a - \mathbf{r}) = q_a \left| S^{l_a}_a(\mathbf{R}_a - \mathbf{r}) \right|^2 \tag{34}$$

where coefficients q_a are atomic charges, and:

$$S^{l_a}_a(\mathbf{R}_a - \mathbf{r}) = \frac{(2\zeta_a)^{l_a + \frac{1}{2}}}{\sqrt{4\pi(2l_a)!}}(\mathbf{R}_a - \mathbf{r})^{l_a - 1} e^{-\zeta_a |\mathbf{R}_a - \mathbf{r}|} \tag{35}$$

The radial power term l_a is taken as the row number of atom a in the Periodic Table or, what is nearly the same, the number of maxima in the radial distribution. Exponents ζ_a are taken to exactly reproduce free atom self-similarity values.

A second strategy to enhance atomic densities defines $\rho^a_{EASA}(\mathbf{R}_a - \mathbf{r})$ as a superposition of l_a STO shells in the form:

$$\rho^a_{EASA}(\mathbf{R}_a - \mathbf{r}) = \sum_i^{l_a} m_i \left| S^i_a(\mathbf{R}_a - \mathbf{r}) \right|^2 \tag{36}$$

Occupations m_i are the number of electrons commonly associated with the atomic electronic configurations. The set of exponents used are those of Clementi et al.[18] for spherical orbitals.

Similarity measures of a set of fluoro- and chloro-substituted methanes, whose ab initio HF/6-31G** values were already known,[8] will be reviewed to illustrate the performance of these two empirical approaches. Table 5 presents the similarity values, the ab initio ones in bold, those computed with functional approach (Eq. 34) in italics, and, those with the approach of Eq. 36 in normal type. Results in the

Table 5. QMSM for Fluoro- and Chloro-Substituted Methanes[a]

	CH_4	*CH_3F*	*CH_3Cl*	*CH_2F_2*	*CH_2Cl_2*	*CHF_3*	*$CHCl_3$*	*CF_4*	*CCl_4*
CH_4	**31.84**	**58.78**	**144.22**	**58.83**	**144.23**	**58.89**	**144.23**	**58.92**	**144.23**
	30.56	55.69	128.49	55.75	128.50	55.80	128.50	55.84	128.50
	37.46	*68.86*	*170.32*	*68.66*	*169.55*	*68.44*	*168.89*	*68.28*	*168.31*
CH_3F		**151.11**	**316.69**	**150.37**	**316.87**	**148.25**	**317.03**	**146.60**	**317.15**
		142.37	281.97	141.94	282.16	140.24	282.34	138.97	282.49
		163.54	*360.33*	*161.27*	*359.59*	*159.30*	*358.60*	*157.66*	*357.37*
CH_3Cl			**1028.15**	**318.81**	**1027.71**	**319.24**	**1027.48**	**319.41**	**1028.04**
			878.59	284.33	878.33	284.80	878.27	284.97	878.92
			1091.08	*364.80*	*1086.24*	*367.84*	*1082.27*	*368.41*	*1079.13*
CH_2F_2				**270.43**	**319.08**	**258.32**	**319.55**	**249.98**	**319.93**
				254.23	284.61	243.77	285.18	236.53	285.67
				289.48	*356.80*	*286.04*	*355.74*	*283.38*	*354.57*
CH_2Cl_2					**2024.52**	**319.49**	**1738.55**	**319.78**	**1401.47**
					1726.63	284.91	1489.62	285.29	1199.69
					2126.12	*354.98*	*2093.98*	*353.93*	*2040.88*
CHF_3						**389.77**	**321.44**	**386.70**	**322.01**
						366.13	287.24	363.66	287.96
						414.85	*353.34*	*412.76*	*352.18*
$CHCl_3$							**3020.95**	**321.74**	**2694.00**
							2574.67	287.56	2303.79
							3146.32	*352.25*	*3109.11*
CF_4								**509.07**	**324.14**
								478.04	290.38
								539.88	*350.53*
CCl_4									**4017.38**
									3422.69
									4155.16

Note: [a] Ab initio HF/6-31G* values are in bold, the empirical ASA values using one STO per atom in italics, and EASA with a STO per shell and atom values in medium type.

first approach, with a single nS STO function per atom, show a good agreement with ab initio values in case of self-similarities, having a 6% error for CH_4–CH_4 or a less than a 5% for the CCl_4–CCl_4 measure, while errors in cross-similarities are larger than 10%. The reason for having more accurate values in self-similarities can be found in the fact that when computing self-similarities there is a perfect matching between the two molecules being compared, which are the same. In this case, the main contribution comes from atoms perfectly superimposed, while contributions from atoms not superimposed are negligible because they are separated by large distances. Given that the exponents in Eq. 34 are taken to reproduce atomic auto-similarities, one can already expect a good result for this case. On the other hand, when dissimilar molecules are compared, one is likely to find pairs of atoms not completely superimposed. These atom pairs are primarily responsible for the greater errors found in this case. The similarity additivity of Eq. 34 is also reflected in the overestimation of all the similarity values, indicating a lack of diffuseness of the atomic densities in molecular environments that this model presents. To better understand this point, one can check that the similarity integral (Eq. 3) increases if charge distribution is concentrated in small areas, being infinity in case of densities collapsed into Dirac deltas. The other approximation, when the density functional form is given by Eq. 36, does not improve the similarity measures in all cases, probably due to the use of nonoptimal exponents to span densities.

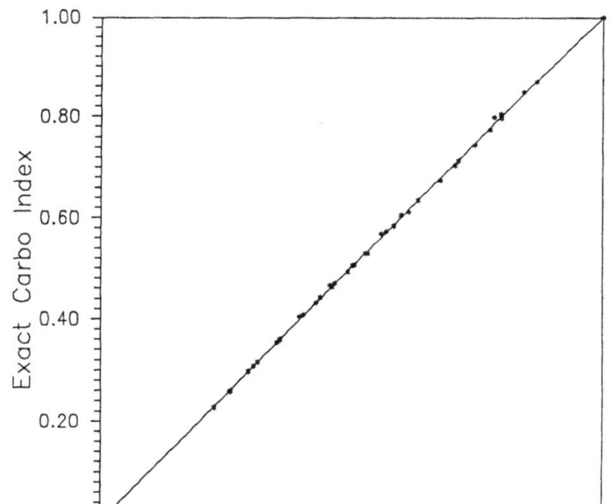

Figure 2. HF/6-31G** versus empirical Carbó indices for the fluoro- and chloro-substituted methanes.

Carbó indices derived from these empirical similarity measures present a better correlation with ab initio values as Figure 2 reveals. This agreement can be explained by the systematic deviation which cancels errors in the index computation.

A third strategy using a single 1*S* GTO function per atom[19] has also been tested with the aim of speeding up similarity maximization. Results are only qualitative and will be presented in next section.

III. SIMILARITIES IN THE ATOMIC SHELL APPROXIMATION

In this section, the performance computing overlap QMSM of several introduced ASA will be analyzed. QMSM for *rigid* molecules are six variable functions, with three of them indicating relative translations and the other three indicating relative orientation. Fixing one of the molecules, molecule *A*, the similarity function is expressed by,

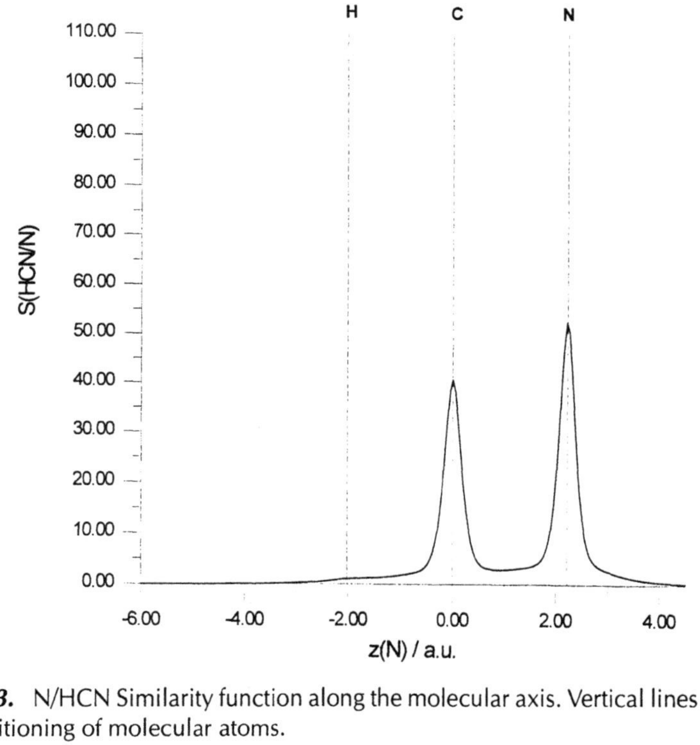

Figure 3. N/HCN Similarity function along the molecular axis. Vertical lines indicate the positioning of molecular atoms.

$$z_{AB}(\Omega) = \int \rho_A(\mathbf{r})\rho_B(\mathbf{r};\Omega)d\mathbf{r} \tag{37}$$

with Ω standing for all six variables. Inside the ASA, similarity measures appear as a sum of isotropic atom–atom contributions, i.e.,

$$z_{AB}(\Omega) = \sum_{ab} z_{ab}(r_{ab}(\Omega)) \tag{38}$$

where the similarity for atomic pairs is given by:

$$z_{ab}(r_{ab}(\Omega)) = \sum_{i \in a} \sum_{j \in b} n_i n_j \int S_i(\mathbf{R}_a - \mathbf{r}) S_j(\mathbf{R}_b(\Omega) - \mathbf{r}) d\mathbf{r} \tag{39}$$

Expression 39 enables a global maximization scheme whose first principles are given in Ref. 8. This scheme is used in all similarity optimizations contained in the present work. Therefore, in Section III.A the similarity function between atomic nitrogen and two linear molecules will be computed at the ab initio MP2/6-311G** level of theory, and differences with the approximate functions will be displayed in order to have a vision of the behavior of the ASA atom–atom contributions (Eq.

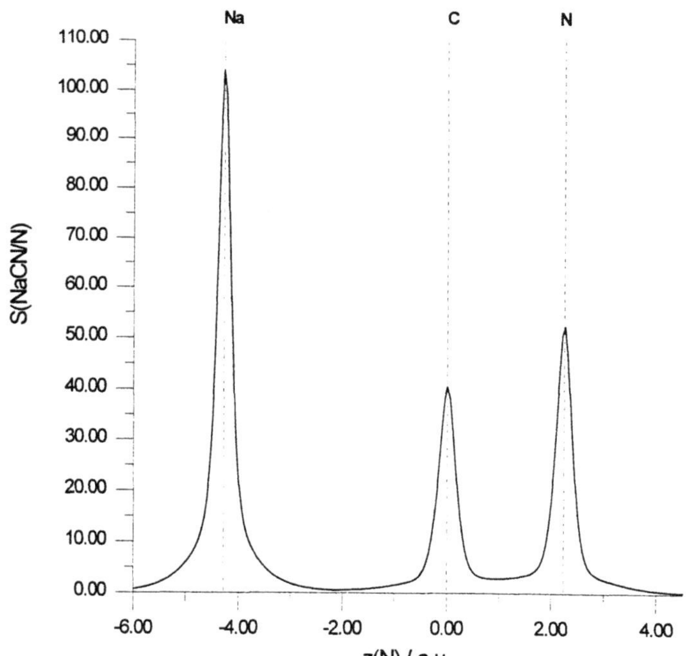

Figure 4. N/NaCN Similarity function along the molecular axis. Vertical lines indicate the positioning of molecular atoms.

39). This will shed some light when, afterwards, in Section III.B the accuracy of the ASA method will be checked in a series of *real* drug design molecules.

Computations of ab initio densities and optimized geometry have been performed using the Gaussian 92 ensemble of programs.[20] Program ExSim[21] has been used to compute ab initio similarities, ASAC[22] for fitting the ab initio densities and computation of their similarities, and MolSimil 95[23] for the empirical computations.

A. HCN/N and NaCN/N Systems

Similarity functions for HCN/N and NaCN/N systems only depend on the coordinates of the nitrogen atom with respect to some fixed frame of axis defining the atomic positions of the cyanide molecule having:

$$z_{XCN,N}(\mathbf{r}_N) = \int \rho_{XCN}(\mathbf{r})\rho_N(\mathbf{r};\mathbf{r}_N)d\mathbf{r} \tag{40}$$

If *XCN* molecules lie along *Z* axis, the pictures of $z_{XCN,N}(0,0,Z_N)$ will be sufficient to show the peculiarities of similarity functions, also present in more complicated

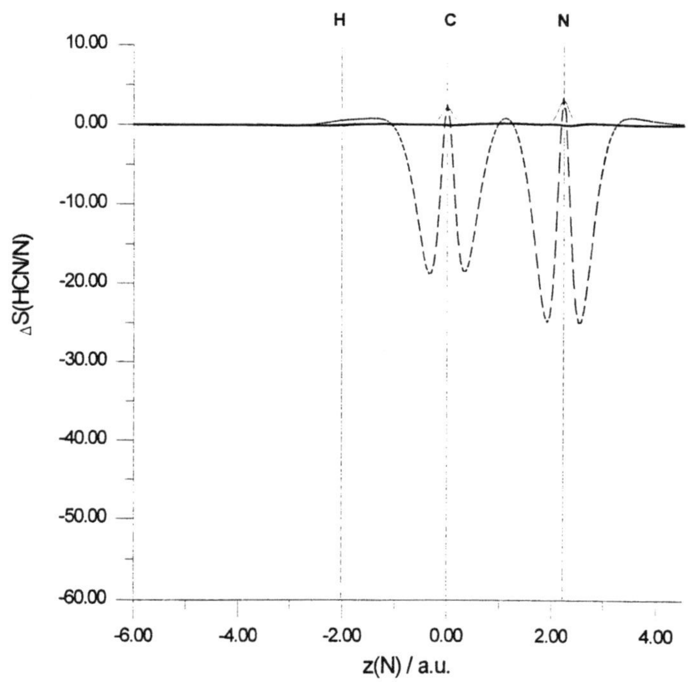

Figure 5. N/HCN ab initio-approximate differences in similarity function. Thick solid line corresponds to ASA computations, fine solid line to Slater empirical approach, and dashed is the empirical Gaussian approach.

systems because of the nearly atom–atom additivity. Figure 3 and Figure 4 represent the similarity function computed at the MP2/6-311G** level of theory for nitrogen vs. hydrogen and sodium cyanide, respectively. The HCN/N function only presents two maxima due to the fact that electron density flows from hydrogen to the electronegative group cyanide. Even if hydrogen was not bonded to an electronegative group, its maximum would appear nearly hidden by the heavier atoms.

The differences with the similarity functions obtained using ASA densities are given for hydrogen and sodium cyanides, respectively, in Figures 5 and 6. Thick lines correspond to the differences between exact and ASA QMSM and are confused with the abscise, showing a nearly complete agreement especially at the maxima. At approximately 1 bohr around carbon and nitrogen coordinates, the maximum difference is found to be 0.2 au in similarity. Fine solid lines correspond to the differences with the empirical function built using Slater-type functions (Eq. 36). They also show a conformity with the exact functions, except at the maxima where they are approximately 10% lower. Dashed lines correspond to the simplest approach analyzed, which consists in a single $1S$ GTO per atom. These functions

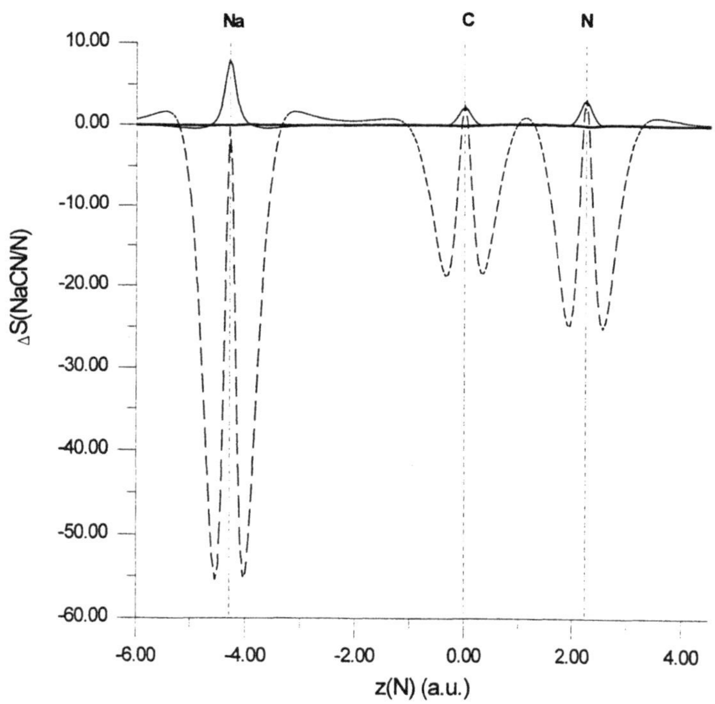

Figure 6. N/NaCN ab initio-approximate differences in similarity function. Thick solid line corresponds to ASA computations, fine solid line to Slater empirical approach, and dashed is the empirical Gaussian approach.

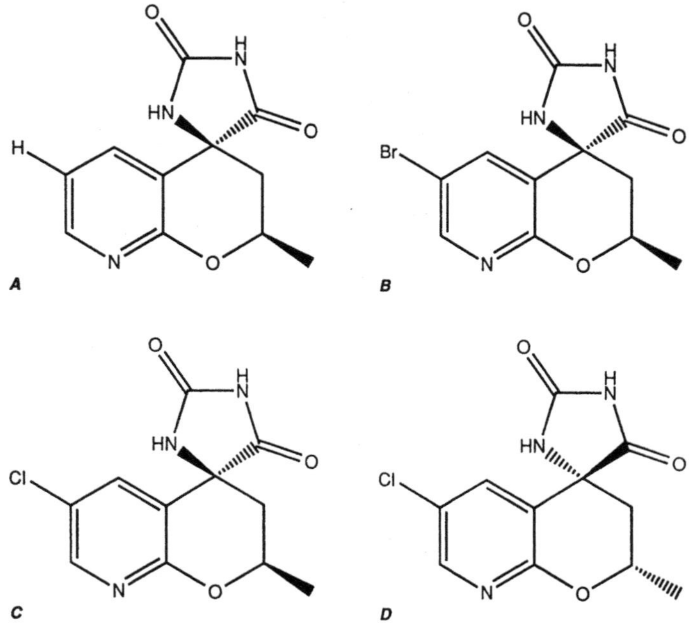

Figure 7. Representation of the four spiro hydantoin aldose reductase inhibitors considered.

are only a qualitative description since a single Gaussian cannot describe simultaneously height and width, thus their use should be restricted to interactive visual matching. Compared molecules usually will be placed at the right maximum arrangement, but the corresponding similarity value will appear highly distorted because of the important errors when nuclei are not perfectly superimposed, the case of most of the nuclei when matching dissimilar molecules.

B. Spiro Hydantoins Comparison

A series of four spiro hydantoin 8-aza-4-chromanones which act as aldose reductase inhibitors[24] has been selected to test the performance of the ASA method in a real case of drug design. Their chemical structure is presented in Figure 7.

Ab initio and ASA similarities have been computed at the fully optimized HF/STO-3G geometry. EASA computations were performed with the set of functions in Eq. 36. Similarities and their derived Carbó indices are presented in Table 6 and Table 7, respectively, the ab initio values appearing in bold, the ASA values in normal type, and the EASA ones in italics. Similarity maximization was only performed using ASA and EASA densities, obtaining in both cases the same maxima with just a negligible difference in the final values of Ω. Then, ab initio punctual similarities were performed at the ASA maxima. In order to easily allow

Table 6. Similarities for Spiro Hydantoins[a]

	A	*B*	*C*	*D*
A	**729.840**	**712.354**	**454.488**	**353.080**
	729.548	713.971	456.296	354.063
	710.737	*630.816*	*446.103*	*346.642*
B		**11053.921**	**3187.972**	**2988.291**
		11051.215	3194.098	2993.832
		8704.890	*2706.978*	*2510.181*
C			**1687.997**	**1294.120**
			1687.574	1294.970
			1558.762	*1177.520*
D				**1687.963**
				1687.541
				1558.773

Note: [a] Ab initio values are in bold, ASA in medium type, and empirical ASA values in italics.

a comparison of the results, exact–approximate differences and percentual errors are presented in Table 8 and Table 9, respectively, while Table 10 and Table 11 give the errors corresponding to the Carbó indices.

Differences in ASA similarities are mainly originated by the atomic sphericity loss since densities for free atoms are excellently reproduced. This deformation, as commented in Section III.A, is more noticeable when nuclei are not completely

Table 7. Carbó Indices for Spiro Hydantoins[a]

	A	*B*	*C*	*D*
A	1	**0.2508**	**0.4095**	**0.3181**
		0.2514	0.4112	0.3191
		0.2536	*0.4238*	*0.3293*
B		1	**0.7380**	**0.6918**
			0.7396	0.6933
			0.7349	*0.6814*
C			1	**0.7667**
				0.7674
				0.7554
D				1

Note: [a] Ab initio values are in bold, ASA in medium type, and empirical ASA values in italics.

Table 8. Similarity Differences, Ab Initio-Approximate, for Spiro Hydantoins[a]

	A	*B*	*C*	*D*
A	0.292 *19.103*	−1.617 *81.538*	−1.808 *8.385*	−0.983 *6.438*
B		2.706 *2349.031*	−6.126 *480.994*	−5.541 *478.110*
C			0.423 *129.235*	−0.850 *116.600*
D				0.422 *129.190*

Note: [a] ASA values are in medium type and empirical ASA values in italics.

superimposed, having in the previous examples maximum differences of 0.2 au in similarity for carbon and nitrogen atoms. Extrapolating these differences to the present example, one can easily understand the different behavior of self- and cross-similarities, the first ones being more accurate. This also explains, for instance, why z_{BC} has the greatest absolute error, while z_{CD} has the precision of a self-similarity (see Table 8). In the first case the arrangement maximizing the electron density overlap superposes bromine and chlorine atoms, whereas all other atoms appear displaced. Figure 8 shows the molecular superposition for the *B–C* pair. By contrast, molecules *C* and *D*, pictured in Figure 9, completely match except for the methyl group and the ring attached at chiral carbons. Nevertheless, the change in chirality

Table 9. Percentage Similarity errors, Ab Initio-Approximate, for Spiro Hydantoins[a]

	A	*B*	*C*	*D*
A	−0.040 *−2.688*	0.226 *−12.926*	0.396 *−1.880*	0.278 *−1.857*
B		−0.024 *−26.985*	0.192 *−17.769*	0.185 *−19.047*
C			−0.025 *−8.291*	0.066 *−9.902*
D				−0.025 *−8.288*

Note: [a] ASA values are in medium type and empirical ASA values in italics.

Table 10. Carbó Index Differences, Ab Initio-Approximate, for Spiro Hydantoins[a]

	A	*B*	*C*	*D*
A	0	−0.001 *−0.003*	−0.002 *−0.014*	−0.001 *−0.011*
B		0	−0.002 *0.003*	−0.001 *0.010*
C			0	−0.001 *0.011*
D				0

Note: [a] ASA values are in medium type and empirical ASA values in italics.

clearly separates these groups, making the overlap contribution of the relevant atoms negligible.

In the case of EASA similarities, errors obviously come from a poor description of electron densities, which is especially evident for the measures involving the bromine-substituted molecule. However, this simple picture of molecular densities places these molecules at the proper maximum arrangement and gives Carbó indices correctly in one decimal figures.

Regarding the possible application of QMSM in QSAR studies, it is interesting to make a qualitative comparison between the activity values for this set of molecules and some of the QMSM values obtained. Thus, it can be seen that, while *B* and *C* are the most active molecules, the Carbó index is higher for the *C*–*D* pair than for the *B*–*C* pair in all the approximations considered, with *D* being an inactive molecule. This result is, at first sight, quite surprising because *B* and *C* share the

Table 11. Percentual Carbó Index Errors, Ab Initio-Approximate, for Spiro Hydantoins[a]

	A	*B*	*C*	*D*
A	0	0.239 *1.104*	0.413 *3.374*	0.313 *3.401*
B		0	0.216 *−0.422*	0.216 *−1.526*
C			0	0.091 *−1.496*
D				0

Note: [a] ASA values are in medium type and empirical ASA values in italics.

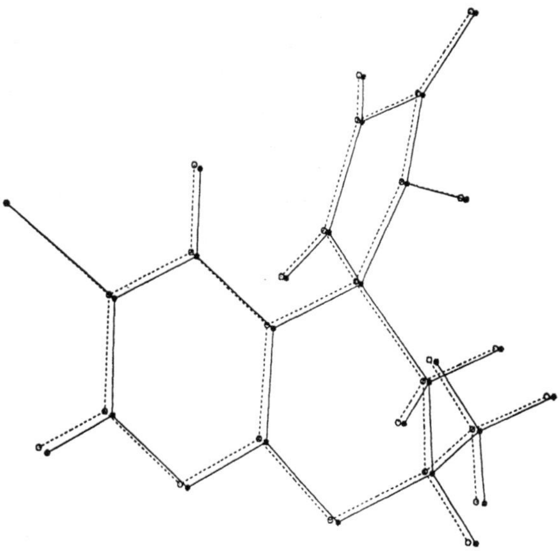

Figure 8. Superposition of the bromine-substituted spiro hydantoin (*B*) with the chloro-substituted (*C*). Pictured by MolSimil 95.

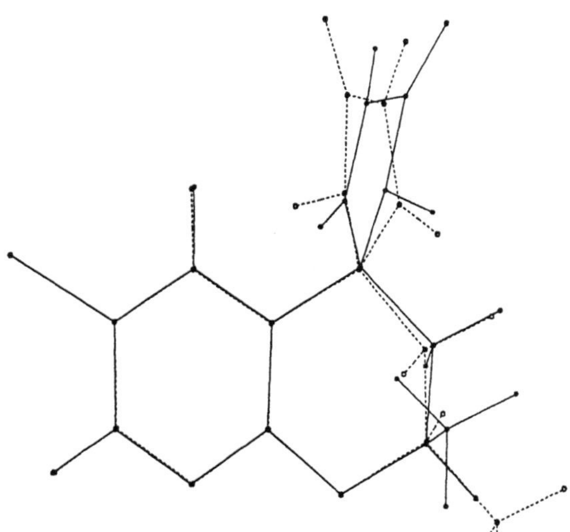

Figure 9. Superposition of the chloro-substituted spiro hydantoins (*C*) and (*D*). Pictured by MolSimil 95.

same structure and differ only in the halogen, while *C* and *D*, although having different halogens, seem to be structurally more different because its five-membered ring cannot be superposed due to the different chirality of the two molecules. However, the low value for the *B–C* pair can be attributed to the shifting of the large common substructure slightly out of the maximal superposition, as can be seen in Figure 8. This is forced by the superposition of Br and C1 and because the C–Br and C–C1 distances are slightly different. This arises not from the ASA fitting but rather from the theoretical background consisting in using electronic densities which do not take into account the vibrational motion of atoms.

IV. CONCLUSIONS

The main conclusion of the present work indicates that QMSM based on electron distributions can be accurately computed, even for large molecules. The purpose of this work has been to assess a fast and correct methodology to quantify molecular similarities based on first-order electronic distributions. The ASA, due to its simplicity, brings not only the means to perform fast QMSM computations, but also possible ways of modeling molecules and defining local similarities. Future work will allow nuclear movements and the averaging of electronic distributions by considering harmonic nuclear displacements, thus giving a more real picture of molecules. We expect that within this framework it will be possible to obtain better correlations between QMSM and biological activities in cases such as the spyro hydantoins considered in section III.B. Furthermore, the concept of local similarities could be valuable in the localization of active centers or common patterns in sets of molecules.

ACKNOWLEDGMENTS

P.C. has benefitted from a CIRIT OA/au BQF93/24 fellowship, and L.A. from a "Ministerio de Educación y Ciencia" fellowship. P.C. thanks Dr. M.D. Pujol from the Pharmacological Chemistry Department at the University of Barcelona for her help in selecting an appropriate set of active molecules.

REFERENCES

1. (a) Löwdin, P.O. *Phys. Rev.* **1955**, *97*, 1474–1489. (b) McWeeny, R. *Proc. Roy. Soc. London* **1959**, *A253*, 242–259.
2. Bader, R.F.W. *Atoms in Molecules: A Quantum Theory*; Clarendon Press: Oxford, 1990.
3. (a) Cioslowski, J.; Mixon, S.T. *J. Am. Chem. Soc.* **1991**, *113*, 4142. (b) Cioslowski, J.; Mixon, S.T. *J. Am. Chem. Soc.* **1992**, *114*, 4382. (c) Cioslowski, J.; Mixon, S.T. *J. Am. Chem. Soc.* **1993**, *115*, 1084.
4. (a) Carbó, R.; Leyda, L.; Arnau, M. *Int. J. Quantum Chem.* **1980**, *17*, 1185–1189. (b) Carbó, R.; Calabuig, B. *Int. J. Quantum Chem.* **1992**, *42*, 1681–1693. (c) Carbó, R.; Calabuig, B. *Int. J. Quantum Chem.* **1992**, *42*, 1695–1709. (d) Carbó, R.; Calabuig, B.; Vera, L.; Besalú, E. *Adv.*

Quantum Chem. **1994**, *25*, 253–313. (e) Besalú, E.; Carbó, R.; Mestres, J.; Solà, M. *Topics in Current Chemistry* **1995**, *173*, 31–62.

5. Cioslowski, J.; Fleischmann, E.D. *J. Am. Chem. Soc.* **1991**, *113*, 64–67.
6. Good, A.C.; Richards, W.G. *J. Chem. Inf. Comput. Sci.* **1992**, *33*, 112–116.
7. (a) Mestres, J.; Solà, M.; Duran, M.; Carbó, R. *J. Comp. Chem.* **1994**, *15*, 1113–1120. (b) Carbó Ed. *Molecular Similarity and Reactivity: From Quantum Chemical to Phenomenological Approaches*; Kluwer Academic: Netherlands, 1995.
8. Constans, P.; Carbó, R. *J. Chem. Inf. Comput. Sci.* **1995**.
9. Unsöld, A. *Ann. Physik* **1927**, *82*, 355–393.
10. (a) Coppens, P.; Pautler, D.; Griffin, J.F. *J. Am. Chem. Soc.* **1971**, *93*, 1051–1058. (b) Schwarz, W.H.E.; Lagenbach, A.; Birlenbach, L. *Theor. Chim. Acta* **1994**, *88*, 437–445.
11. Walker, P.D.; Arteca, G.A.; Mezey, P.G. *J. Comp. Chem.* **1991**, *12*, 220–230.
12. (a) Paoloni, L.; Giambiagi, M.S.; Giambiagi, M. *Estratto da Atti della Società dei Naturalisti e Matematici di Modena* **1969**, C, 89–105. (b) Frost, A.A. *J. Chem. Phys.* **1967**, *47*, 3707. (c) Moncrieff, D.; Wilson, S. *Molecular Physics* **1994**, *82*, 523–530.
13. Reeves, C.M.; Harrison, M.C. *J. Chem. Phys.* **1963**, *39*, 11–17.
14. (a) Ruedenberg, K.; Raffeneffi, R.C.; Bardon, D. *Proceedings of the 1972 Boulder Conference on Theoretical Chemistry*; Wiley: New York, 1973, p. 164. (b) Schmidt, M.W.; Ruedenberg, K. *J. Chem. Phys.* **1979**, *71*, 3951–3962. (c) Feller, D.F.; Ruedenberg, K. *Theoret. Chim. Acta* **1979**, *52*, 231–251.
15. (a) Politzer, P.; Parr, R.G. *J. Chem. Phys.* **1976**, *64*, 4634–4637. (b) Proft, F.; Geerlings, P. *Chem. Phys. Lett.* **1994**, *220*, 405–410.
16. (a) Huzinaga, S. *J. Chem. Phys.* **1965**, *42*, 1293. (b) Huzinaga, S. *J. Chem. Phys.* **1977**, *67*, 5973–5974.
17. Pauling, L. In *The Nature of the Chemical Bond and the Structure of Molecules and Crystals*; Cornell University Press: New York, 1960.
18. (a) Clementi, E.; Raimondi, D.L. *J. Chem. Phys.* **1963**, *38*, 2686. (b) Clementi, E.: Raimondi, D.L.; Reinhard, W.P. *J. Chem. Phys.* **1967**, *47*, 1300–1302.
19. Besalú, E.; Carbó, R.; Lobato, M. *Sci. Gerund.*, in press.
20. Frisch, M.J.; Trucks, G.W.; Head-Gordon, M.; Gill, P.M.W.; Wong, M.W.; Foresman, J.B.; Johnson, B.G.; Schlegel, H.B.; Robb, M.A.; Replogle, E.S.; Gomperts, R.; Andres, J.L.; Raghavachari, K.; Binkley, J.S.; Gonzalez, C.; Martin, R.L.; Fox, D.J.; Defrees, D.J.; Baker, J.; Stewart, J.J.P.; Pople, J.A. *Gaussian 92*, Revision B, Gaussian, Inc., Pittsburgh PA, 1992.
21. Constans, P. *ExSim Program* version 1.0 (CAT, 1995).
22. Constans, P.; Carbó, R. *ASA Calculations* version 2.0 (CAT, 1995).
23. Amat, Ll.; Besalú, E.; Carbó, R. *MolSimil 95* (CAT, 1995).
24. Sarges, R.; Goldstein, S.W.; Welch, W.M.; Swindell, A.C.; Siegel, T.W.; Beyer, T.A. *J. Med. Chem.* **1990**, *33*, 1859–1865.

AUTOMATIC SEARCH FOR SUBSTRUCTURE SIMILARITY:
CANONICAL VERSUS MAXIMAL MATCHING; TOPOLOGICAL VERSUS SPATIAL MATCHING

Guido Sello and Manuela Termini

Advances in Molecular Similarity
Volume 1, pages 213–241

ISBN: 0-7623-0131-7

ABSTRACT

In the past few years we became interested in studying a system for the evaluation of the similarity of (sub)structures using an empirical method for the calculation of electronic energy. After having verified its applicability to structures of different complexity we were faced with the need to automate the matching in order to extend the dimension and the number of the analyzed compounds.

To operate a canonical matching we needed a sequencing methodology that was univocal, reliable, and connected to the calculation system. Subsequently we used the obtained sequences to effect the matchings. To increase the accuracy of the automatic comparison we introduced different methods to improve the matchings. The matchings concerned both topological and three-dimensional molecular representations.

The resulting method has been applied to a series of compounds and the results will be discussed taking into account the differences of the maximal and canonical and the topological and spatial analyses.

I. INTRODUCTION

A. Similarity Measure

Analogy is a cognitive process[1] that plays a fundamental role in our perception of the external world. In everyday life it represents the logical link between situations and events; therefore it's definitely a natural and instinctive process for the human brain. Reasoning by analogy allows us to explain unknown events starting with their resemblance to known facts. Therefore it's an indispensable step in knowledge progress (even in scientific fields) though the increase of knowledge is only probable and not guaranteed.

Analogy is a relationship of likeness that links distinct objects; it could be defined only by similarity of objects or by partial identity of their qualities, features, appearance, and so on.[2] When performing an analysis by analogy we quantify object similarity,[a] or rather we estimate the quality and the quantity of the characteristics common to the objects of an analyzed set. In a scientific field, particularly in chemistry, the availability of clear criteria allowing one to specify the similarity of a set of molecules provides a useful tool for predicting reactivity, activity, molecular properties, and in general, molecular behavior.

To measure the similarity or the dissimilarity between two objects we must first define some representative features of the objects and the criteria that permit one to establish if the objects share any peculiarity. To have reproducible results, object representation and analysis criteria must be clear. However, since resemblance is an attribute that we arbitrarily assign to objects, and that will depend on the particular analysis criteria we choose, similarity is inherently subjective.

Its usefulness, its instinctive use, and the flexibility of its measure and quantification, on the one hand, and the development of computer science, the capability of computers to process large amounts of data, and the necessity of making the methodology objective, on the other hand, are some reasons that have led to the development of several computer methods based on similarity.

The first difficulty one has to face performing an automatic similarity analysis is represented by the problem of chemical structure perception by a computer. Namely, it is necessary to look for a suitable molecule description that the computer can handle. All those descriptors that can be correlated to physical or physicochemical properties of molecules will be suitable. The attribution of similarity as well as the choice of molecule representation is subjective and dependent on the particular criteria of the user, and thus it will be peculiar to each method. Many different kinds of molecule representation based, for example, on electron density surfaces,[3,4] steric volumes,[5] molecular surfaces,[5] chemical graphs,[6] topological indices,[7] have been described in the literature. In the present approach, the molecular description used is the electronic energy calculated by an empirical equation.[8]

After having found a good molecular description it must be decided which features to compare or which criteria to use in order to evaluate the similarity of objects. The mathematical form of the molecular description leads the method of comparison according to the manipulations to which it can be subjected. The manipulation of the representations is the key to obtaining a data organization on similarity where the objects can be grouped and ordered.

To better explain this item let's use a trivial example, namely a simple continuous mathematical function, which is derivable in the problem interval, as the descriptor of the property of interest. Let's also choose the function values in its maximum and minimum points as the measure of the similarity between the studied objects. From the derivative (manipulation) of the function we can then get the values of the variables at the extremal points and, as a consequence, the corresponding values of the function (similarity measure). At this point we can order the objects following the calculated values of the function at the extremal points, and we can state that two objects are more similar (at least concerning our similarity measure) the more similar are the calculated values. In this way we have fixed a similarity hierarchy between our objects. Let's notice that the similarity link between the objects and the descriptor is the hypothesis of our analysis; moreover the link between the objects and the property we are describing is also known (in fact the property must

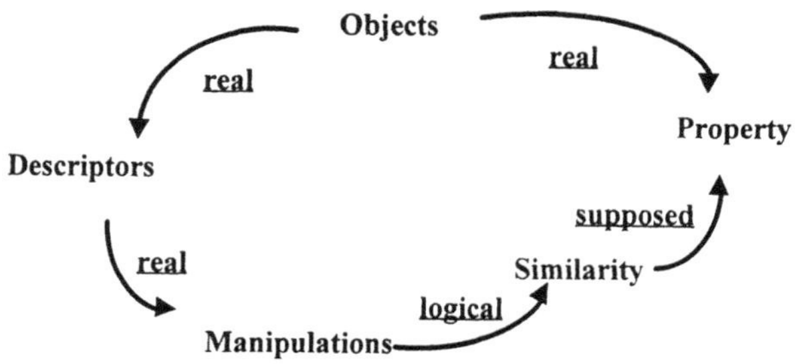

Scheme 1. The links between known, calculated, hypothesized, and logical items of a similarity analysis.

be measurable). By contrast, the link we would like to demonstrate (i.e. our thesis) is the link existing between the similarity measure and the physical or physico-chemical property. We can thus build a graph of links (Scheme 1).

This represents an example that, even if lacking any physical meaning, explains the links existing between similarity, objects, and properties, and gives an idea of the possibility of building a method of similarity analysis.

B. Comparison Methods

A different but equally important phase of similarity analysis concerns the comparison method. In fact, although each method of analysis is strictly connected to the use of molecular description, we can point out some general features that are independent from any molecular description; i.e. the same comparison method can be used for different similarity descriptions appropriately fitting some working features. As an example we recall the proximity measures; these, independently from the molecular description, can be described as the expression of the amount of the affinity between two objects. The way of exactly calculating this amount in different cases depends on the choice of the molecular description. This method is widely used when the mathematical form of the molecular description is vector- or function-like, as in the case of topological indices or electron density surfaces. However, it is applicable in other cases as well.

Of further interest is another method of analysis that works by matching through superimposition. This method can be generally defined as an indication of the share, in terms of similarity, of the structure that two molecules have in common when represented by a particular molecular description and superimposed in such a manner as to join the maximum number of points. The application of this method is particularly convenient when the mathematical form of the molecular description is, as in the present case, represented by sequences.

The method that we developed uses two different ways of superimposition: the first one considers the structures as connected points placed on a plane surface (i.e. considering the bonds between atoms) and the second one also takes into account the relative positions of the points of the molecule in a three-dimensional space. This second type of comparison is not necessarily more restrictive than the first one; it simply gives information on the similarity of atoms not directly connected.

There are various kinds of matching but we will focus our attention on two of them: the maximal and the canonical matchings. The first one looks for the maximal matching of points of two structures, taking into full consideration the bonds, the interatomic distances, the conformation, and the configuration of molecules. This kind of analysis can be expensive in terms of time and memory when the size of the molecules being compared becomes large. This is mainly due to the increase of the number of ways in which we can combine points. A "canonical" match reduces the number of calculations to perform by introducing an approximation. The canonical method is thus an approximation of the exhaustive method where the introduced rules (the "canons") aim to maintain its efficiency, assuring the reproducibility and minimizing the negative aspects of time costs. First of all we must distinguish between canonical method and "canonical analysis." All the methods based on similarity could be named canonical in the sense that they fix rules to represent and compare molecules. By contrast, we want to emphasize that we perform a canonical analysis[9] in the sense that we introduce a limitation on the number of matchings to calculate and that this approximation is strictly regulated; the links introduced assure the validity and the reproducibility of the method. Introducing an approximation on the number of calculations has an important limit because we can't be as sure to get the maximal similarity as we could by using an exhaustive comparison. Therefore, on the one hand, we have the potential loss of information that could invalidate our similarity analysis but, on the other hand, we have a reduction in the computation cost. We must look for the best compromise between these aspects in order to minimize the information loss. Our method tries to solve the problem by keeping its balance between an entirely exhaustive method and a rigorously canonical one. Further explanations on this subject are covered later in the method description section.

In general, we can make an analysis canonical by sequencing the elements of the molecular description and performing the match only among corresponding positions of the sequences. Therefore a delicate step in a canonical analysis is the building of sequences; they must be well-defined in order to guarantee the reproducibility of the results and, at least theoretically, they must reach the maximum of similarity with a single-step match. In fact, if the measuring system would permit a sequentiation absolutely unique, the canonical comparison would be completely equivalent to the maximal comparison. For example, if we take two sets of integers and decide to sequence them by their absolute value, the search for the subsets containing equal numbers gives an exact result in one step, comparing corresponding positions. By contrast, when the measuring system has some uncertainty and/or the grouping rule is ill-defined, the sequentiation, even if unique and predictable, can be not fully representative (as is the case in similarity analyses). For example, if the two numerical sets are first sequenced by approximating subsets and then grouped by differences between number pairs, the result is not necessarily the absolute maximum.

To decrease the number of calculations some authors keep some descriptors fixed and it is possible to consider this solution an alternative to the canonical match;[10] it is not the purpose of this paper to compare our method to others, but we would like to present a rigorously canonical analysis as used in a similarity study.

II. BACKGROUND

A. Similarity Measure

In order to make the discussion easier a short summary of our approach is helpful. The aim of our choice of similarity for a representation of chemical structures is the generation of an effective tool to correlate structures to activity; we are especially interested in predicting the activity of particular portions of a chemical structure once we know its relation to other compounds with known activity. As a consequence, the approach must be able to describe small portions (as small as single atoms) of a structure: it must be a point descriptor. A good choice would fulfill both conditions: the accuracy of the description, and the easy connection between the descriptor and the chemical behavior. We selected the electronic energy of atoms;[8] more precisely, the variation of atomic electronic energy generated by the molecular environment (ED = energy difference). ED is a good descriptor because it is characteristic of each atom in a particular environment, i.e. it is representative of the atomic response to the environment perturbation. The use of ED as a similarity measure is thus straightforward.

B. Electronic Energy

In principle, any kind of energy calculation could be used but, because of simplicity and calculation speed, we adopted an empirical method.[11] It uses the well-known relation between electronic energy and electron density (shell occupation) where the electronic energy is calculated by the integral of the chemical potential. The energy can be calculated considering the molecules as pure topological objects or as three-dimensional objects. These two alternatives give different results; in fact, in the first case each atom "feels" only its connected sphere, while in the second case each atom "feels" all the near neighbors (Figure 1).

C. Results

By using this approach we obtained some exciting results. It was possible to introduce a new general definition of functional group[12] where the identification is driven by EDs and ED gradients. And in the similarity area, the possibility to compare structures and substructures has shown its power. In fact, we could compare different molecular situations: from simple atomic groups to whole structures (Figure 2).[8]

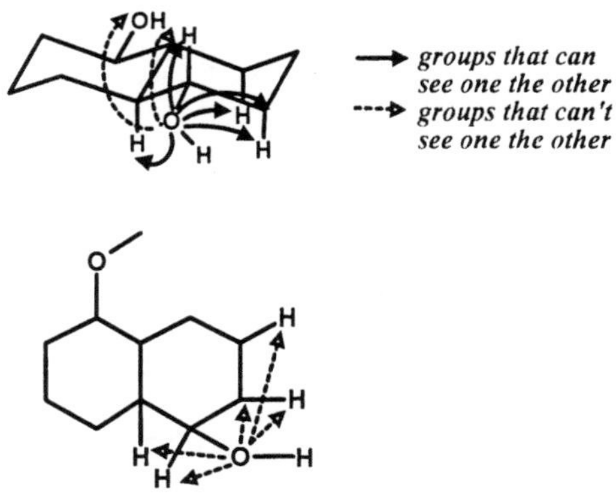

Figure 1. The different influence of the atom environment in topological and spatial calculation of electronic energy.

We also introduced two ways of comparison, the first directly using the EDs, the second using the ED variations along atomic chains that we called trend comparison (Figure 3). As an example, we can compare substituted benzenes by ED and group them by substituent electronic effects, or we can compare them by ED trends and group them by substituent positions (Figure 4).

Finally, the possibility of using different calculations of ED (topological or three-dimensional) offers another chance of getting different results by affecting the similarity measure (Figure 5).

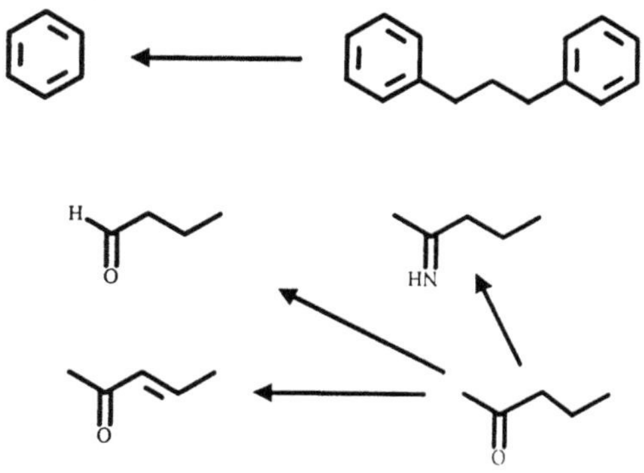

Figure 2. Examples of calculated similarities: from functional groups to substructures.

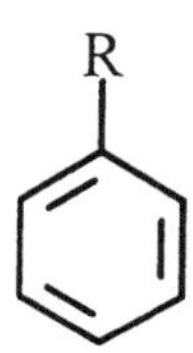

Group 1	**R = OH, OMe, NMe_2, F**
Group 2	**R = NH_2, SH, SMe, Me, Br, I**
Group 3	**R = Cl, DB, Ph**
Group 4	**R = COOH, COOMe, CHO, NO_2**

Figure 3. Monosubstituted benzenes: grouping by ED similarity.

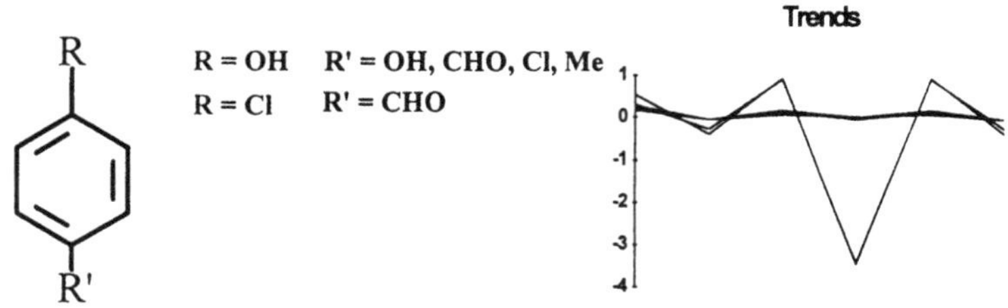

Figure 4. 1,4 disubstituted benzenes: grouping by ED trend similarity. On the right is the graph of the trends.

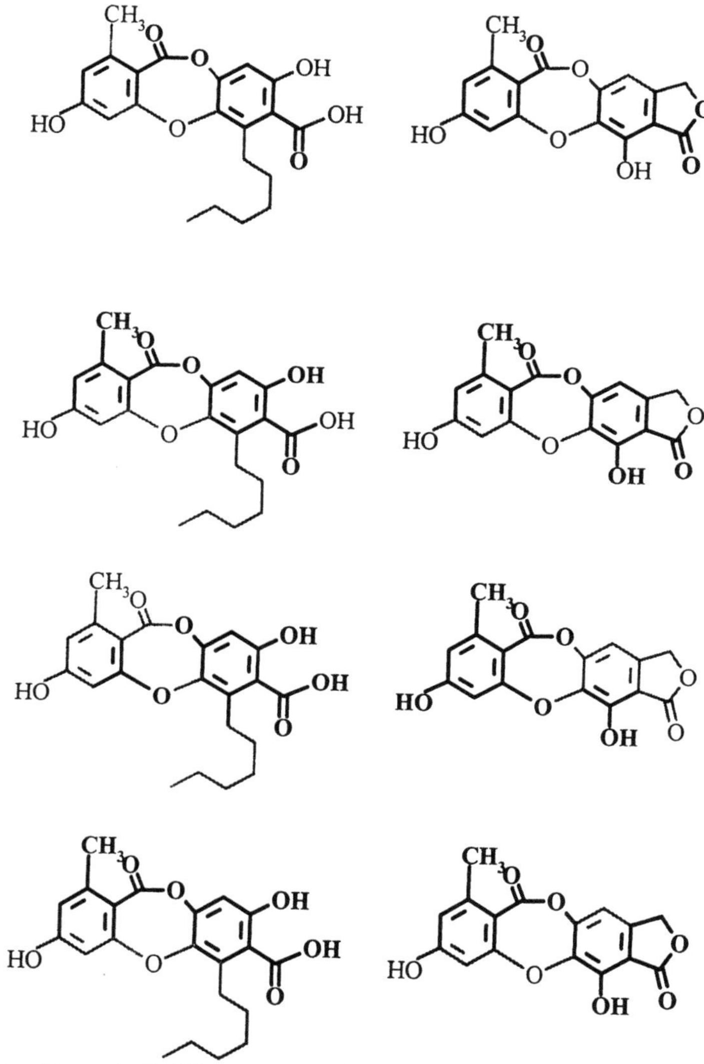

Figure 5. *Upper half*: differences in substructure ED similarity considering spatial influences (*upper example*) and ignoring spatial influences (*lower example*). *Lower half*: differences in substructure ED trend similarity considering spatial influences (*upper example*) and ignoring spatial influences (*lower example*).

D. Investigation Methodology

All the obtained results implicitly refer to point-to-point comparisons. In fact, the atomic EDs and ED trends are a representation of the molecule as an ordered collection of objects. The choice of an investigation methodology is therefore quite natural. From the different possibilities we chose matching by superimposition that represents a classical point-to-point matching. As already mentioned, the matching could be in principle effected by an exhaustive search (maximal matching), but time saving requires a different approach, particularly when doing many comparisons on complex compounds.

III. SEQUENTIATION

Sequentiation is fundamental for a canonical search by superimposition, therefore it is very important to fix the rules that must give reproducible and reliable results. However, the most important characteristic is the connection between the sequence and the measure (and consequently the chemical property) that must be clear and

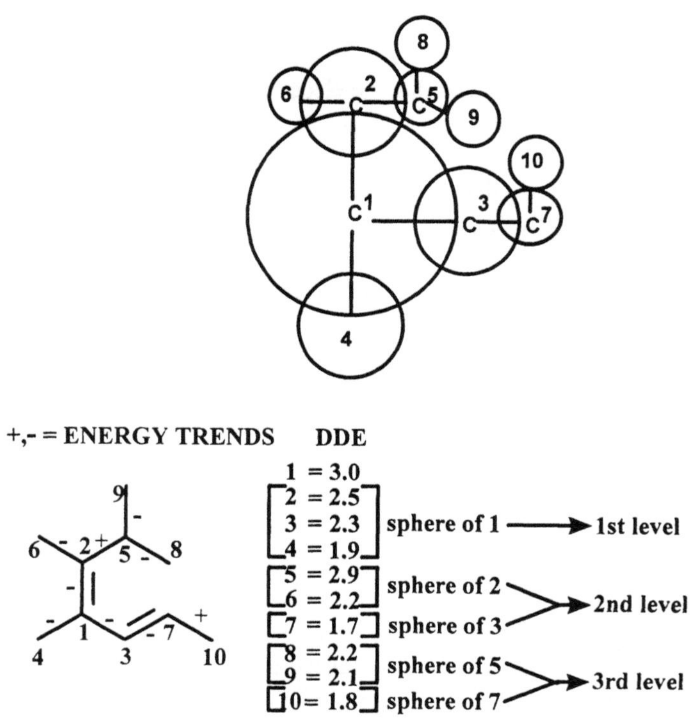

Scheme 2. Defining atomic sequences: the growth by sphere based on ED weight.

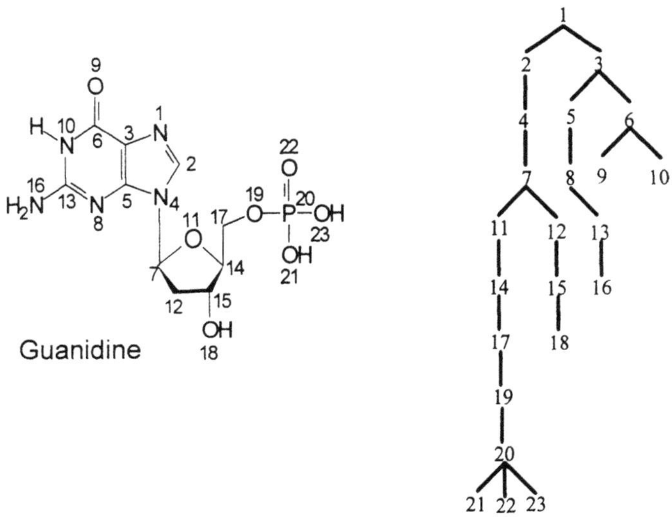

Scheme 3. Defining atomic sequences: the example of Guanidine. Tree representation of the sequence levels.

meaningful. In our case it is important to sequence a structure in a way that favors the most important atoms concerning the electronic energy. The sequence will be built following the rules:

1. place first the atoms that have the highest ED,
2. add first to the sequence the atoms linked to those already in the sequence, and
3. grow the sequence by spheres of connected atoms.

A simple example illustrated in Schemes 2 and 3 may make the concept clear.

The result is a sequence that represents the corresponding structure as a tree of connected atoms ordered by decreasing ED in each sphere. This allows a canonical comparison where the points are compared following their importance in the structure.

IV. TOPOLOGICAL MATCHING

We will first discuss matching with respect to the topology of structures (along molecular bonds). We have already introduced the possibility of making a match using EDs or ED trends. Besides this possibility we can also envisage the opportunity of making both inter- and intramolecular comparisons. In the second case the necessity of comparing different substructures of the same molecule implies the corresponding creation of a second sequence. This will be created by the described method but starting from the most distant atom with a comparable ED. In the case when an atom with this characteristic (comparable ED) does not exist, the matching will be deemed impossible, therefore a new sequencing phase will begin from the second most important atom and the loop is repeated until either a suitable secondary starting point is found or no matching is possible (empty set). Once two sequences are available (either coming from two different molecules or from only one molecule) the matching can start without considering the origin of the sequences.

When doing a match by EDs the algorithm is as follows:

1. starting from the maximum ED, search for the first atom with similar ED ($< \Delta$) in the other sequence;
2. then compare the atoms on the spheres of those selected at point 1 and include the similar ones in the similarity set (ASS = atom similarity set); and
3. continue the search until no new entry is present in the ASS.

At this point we have two sets of atoms that are connected and similar, i.e. two connected similar substructures. (We only save sets containing at least four atoms.) If the number of atoms not yet examined is ≥ 4 then the search is restarted and, possibly, new similar substructures are found.

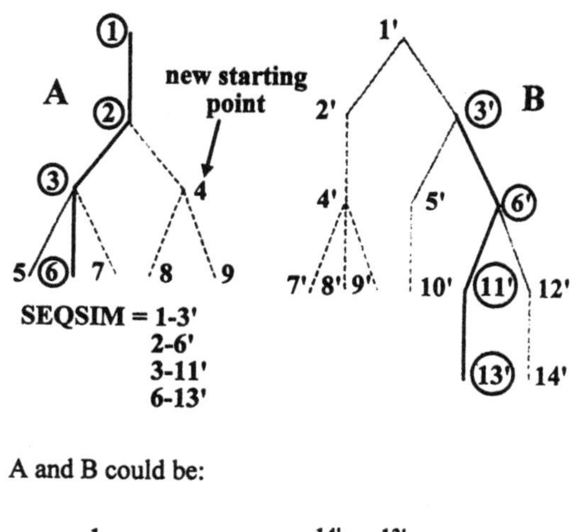

A and B could be:

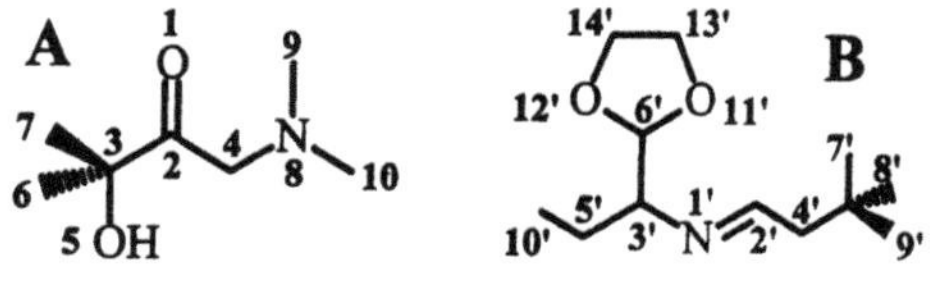

Figure 6. Topological matching mechanism. Atom 1 is a primary starting point; atom 4 could be a secondary starting point.

Figure 6 shows a typical example where atom 1 and 1′ are not similar (thus discarding atom 1′) and the similar substructures start from atom 1 and atom 3′, respectively. After the first two substructures have been determined the search could start again from atom 4 and atom 2′.

A second case where more than one search can be helpful appears when the two molecules being compared are different in dimension, i.e. the smallest one can be found similar to more than one substructure of the largest one;[b] a typical example being a molecule that is the monomeric component of a polymeric compound.

In Figure 7 some examples of matchings are shown.

The algorithm used in the case of ED trend comparison is slightly different. It follows the rules:

1. only atoms at the same level in the sequence are compared;
2. only atoms with the same number of connections can be similar; and
3. only atoms with the same ED trend are added to the similarity set.

This comparison is more restrictive than the previous one concerning the superimposition and less restrictive concerning the ED similarity. Here, again, it is possible to repeat the search starting from atoms not yet used if needed.

The example illustrated in Figure 8 is self-explanatory. Once the first two substructures are found the search is restarted from atoms 6 and 2′ with the corresponding reset of the sphere levels.

The method just described works nicely and gives interesting results, but one problem still remains: we cannot be sure we are getting the maximum similarity because we are using a canonical, one-shot match. For the sake of completeness we then introduce another mechanism to increase our confidence in the method—let's

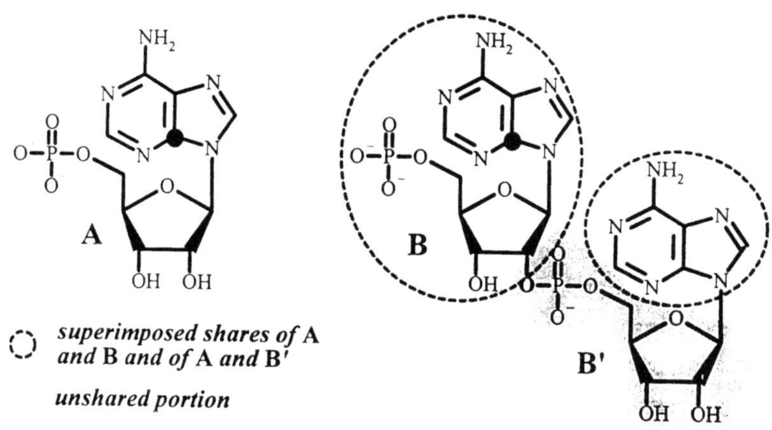

Figure 7. Topological matching: the example of a monomer compared to its dimer. The dotted atom is the sequence starting point; the grey portion of the dimer is not found similar because of the sequencing mechanism.

call it "Jumping Jack" (JJ). What does JJ do? In principle, it is a repetition of the standard mechanism but using a different sequence. It works as follows (let's justify the JJ name):

1. the first search is standard;
2. then one of the two structures is sequenced starting from another atom with ED similar to the primary starting point but not connected to it;
3. the search is repeated and the best result saved;

(a)

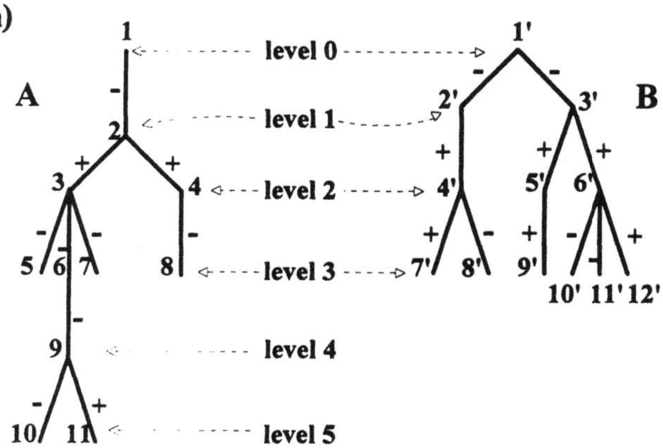

Pair	level?	link?	trend?	Similar
1-1'	x			
2-2'	x		x	
2-3'	x	x	x	x
3-5'	x		x	
3-6'	x	x	x	x
5-10'	x	x	x	x
6-11'	x		x	
7-12'	x	x	x	x
4-4'	x		x	
4-5'	x	x	x	✓
8-9'	x	x		

(continued)

Figure 8. Topological matching mechanism using ED trends (**a**,**b**). After the first comparison the level of the first compound are reset. Two substructures are found similar. The substructures starting from atoms 4 and 5′ are too short to be considered.

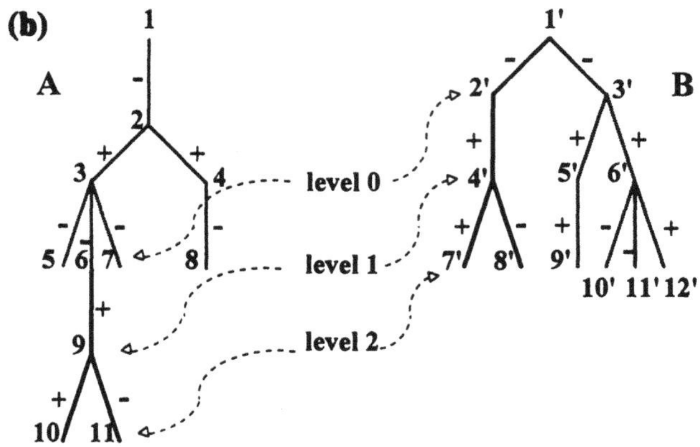

Pair	level?	links?	trend?	Similar
6-2'	×	×		×
9-4'	×	×	×	×
10-7'	×	×	×	×
11-8'	×	×	×	×

Sequence of similar atoms:

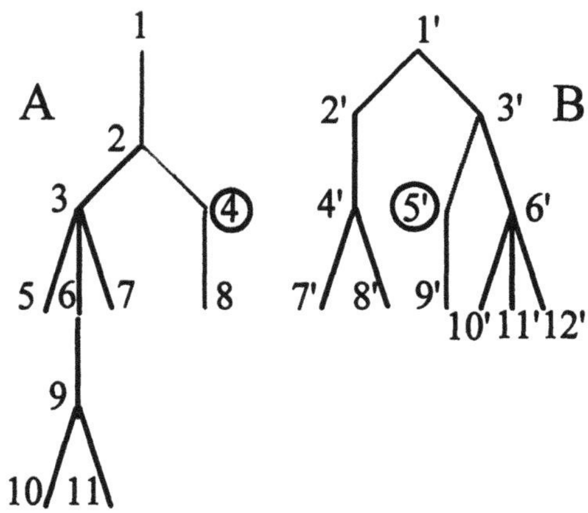

Figure 8. (Continued)

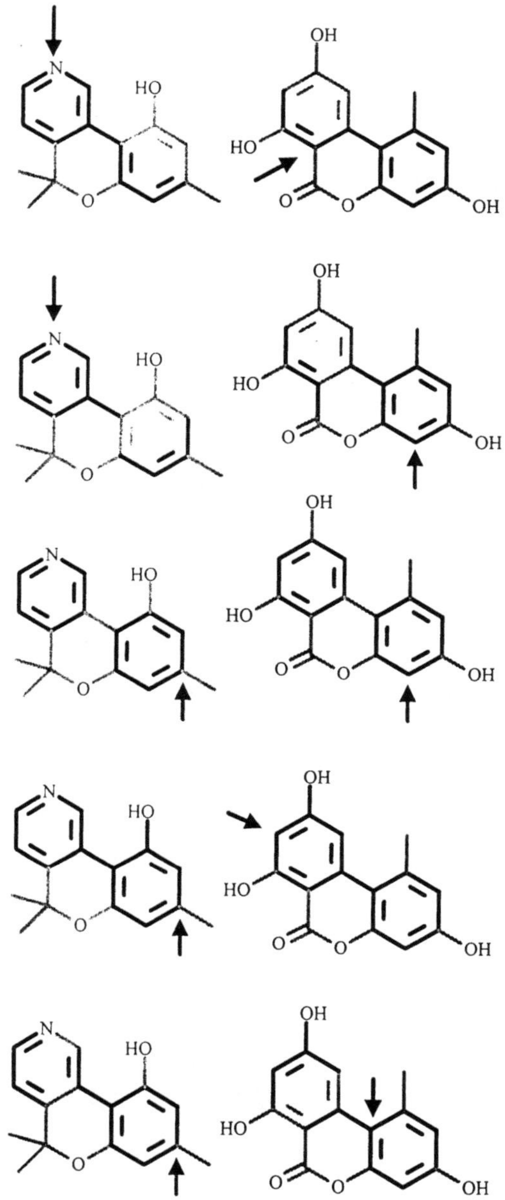

Figure 9. Jumping Jack topological matching mechanism. Arrows point to sequence starting points. They change at each matching until the increase in similarity stops.

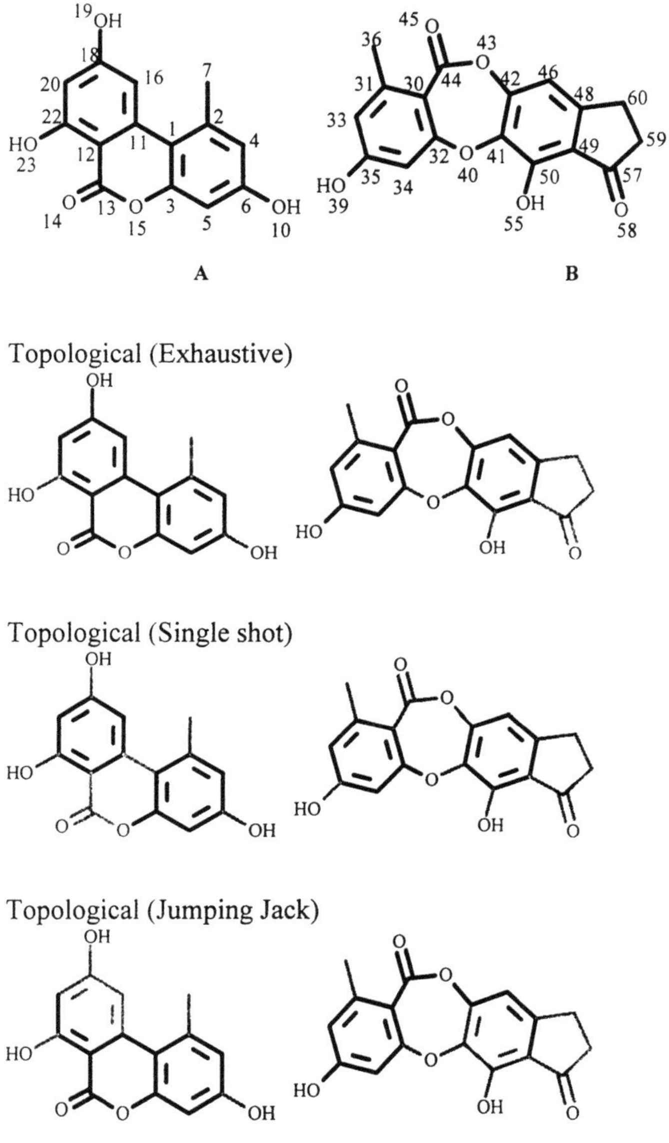

Figure 10. ED topological matching: comparison between exhaustive, single shot, and Jumping Jack search.

4. a third search is done after resequencing of the second molecule;
5. then if one of the two last searches has given a result better than the first search the procedure is repeated sequencing again the structure that was not resequenced in the accepted search; and
6. the process continues until either the new search is less effective than the previous one or there are no more new potential starting points (with ED similar to the primary starting point).

Jumping Jack thus allows a deeper search of the absolute maximum while still following the general rules of canonicity (sequence and matching). It is worth noting that JJ finishes its work in a finite number of steps (usually less than 10). The example in Figure 9 clearly shows the gain in similarity obtained by JJ.

A. Results

The first result that we will discuss concerns the comparison between exhaustive and canonical search. The two structures shown in Figure 10 have, evidently, many

Table 1. Sequences and Similar Substructures[a]

A	*B on A*	*B*	*A*	*B on A*	*B*
5	**30**	**30**	**12**	**30**	**30**
6	**31**	**31**	**11**	**31**	**31**
3	**32**	**32**	**22**	**32**	**32**
4	**33**	44	**13**	44	**44**
10	—	**33**	16	—	**33**
1	**34**	36	**1**	33	36
15	**40**	**34**	**20**	**34**	**34**
2	**35**	**40**	23	—	40
11	**41**	45	**14**	45	**45**
13	—	43	**15**	43	**43**
7	—	**35**	18	—	**35**
12	**50**	**41**	**2**	**35**	41
16	**42**	**42**	**3**	**42**	**42**
14	—	39	19	—	39
22	**49**	**50**	**4**	**49**	50
18	**46**	**46**	7	—	**46**
20	**48**	**49**	**5**	**46**	**49**
23	—	55	**6**	48	55
19	—	**48**	10	—	**48**
		57			57
		60			60
		58			58
		59			59

Note: [a] Bold numbers are atoms in the similarity set.

similar atoms and, consequently, substructures. In Table 1 the sequences coming from the two structures are reported together with the similar substructures found by exhaustive canonical–single-shot and canonical JJ matchings. The following comments apply:

1. The exhaustive matching that has been done by hand follows the energy rules of the canonical matching, i.e. atoms are energetically similar if the difference between their EDs is within a threshold and only sequences of at least four atoms are accepted.
2. The longest sequence of similar atoms contains 16 atoms and, in the case of alternariol, is made by all but 3 atoms.

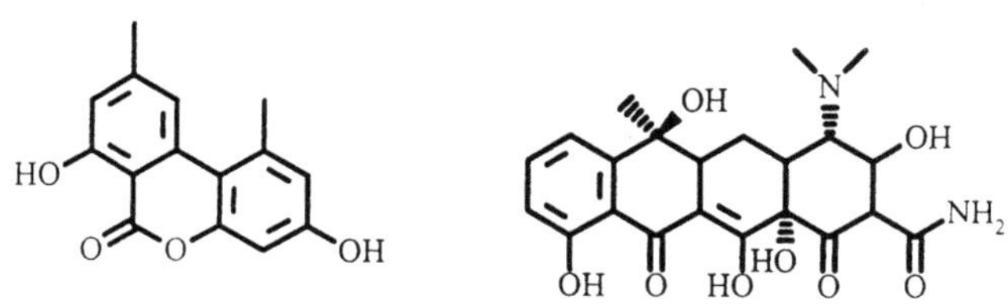

Alternariol -Tetracycline (0.5033)

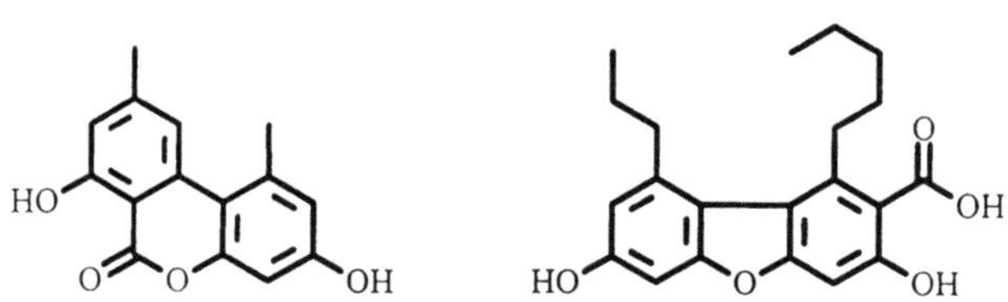

Alternariol - Didymic acid (1.0020)

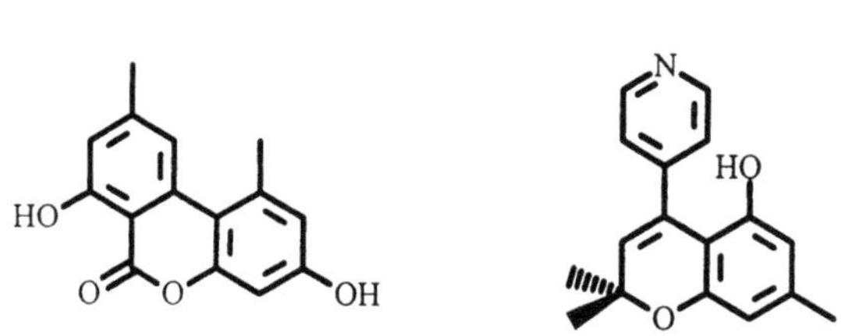

Alternariol - Cannabinol der. (1.4368)

Figure 11. Alternariol used as a probe: numbers (calculated by Eq. 1) in parenthesis assign hierarchy.

3. Both single-shot and JJ procedures found the same number of similar atoms[13] that is smaller than the maximum. The main difference is that, in the first case, the atoms are put into two separate sequences while, in the second, they are part of the same sequence. This second case, therefore, represents a better result, at least in terms of substructure search.
4. It is worth noting that the chosen example is highly critical because the compounds contain a high number of atoms with very similar EDs that have, as a consequence, a high probability of sequencing the two structures differently. (In fact, the most important atom can be chosen from several alternatives.)

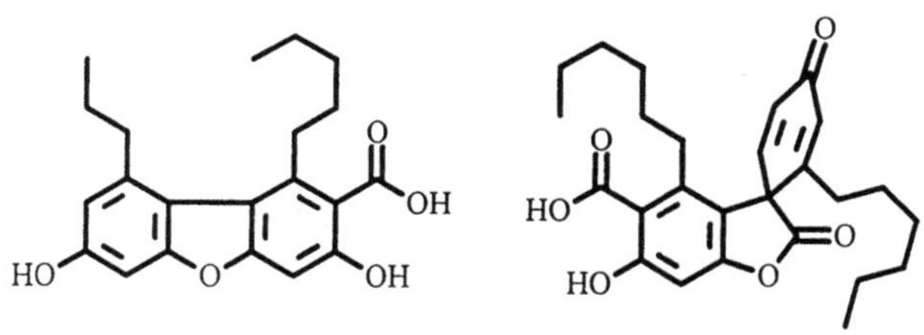

Didymic acid - Picrolichenic acid (0.4183)

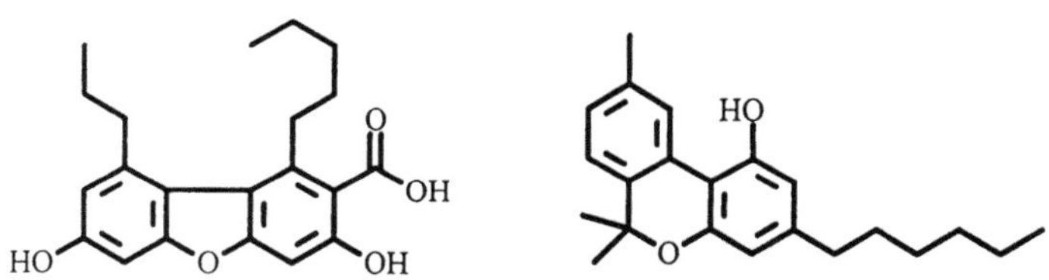

Didymic acid - Cannabinol (0.7212)

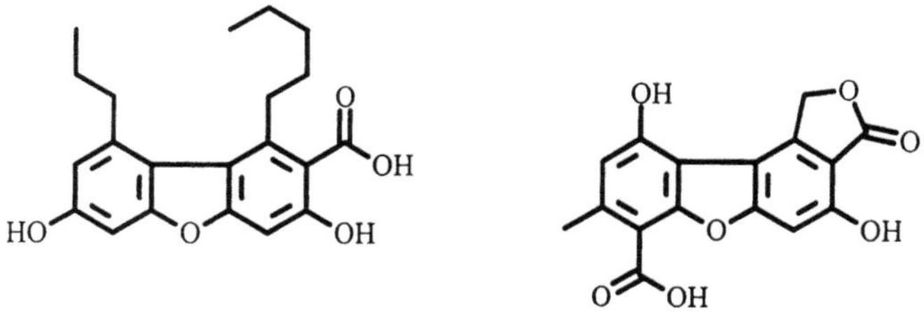

Didymic acid - Porphyrilic acid (1.1472)

Figure 12. Didymic acid used as a probe: numbers (calculated by Eq. 1) in parenthesis assign hierarchy.

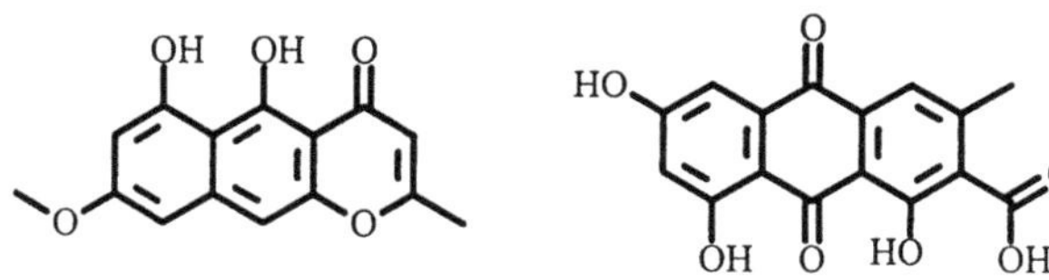

Rubrofusarin - Endocrocin (1.2152)

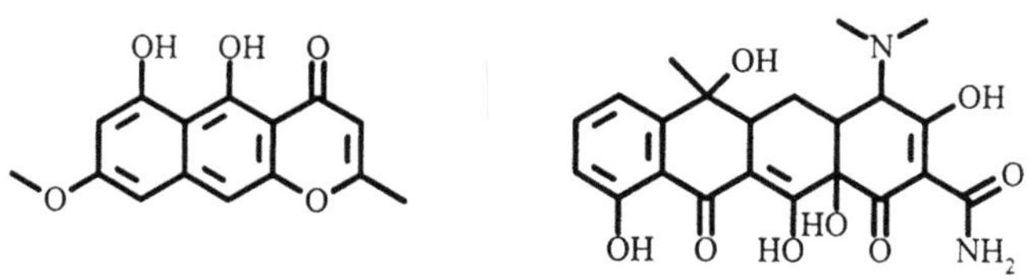

Rubrofusarin - Tetracycline (1.1375)

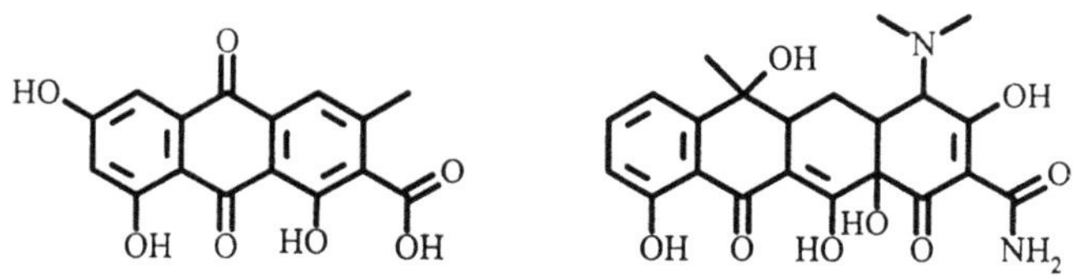

Endocrocin - Tetracycline (0.5978)

Figure 13. The values calculated for Rubrofusarin are not transferable to compare endocrocin to tetracycline.

5. The JJ analysis shows its importance by two aspects: (a) the found sequence is longer; and (b) This result is achieved by sequencing using an atom of a different aromatic ring as starting point. It is clear that the presence of many aromatic carbon atoms is the fundamental reason for inaccuracy.[c]

The second result we will present concerns a potential expansion of the use of the similarity matchings. In Table 2 the results of several matchings between two compounds, used as probes, and a set of molecules chosen from a single biogenetic path are reported. The effectiveness of the matching is represented by an index that weights the similarity of each pair of compounds,

$$I = N \times (A + B)/A \times B \tag{1}$$

where N is the number of atoms found to be similar, A and B are the numbers of significant[d] atoms in molecules A and B. The calculation gives a list of molecules ordered against the probe. In principle this is exactly what is expected from a

Table 2. Similarity Ordering Obtained by Equation 1 Using Alternariol or Didymic Acid as Probes

Molecules[a]	*I*	*Molecules*[a]	*I*
ALTE-AUR	1.2030	DIDY-ALTE	1.0020
ALTE-CAN	0.7544	DIDY-AUR	0.8608
ALTE-CIC	0.5033	DIDY-CAN	0.7212
ALTE-CRO	1.1523	DIDY-CIC	0.4183
ALTE-DCA	1.0819	DIDY-CRO	0.9833
ALTE-DIDY	1.0020	DIDY-DCA	0.8600
ALTE-FUC	0	DIDY-FUC	0
ALTE-GRI	0.5658	DIDY-GRI	0.6410
ALTE-IMC	1.3248	DIDY-IMC	1.0385
ALTE-MICE	0.6158	DIDY-MICE	0.6192
ALTE-MOR	1.1770	DIDY-MOR	0.5035
ALTE-NDC	0.7083	DIDY-NDC	0.5558
ALTE-NIC	0.6484	DIDY-NIC	0.4808
ALTE-NOCE	0.6866	DIDY-NOCE	0.5035
ALTE-PDC	1.4368	DIDY-PDC	0.8845
ALTE-PHY	1.2368	DIDY-PHY	0.8512
ALTE-PIC	0.5033	DIDY-PIC	0.4183
ALTE-POR	0.7689	DIDY-POR	1.1472
ALTE-RUB	0.8211	DIDY-RUB	0.9731
ALTE-VAR	1.2494	DIDY-VAR	0.9013

Note: [a] Acronyms correspond to the names of the molecules in the test set (see Abbreviations).

similarity analysis. From Figures 11 and 12 it is possible to see that the proposed ordering is quite natural and, as much as possible, expected. The use of EDs for the comparison gives good results even for atoms of different types (e.g. N and C in a alternariol–cannabinol derivative comparison). On the other hand, the results are not transferable as clearly shown in Figure 13, where a rubrofusarin probe cannot be used to compare endocrocin to tetracycline.

B. Conclusion

We hope to have demonstrated that canonical matching, especially the JJ version, can be fruitfully used to automatically compare structures. The results obtained are satisfactory and the problem of local minima is mostly solved.

Moreover, the matchings give interesting hints concerning the similarity between molecules. The use of the maximal matching approach would obviously give the best result, but considering its cost we recommend the choice of the canonical alternative for routine work.

V. SPATIAL MATCHING

A different approach to similarity matching concerns the comparison of molecules in three-dimensional space. In this case the information gained will be different because a second aspect, the relative space position, comes into play and influences the similarity evaluation. The importance of the spatial position of atoms and groups in chemical activity is well known and very often has a fundamental role. There are

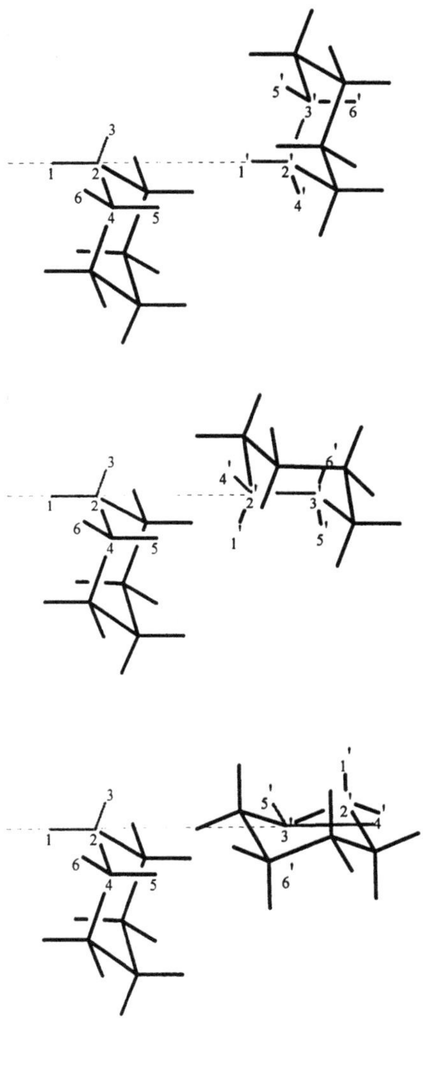

Figure 14. Three subsequent orientations obtained using triples of atoms from sequences. The first structure is kept fixed.

application areas, such as drug–receptor interaction, that are heavily dependent on the geometry of both partners. It is therefore natural to extend our approach to three-dimensional space considerations.

The comparison of two molecules in space emphasizes the problem of maximal matching. In fact, the number of alternatives in point matching grows rapidly because the comparison must ignore the bond frame of the molecules. It is thus even more important to adopt a canonical method to matching. In our view, the sequence search must remain identical in order to guarantee a consistent set of results. But in the spatial approach it is necessary to also have a canonical way to orient molecules in space because we need a unique result for all matchings. The problem of molecular orientation has been analyzed by several authors,[13] but it still remains unsolved. In fact, in our opinion, the orientation must be related to the descriptor used, i.e. it is impossible and unwise to have only one orientation methodology. Again, for the sake of complete self-consistency, we chose ED as the reference descriptor for positioning structures. The method is as follows:

1. take one molecule fixed with the first atom in the sequence placed to the origin 0,0,0; the second atom along the X axis (positive); the third in the XY plane (positive Y);
2. align the second molecule with the same orientation;

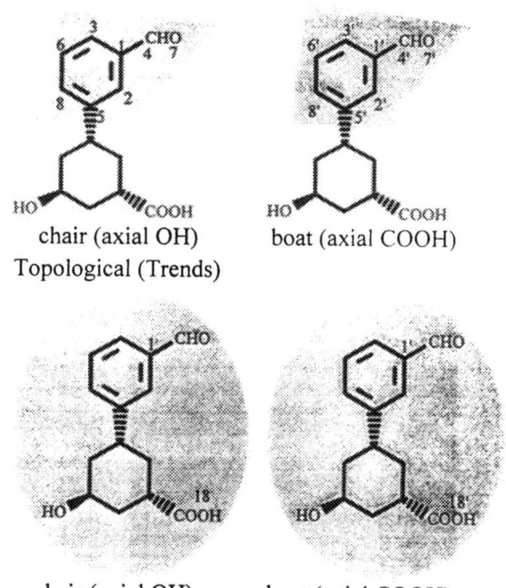

Figure 15. ED and Ed trend similarities calculated topologically with spatial EDs.

3. compare all the atoms and include in the ASS those atoms that have a difference in ED within a threshold and that are near, i.e. at a distance shorter than another threshold;
4. reorient the second molecule using the next triple of atoms in the sequence and repeat the matching; repeat until all the possible triples are used; and
5. reorient the first molecule as described at point 4 and repeat from point 2.

This methodology is similar to an exhaustive search, but recall that we are only using atoms ordered in the sequence. Figure 14 shows the first three steps of the orientation procedure.

A. Results

The first example in Figure 15 shows the application of the procedure to a simple case. The bicycle structures sketched are two conformations (boat and chair) of the

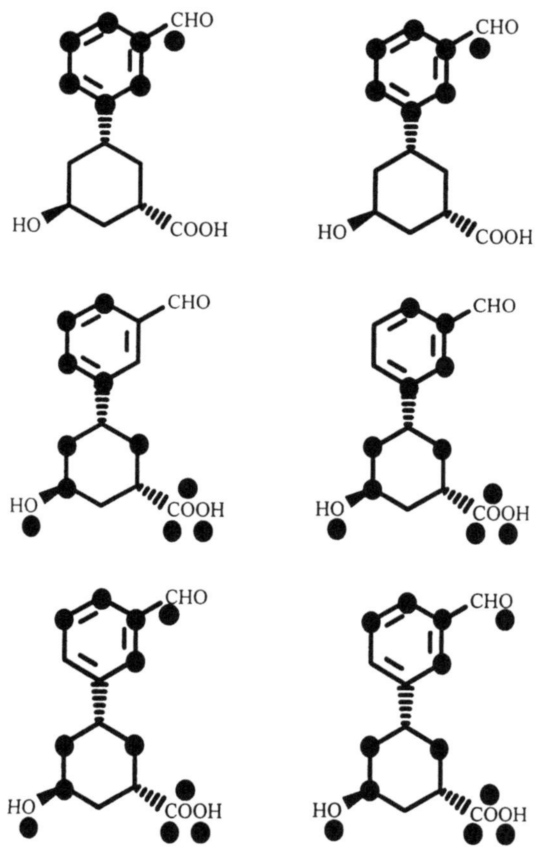

Figure 16. Some examples of ED similarities calculated in three dimensions with sequence dependent orientations.

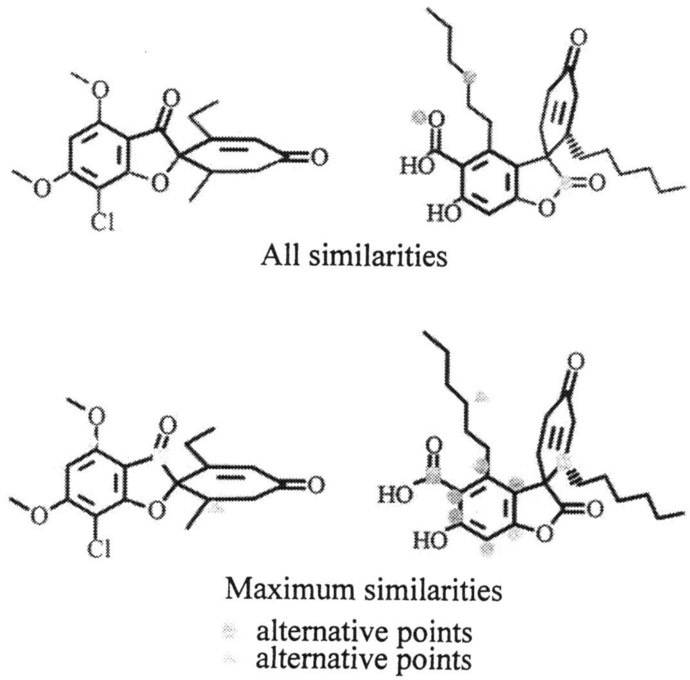

Figure 17. Spatial similarities between Griseofulvin and Picrolichenic acid. The combination of all similarities (*upper example*) and the biggest substructures (*lower example*) with alternative points.

(a) **CANONICAL VERSUS MAXIMAL MATCHING**

Maximal matching

☺ Very accurate result (Absolute maximum)

☹ Great number of solutions

Canonical matching

☹ Less accurate result (Local minima problem)

☺ Small number of solutions

Canonical matching & Jumping Jack

☺ Quite accurate result (Escaping from local minima)

☺ Small number of solutions

Scheme 4. Positive and negative aspects of matching methods (**a**,**b**).

(b) **TOPOLOGICAL VERSUS SPATIAL MATCHING**

Topological matching

Using DDE

- Keeps structural information
- Is independent from conformational problems
- Is a punctual similarity
- Gives substructural similarities with evident chemical meaning

Using Trends

- Keeps structural information
- Is independent from conformational problems
- Is a path similarity
- Gives substructural similarities with a different meaning

Spatial matching

- Looses some structural information (bond connectivities)
- Depends on conformations
- Is more exhaustive
- Is a punctual similarity
- Gives spatial similarities between unconnected atoms

Scheme 4. (Continued)

same molecule where the hydroxyl and the carboxyl groups are either axial–equatorial or vice versa.

The topological searches (EDs and ED trends) give two apparently different results because in the ED search the OH and COOH groups that are composed of less than four atoms are not saved as sequences. Thus the common result is a complete equality of the two conformations as expected. If the spatial search is applied to the problem we get different ASSs depending on the relative orientation of the two structures. For example in the first result shown in Figure 15 (with the two structures equally oriented) the common substructures containing the aromatic ring is found, whereas the other two groups (OH and COOH) are missing because of their different positions in space. When the two molecules are differently positioned, the result changes and a subset of them is given in Figure 16.

A second example is illustrated in Figure 17. In this case the two molecules are different and the results can be summarized as follows: If we add all the ASSs together we can see all the possible similarities between the atoms of the two compounds (15 atoms) and the largest sets of similar atoms (7 atoms) found in one comparison. It is worth noting that in the last result we can easily point out atoms that can represent alternatives in similar activity (e.g. the carbonyl carbon of compound *A* and either the carboxyl carbon or the alkenic carbon of compound *B*).

Finally, if we compare the results of the topological ED, topological trend, and spatial matchings (all canonical), we can note the different aspects that are furnished by each methodology. (It is evident that each one can be helpful in its own application, none being clearly superior.)

B. Conclusion

Concluding this section we would like to point to some characteristics of each matching. These are shown in sketch Scheme 4.

VI. FINAL CONCLUSION

In this review we have faced the problem of automatically matching of molecules according to their similarity. We were particularly interested in discussing the problem in connection with our approach to similarity. The addition of a calculation considering spatial position of atoms to the previous achievements completed the potential applicability of the method. The usefulness of a canonical search compared to a classical search were pointed out and the consequent needs of sequencing and canonical matching were solved. The introduction of an expansion to rigid, one-shot matching was discussed and showed an improvement in the performance of the method. Finally, the possibility of canonical matching in space was presented. All the points were discussed with examples and compared.

Our conclusion is that the use of a canonical approach to solve the automatic matching problem in the similarity area is worthy of consideration. In particular the consistent use of a methodology connected to the molecular representation used is a guaranty of canonicity and understandability.

Recalling the introductive notes, we have fully achieved the objectives of our hypothesis and we can now begin to study the possibility of demonstrating our thesis. The first attempt in this direction is presented elsewhere in this volume.

ACKNOWLEDGMENTS

The authors gratefully thank the organization of the "Summer School and 2nd Girona Seminar on Molecular Similarity" for supporting and granting their participation in the congress. Partial funding by Italian M.U.R.S.T. and C.N.R. is acknowledged.

ABBREVIATIONS

ALTE	Alternariol
AUR	Aureosidin
CAN	Cannabinol
CIC	Tetracycline
CRO	Endocrocine
DCA	Cannabinol derivative (1)
DIDY	Didymic acid
FUC	Fuchsin
GRI	Griseofulvin
IMC	5-hydroxy-2-methyl-chromone
MICE	Citromycetin
MOR	Morin
NDC	Cannabinol derivative (2)
NIC	Usnic acid
NOCE	Monocerin
PDC	Cannabinol derivative (3)
PHY	Physodic acid
PIC	Picrolichenic acid
POR	Porphyrilic acid
RUB	Rubrofusarine
VAR	Variolaric acid

NOTES

[a]We must be careful using the words analogy and similarity because they don't have the same meaning. Analogy is the relationship that exists among objects; similarity concerns the (common) qualities of objects linked by the relationship of analogy.

[b]The difference in dimension between the two molecules must be ≥ 4 atoms, thus potentially allowing the generation of another ASS.

[c]All the atoms that have similar environments also have similar ED; this situation is quite common in aromatic rings.

[d]Only atoms whose ED is greater than a fixed threshold are considered and they are defined "significant."

REFERENCES

1. Rouvray, D.H. *J. Chem. Inf. Comput. Sci.* **1994**, *34*, 446–452.
2. *Vocabolario della lingua italiana*; Zingarelli: Milano, 1990.

3. Carbó, R.; Calabuig, B. *Int. J. Quantum Chem.* **1992**, *42*, 1681–1693, 1695–1709.
4. Mezey, P.G. *J. Chem. Inf. Comput. Sci.* **1992**, *32*, 650–656.
5. Dean, P.M.; Perkins, T.D.J. *Trends QSAR Mol. Modell. 92, Proc. Eur. Symp. Struct.-Act. Relat.: QSAR Mol. Modell., 9th 1992*; Wermuth, 1993, pp. 207–215.
6. Wochner, M.; Brandt, J.; von Scholley, A.; Ugi, I. *Chimia* **1988**, *42*, 217–225.
7. Randić, M. *J. Math. Chem.* **1991**, *7*, 155–168.
8. Leoni, B.; Sello, G. In *Molecular Similarity and Reactivity: from Quantum Chemical to Phenomenological Approaches*; Carbó R., Ed.; Kluwer Academic: Dordrecht, The Netherlands, 1995, pp. 267–289.
9. Maggiora, G.M.; Johnson, M.A. *Concepts and Applications of Molecular Similarity*; Maggiora, G.M.; Johnson, M.A., Eds.; Wiley Interscience: New York, 1990, p. 4.
10. Carbó, R. *Concepts and Applications of Molecular Similarity*; Maggiora, G.M.; Johnson, M.A., Eds.; Wiley Interscience: New York, 1990, pp. 147–172.
11. Baumer, L.; Sello, G. *J. Chem. Inf. Comput. Sci.* **1992**, *32*, 125–130.
12. Sello, G. *J. Am. Chem. Soc.* **1992**, *114*, 3306–3311.
13. Moock, T.E.; Henry, D.R.; Ozkabak, A.G.; Alamgir, M. *J. Chem. Inf. Comput. Sci.* **1994**, *34*, 184–189. Hurst, T. *J. Chem. Inf. Comput. Sci.* **1994**, *34*, 190–196.
   Clark, D.E.; Jones, G.; Willet, P.; Kenny, P.W.; Glen, R.C. *J. Chem. Inf. Comput. Sci.* **1994**, *34*, 197–206. Bures, M.G.; Danaher, E.; DeLazzer, J.; Martin, Y.C. *J. Chem. Inf. Comput. Sci.* **1994**, *34*, 218–223.

USING A CANONICAL MATCHING TO MEASURE THE SIMILARITY BETWEEN MOLECULES:
THE TAXOL AND THE COMBRETASTATINE A1 CASE

Guido Sello and Manuela Termini

Advances in Molecular Similarity
Volume 1, pages 243–266

ISBN: 0-7623-0131-7

ABSTRACT

The incomplete understanding of tumor proliferation and the structural complexity of the few natural antitumor agents are impediments to the production of effective synthetic drugs. Polihydroxyphenol derivatives with stilbenic skeleton, such as some combretastatine A1 derivatives, proved to be promising as potential antitumor agents. The possibility of modeling structure and biological activity relationships could allow us to find new drugs to be synthesized more easily and with controlled pharmacological properties, such as activity and selectivity, thus giving great benefits. Knowing the structure and the properties of one of the few antitumor drugs currently available (taxol) and having at our disposal an analysis method to detect similarities, we started a conformational study of the similarity between taxol and some combretastatine A1 derivatives. The aim was to check the possibility of these simple compounds substituting taxol in its biological activity. The results obtained have been compared to those from a modeling program (CHEMX) to test and confirm the correctness of our methodology.

I. INTRODUCTION

The search for new drugs is one of the main goals of medicinal chemistry. The capability of making molecules with specific properties would enable us to strengthen the benefits of a drug, such as effectiveness and selectivity, and to minimize the negative aspects, such as toxicity. In this area the computer-aided drug design techniques represent a useful tool in supporting the chemist's work, allowing the examination of large molecular systems, and determining pharmacological problems at the molecular level.

The action of a drug depends on a wide variety of factors; among the most important there are two of particular interest in the present discussion:[1] (a) affinity to the receptor,[a] and (b) intrinsic activity.[b]

The main role of a theoretical study is based on these two factors, giving rise to two different approaches to the problem of drug design according to the information available:[2]

1. those in which the molecular structure of the receptor is known (based on a);
2. those in which either a set of active compounds or the origin of the activity is known, e.g. in the interruption of a particular biochemical transformation (based on b).

The computational techniques[2] used in the two cases are most often the same, while the application methodology is heavily influenced by the type of problem.

When the structure of a receptor is known, the design of potentially active compounds can appear straightforward; in fact the characteristics and the position of the interacting substructures are easily derived. Thus, the modification of a hypothetical drug, even by sophisticated calculation techniques, can lead to the design of one or more potentially interesting compounds. However, the problems of transport, stability, etc. that can make a compound active "in vitro" and an active drug "in vivo" remain to be solved. For the problems, similarity can be fairly useful, while the management and the accuracy of the calculations modeling the interaction between the macromolecule, the drug, and the environment become essential.

On the other hand, when the receptor is not well known, presently the most popular approach is the selection of a large set of compounds with known activity followed by an attempt to select those common substructures that can be thought of as necessary to provide a particular activity. From these data it is possible to hypothesize new compounds that, having the appropriate chemical and geometrical features, are potentially active with the same mechanism of action.

The same method is also used where it is possible to guess the structure of the molecule at the transition state along a biosynthetic path. Here the goal is the modeling of a molecule that, by imitating the transient structure, can substitute it and consequently inhibit the biosynthetic path.

The role of similarity in this second methodological approach is clear. In fact, the major purpose of the study is the identification of molecules similar to those whose activity is known and where similarity can be interpreted at different levels: from similarity in macroscopic properties (hydrophilicity, hydrophobicity, dipole moment, partition coefficient in H_2O/n-octanol, etc.) to similarity at atomic level (shape and energy of molecular orbitals, electronic population of each atom, etc.).

Generally speaking, an attribute that can be assigned to a molecule (or to its components) in relation to a descriptor is thought to be related to one property (activity).

There are two consequences: first, similarity is completely defined by the descriptor and therefore by its quantification; second, its use at a predictive level is the more limited the more precise is the descriptive model used.[3] Thus, when we examine problems where similarity is relevant, it is necessary to keep in mind both the limits and the approximations of the computational technique, and the level of generalization needed to avoid making trivial predictions. Therefore, in the area of drug design, where the aim is the prediction of new compounds without knowing the structure of the receptor, similarity has particular importance and has been quite often applied.

The study we present here uses similarity-based methodologies for substructural research. We started from two compounds: for the first one, the biological activity and the parts of the structure that are responsible for the activity, are known; for the second one, we know that it shows behavioral analogies with the first one. We have pursued modifications of the second structure that could make it a good substitute for the first one, naturally in accordance to our similarity criterion.

II. BIOLOGICAL ACTIVITY

A. Taxol

The research of tumors has been of primary importance to medicinal chemists for decades. Despite the fact that many techniques for the treatment of tumors are currently available and many others are being tested, until now chemotherapy hasn't been able to give definitive solutions to the problem. There are many difficulties in this research: many substances that show in vitro antitumoral activity but aren't equally active in vivo; the limited understanding of the phenomena involved in the growth and in the proliferation of tumors is a hindrance to the search for new drugs; and only few tests on human tumor cells are currently available. The majority of the effective agents known until now come from natural sources. This fact gives rise to further difficulty because their extraction can provide only small quantities of products. Often, the structures of these compounds are too complex to be produced by synthesis; moreover the presence of stereocenters make their synthesis in significative yield and sufficient purity difficult. Besides that, the few active drugs used today are effective only on tumors of rapid proliferation while active agents on solid tumors are minimal.

Taxol[4] (Figure 1) is a natural compound derived from the leaves of a variety of European yew, *Taxus Baccata*. This substance showed in vitro citostatic antitumoral activity due to an antimitotic mechanism that inhibits the tubulin depolimerization. The polymers of tubulin are the major proteic component of microtubules that, once assembled, give rise to the mitotic spindle. One of the functions of the mitotic spindle is to separate and lead the migration of the duplicate genetic material to the opposite poles of the mother cell that, through a fission mechanism (mitosis), generates daughter cells.

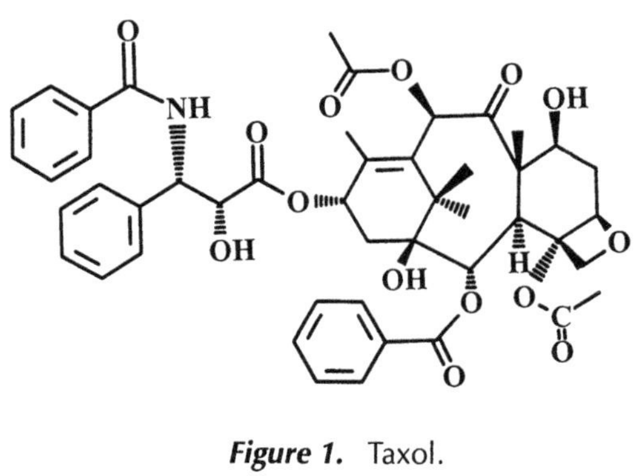

Figure 1. Taxol.

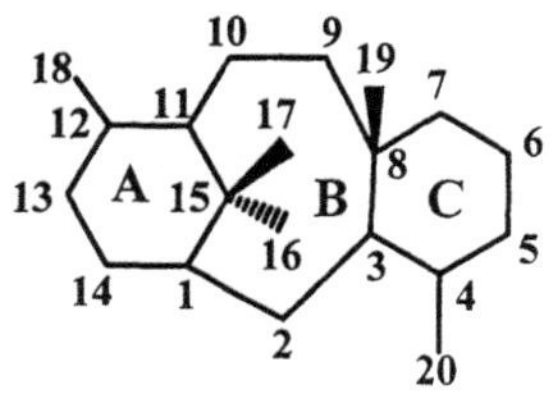

Figure 2. Taxane skeleton.

Stopping the tubulin depolimerization prevents the cell from making the cellular membrane of the daughter cells it can generate by mitosis, i.e. locking the replicative process. This kind of effect is called *citostatic* because it doesn't kill the cell (this would be a "citotoxic effect") but only impairs its reproductive cycle.

The essential functions that allow taxol to exploit its antitumoral activity are known from the literature.[5] The tricyclic portion of the skeleton, called taxane (Figure 2), is fundamental to maintain the rigidity of the molecule that probably assists the correct positioning within the receptor site.

Between the groups connected to the taxane portion, only the benzoyl group at position 2 and the acetyl group at position 4 proved to be essential. Their importance is probably due to the introduction on this part of the structure of a hydrophobic area. The presence of the acetyl group at position 10, of the carbonyl group at 9, and of the hydroxyl at 7 doesn't seem to influence the global activity, thus these groups can be considered unessential. The relative importance of the four-member ring attached to position 4 and 5 of the ring C could be due to the introduction of free hydroxyl groups at those positions following its opening.

By contrast, the lateral chain attached at position 13 of ring A (Figure 3) has proved, in structure–activity relationships (SAR) tests, to be essential for the activity because of its direct involvement in bonding to the receptor site. The importance of the hydrophobic ends is clearly shown by the decrease of activity determined by a primary amine at position 3′. The free hydroxyl at 2′ and the

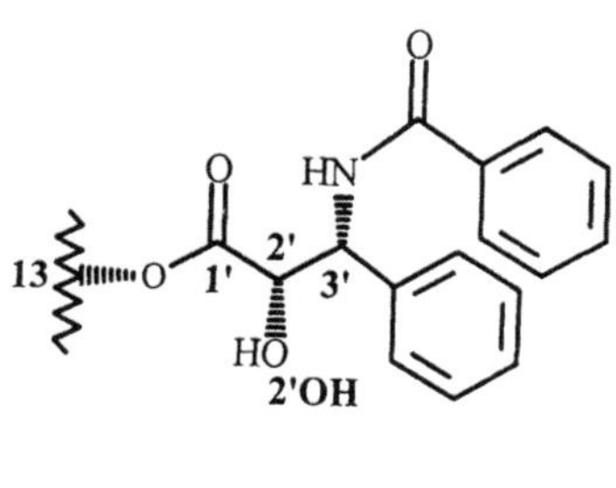

Figure 3. Taxol-like lateral chain.

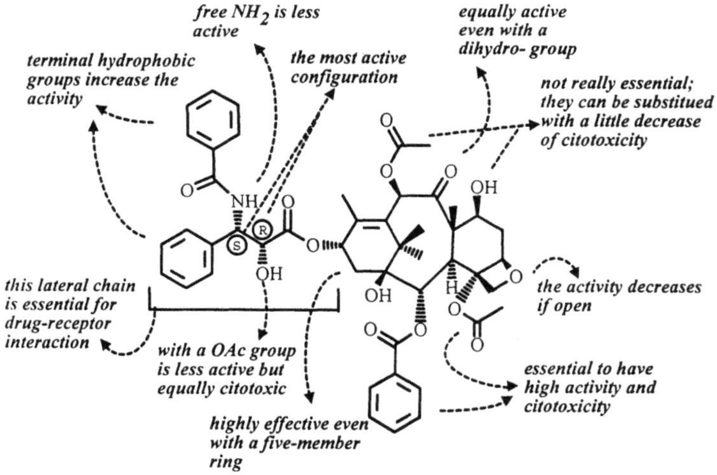

Figure 4. Summary of the essential functions for taxol activity.

absolute configuration of the 2′ and 3′ stereocenters have great importance for the activity (Figure 4).

B. Combretastatine A1

Poliphenols are widespread in nature and the therapeutical properties of many of them have been known for a long time. For example, some derivatives of plants of the genus *Combretum* that live in tropical and subtropical areas are used in the natural medicine of natives.[6]

Particularly interesting for their potential antitumor biological activity are the secondary metabolites with stilbenic polihydroxyphenolic skeleton and their 2′-β-*O*-glucosides coming from the seeds and the leaves of the plant of the species (*C.*) *Kraussii*. Figure 5 shows the principal metabolites (combretastatines) currently undergoing biological tests as potential antitumor agents. Table 1 summarizes the activity data.

Contrary to taxol, the glucoside derivatives of combretastatine A1 showed citotoxic activity that suggests an action mechanism different from that of taxol (that has citostatic activity). From this, we can assume that the glucoside derivatives cannot substitute taxol in its antitumoral activity. By contrast, the corresponding aglycones showed a taxol-like citostatic activity, even though it is less evident than taxol. Further, there is proof that even if the global effect is the same, the antitumoral mechanism of action of aglycone is different from that of taxol (inhibition of the tubulin polymerization instead of inhibition of its depolimerization).[7] From this we can also assume that aglycone cannot substitute taxol. The combretastatine 3′-*O*-glucosilate derivative (compound A), synthesized in our laboratories, represents an

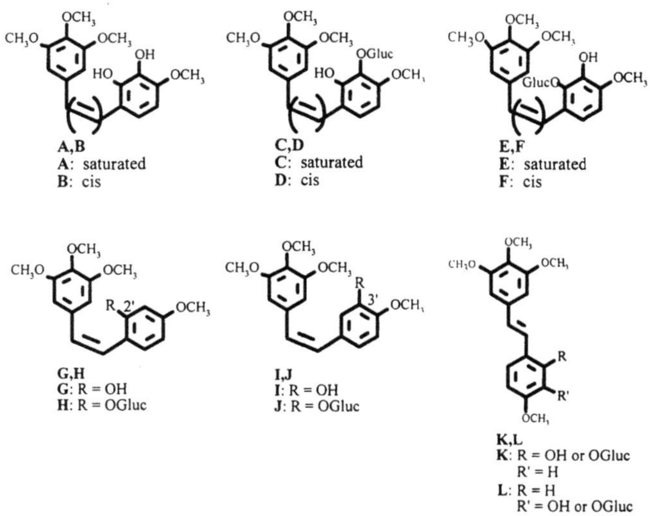

Figure 5. Combretastatine A1 derivatives under test.

exception. In fact, this compound showed citostatic activity by inhibition of the depolymerization of the tubulin protein in contrast with the other glucosides that proved to have either citotoxic or no activity at all. Thus, this glucoside is the only one that could, theoretically, act as a taxol substitute.

The benefits could be extensive if this hypothesis is true. One such benefit is that compound A would be much simpler to synthesize than taxol.

Table 1. Combretastatine A1 Derivatives Activity

Compound	*Natural*	*Citostaticity*
A		–
B		–
C		–
D		+
E	x	–
F	x	+
G	x	–
H	x	–
I	x	–
J	x	–
K	x	–
L	x	–

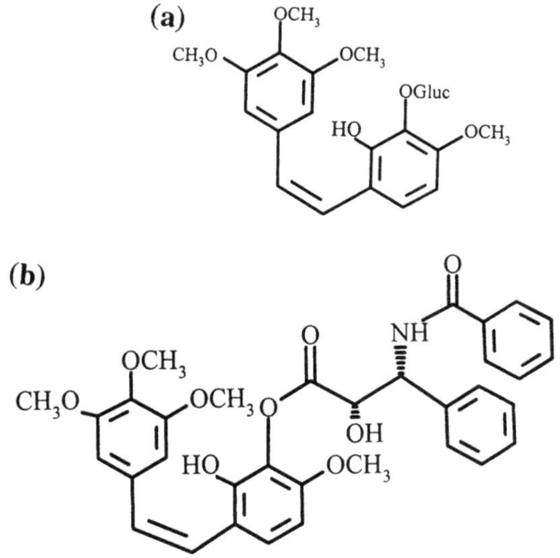

Figure 6. Combretastatine derivates: (**a**) compound A; (**b**) compound B.

These considerations, combined with the experimental data on taxol's essential functions (reported in the literature) that indicate which features potential mimics of taxol must have, have been the motivation to initiate current studies.

Figure 4 shows how the lateral chain of taxol is essential for activity because of its direct involvement in the interaction with the receptor; this gave us the idea to replace the glucoside portion of compound A (Figure 6a) with the taxol lateral chain (compound B; Figure 6b). At least from a theoretical point of view, some conformations of the resulting derivative could behave like taxol.

Having available a similarity-based methodology suitable for this kind of analysis, we began a study of conformational similarity between compounds A, B, and taxol (see Figures 1, 6a, 6b). In the next section we report the results obtained. For details about the methodology we remind the reader to refer to the "Spatial Matching" section included in the chapter titled "Automatic Search for Substructure Similarity: Canonical versus Maximal Matching; Topological versus Spatial Matching" of this volume,[8] of which this study is a practical application.

III. METHODOLOGY

We will cover only the main aspects of the methodology used.

We have an "accurate" measure of similarity that enables us to compare two structures point by point, i.e. to superimpose them. To limit the ways in which the points can be superimposed we define certain rules, a practical consequence of which is the great time-saving in computation. Limiting the number of calculations to be performed implies the ordering of the points to be superimposed to establish priorities and the criteria to lead to the match. The point ordering generates *sequences*; these are built using the same property used to measure similarity. Building the sequences, the connections between points (such as bonds, electron changes due to delocalization or isomerism, long-range interactions, etc.) are taken into account.

The method can perform different types of matching but in the present case (a conformational study) we have used the three-dimensional approach, more suitable than the others for this problem. The structures are handled as rigid entities and oriented within the 3D space using three atoms to locate the axes. To find the best superimposition all the possible orientations generated by the sequences have been tested for both molecules. Only the portions of the molecules positive to the similarity test and occupying the same spatial position when oriented in a certain way can be defined as similar. Because of the nature of the measure, both single points (atoms) or connected substructures can be included in the similarity set.

The problem of orienting the molecules in space is fundamental in our analysis method and we are aware that it can be complex and not easily understandable. For this and any other questions about the methodology, we invite the reader to consult the section "Spatial Matching" referred to earlier.

The spatial position is a necessary condition but it is not sufficient to establish if two points are similar or not; a fundamental condition, but also not sufficient, is a positive result for the similarity test. Let's emphasize the fact that some similar substructures are not connected to others. This fact can be easily understood considering the nature of our similarity measure which can refer to single points as well as to connected substructures. However, the similarity measure doesn't ignore the bounds existing between points; on the contrary, this information is taken into account in the measure, as well as electron delocalization, long-range interactions, and so on.

Our similarity measure is based on an energetic criterion.[9] Similar points or substructures may or may not belong to the same chemical class of functions; however, our aim is to obtain nontrivial answers from the method. Thus, for example, according to our similarity criterion a carbonylic C, a carboxylic C, an amidic C, and an olefinic C are not trivially similar. We want to stress the point that these functions are not similar in an absolute sense, but they are similar only when compared through our particular criterion which, by the way, finds its justification in some reactional behaviors. That is why it is not surprising that an ethereal O could be similar to an aliphatic C or a carbon–carbon double bond to a carbonyl group.

The first impression could be that our method for similarity analysis is contradictory and insensitive. On the contrary, it can measure the influence of a different molecular neighborhood on a particular atom, or of a certain molecular neighborhood on different atoms, and it can identify functional groups. In other words, from this study and from previous ones, we can affirm our measure is fairly sensitive to small perturbations.

The energetic criterion is connected, by an empirical equation, to the occupational level of the atomic shells (that is influenced by the structural neighborhood) and to the chemical potential (that changes with the changes of the electron distribution between an atom interacting with the others in a structure with respect to the same atom hypothetically isolated). This allows us to take the various perturbations into account.

In this conformational study taxol is considered the reference molecule. Its conformation has been derived from X-ray crystallography and is considered fixed and, because of this fact, neither minimized nor modified in the study. By contrast, the conformation of compound B has been changed searching for the best arrangement in which its functional groups assume a particular spatial position to exhibit some taxol functions (possibly those recognized as essential for the activity) with respect to both the similarity measure and the three-dimensional shape.

The conformations of compound B considered in this study have been obtained by rotations by steps of 30° of the angles indicated by 1, 2, and 3 in Figure 7. The conformation of the lateral chain at position 3′ is exactly the same as that of taxol because it is essential for the interaction with the receptor. It is justified to consider the three-dimensional shape obtained from X-ray data as a good approximation of the real conformation.

Considering the lateral chain as fixed, the choice of the angles to be rotated is restricted. We have chosen the rotation around those single bonds that can influence the spatial arrangement of the whole structure. The three angles have been first rotated one by one, and, subsequently, in combination.

For the combretastatine derivatives we defined the "best conformations," in terms of similarity, as those obtained by rotation of each dihedral angle leading to quantitatively greater similarity to taxol. The amount of similarity is measured by

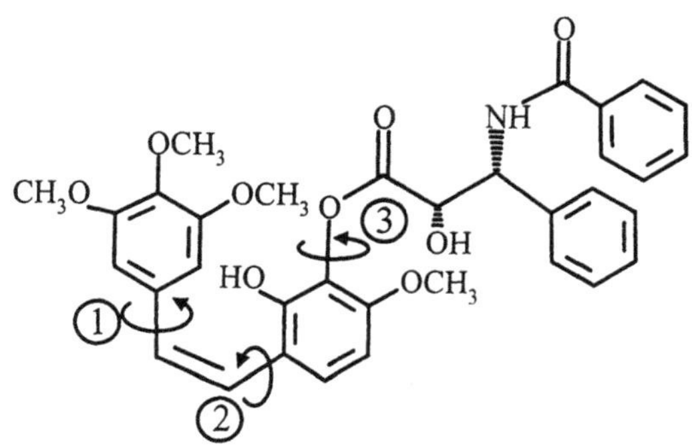

Figure 7. Dihedral angles to be rotated in the conformational study.

the size of the similarity set. For the best conformations small rotations of ±5° were added to the starting angles for testing the sensitivity of the method to small perturbations.

Combinations of the dihedral angles have been chosen among the (local) minima obtained by a modeling program using (CHEMX) molecular mechanics calculations. Some other ones have been selected from rotational combinations coming from the "best" results of the one by one rotations of the single dihedral angles. Finally, we have also compared taxol with a conformation of compound A obtained by minimization with a molecular mechanics calculation performed by the CHEMX program.

IV. CHEMX PROGRAM[10]

The results obtained by superimposition of the molecules in accordance to our similarity-based methodology, have been compared to some fittings performed by CHEMX.

Among the possibilities available we have considered four types of fittings:

1. Automatic;
2. Flexible Torsion;
3. Flexible XYZ; and
4. User Selected Rigid.

When an *automatic* superimposition is performed, the program needs first to identify suitable templates. So it generates a number of pharmacophore templates obtained by a three-center interaction (in the present case; another possibility is a four-center interaction) arranged at fixed distances. Three is the minimum number of centers that defines a three-dimensional binding site. Subsequently the program rigidly superimposes the structures on the generated templates. The tolerance for the center match and for the generation and selection of the best templates to perform the superimposition is controlled. The process is completely automatic.

The *flexible* fittings involve a minimization step of the distortion energy. Once an approximated fitting is rigidly obtained, a process of energy minimization starts automatically to improve the superimposition. An optimization step is required to achieve the best complementarity, eventually in accordance to some geometric restraints previously chosen. The standard minimization used is that obtained by a molecular mechanics calculation. This kind of fitting can be connected to two different types of geometrical freedom: rotational freedom of the dihedral angles (*flexible torsion* fitting), or total freedom for the atoms restricted only by the constraints (*flexible XYZ* fitting). Despite the fact that it is possible to add limitations during the minimization step so that the structures are forced to adopt a certain conformation in the superimposition, in our analysis only flexible fittings without any additional restraint have been used.

In the *manual rigid* fitting case, the user can choose any reference point for the superimposition or, in a simpler way, the molecule is used as fixed reference, but, in any case, the superimposition is performed rigidly without minimization.

V. RESULTS AND DISCUSSION

As mentioned in the methodology section, all the conformations[c] of compound B derived from the rotation by steps of 30° of the dihedral angles 1, 2, and 3 have been examined; for each conformation the level of similarity to taxol in the conformation derived from the X-ray diffraction has been evaluated through our method. For each dihedral angle, the conformations whose real existence is doubtful because of the interaction of some groups have been dropped from the study. Let's first examine the results obtained by rotating the three angles separately.

A. Rotation of Dihedral Angle 1

For this angle the conformations of compound B with dihedral 1 at 60°, 90°, 210°, and 240° with respect to the starting angle (considered as rotation "zero" with respect to any angle) have been left out for the reasons discussed above. Combretastatine at rotation "zero" is the standard conformation provided by the graphical builder of CHEMX for compound B, with the taxol-like lateral chain attached at position 3′ kept fixed in the same conformation as in the original molecule.

Figure 8 shows an example of the result that our similarity-based program has given for a particular rotation of dihedral angle 1. The highlighted portions are the parts accepted as similar by our program; Figure 8 summarizes the global result achieved combining all the possible superimpositions obtained from the different orientations in the space of the two molecules.

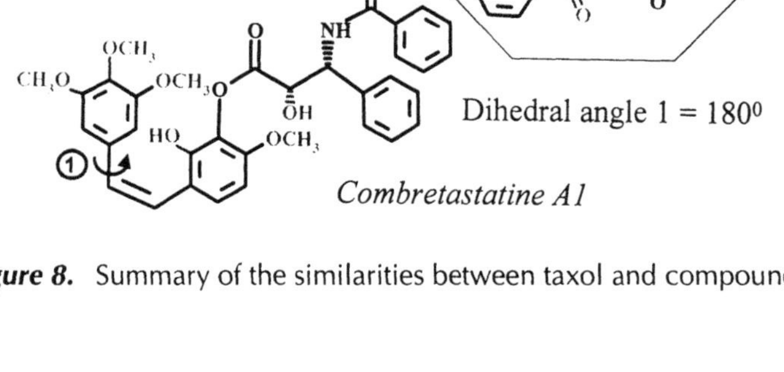

Figure 8. Summary of the similarities between taxol and compound B.

The two molecules are highlighted differently because the superimposition hasn't been reached in a single iteration. Taxol, for example, has three aromatic rings (all of them highlighted; that means all are recognized as similar to parts of compound B), while compound B has four rings, all also highlighted.

This result, apparently contradictory, is in fact logical if we take into account that the different aromatic rings change their spatial position in the different orientations of the two molecules and, because of this fact, don't coincide necessarily in all the orientations of the molecules. That means two rings occupying the same spatial position in a particular orientation can be distant; that is, not similar to one another (Figures 9a,b).

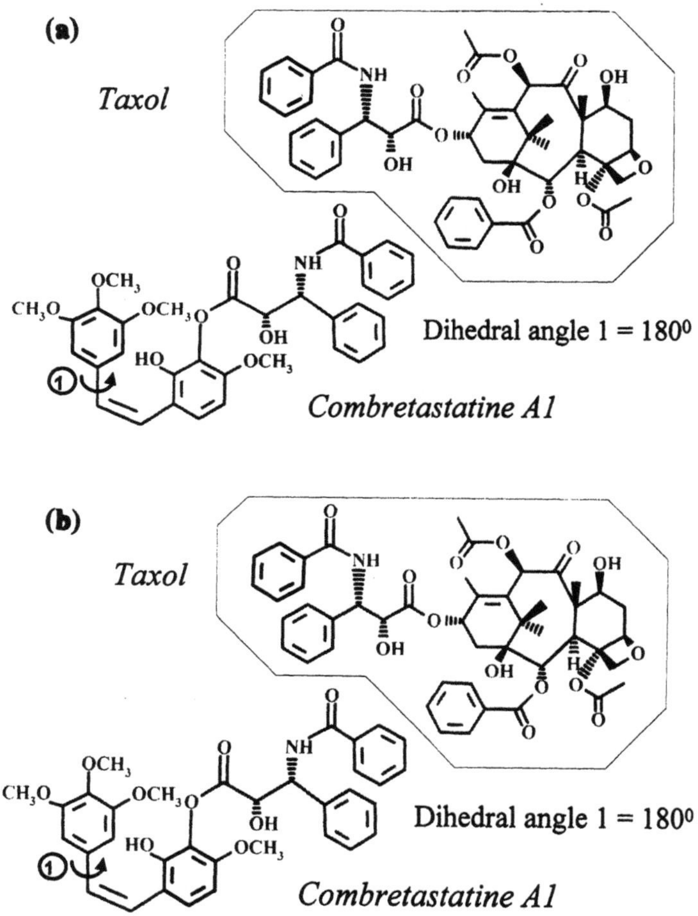

Figure 9. Similarities between compound B and taxol in 2 subsequent iterations: **(a)** 1st iteration; **(b)** 2nd iteration.

For all the rotations of dihedral angle 1 we found, as a general result, the complete superimposition of the lateral chain and of some other points or substructures of the rings A, B, and C of taxol, or of the functions connected to them, and the stilbenic portion of compound B.

From Figure 8 the superimposition derived from the rotation of dihedral angle 1 equal to 180° seems to be fairly satisfactory because compound B approximates, more or less, all the essential functions of taxol. The result is not as good if we consider that in a single iteration (namely for a single orientation arising from that conformation) similar sets containing less than 15 atoms, for molecules of 67 (taxol) and 44 atoms (compound B) each, have been found. The hydrogen atoms can be ignored because when highly perturbable atoms are present in the molecule they don't give a great contribution to the search for similarities.

Analogous results can be obtained for each rotation of dihedral angle 1. In conclusion we can say this angle seems to be scarcely important for improving the level of approximation of compound B to taxol. But this is easily understandable because the rotation around this dihedral angle doesn't substantially influence the three-dimensional shape of compound B.

The largest substructure obtained in a single iteration has been derived for the starting conformation (called conformation "zero" where the values of dihedral angles 1, 2 and 3 are the starting ones) and is composed of 20 atoms.

B. Rotation of Dihedral Angle 2

In this case the excluded conformations are at 270° and 300° rotations of dihedral angle 2. All the general considerations previously presented in the section about the methodology and the results obtained remain equally valid, and we can only derive few additional indications. For example, the largest substructure has been obtained when dihedral angle 2 is equal to 120°, but there is no orientation of compound B

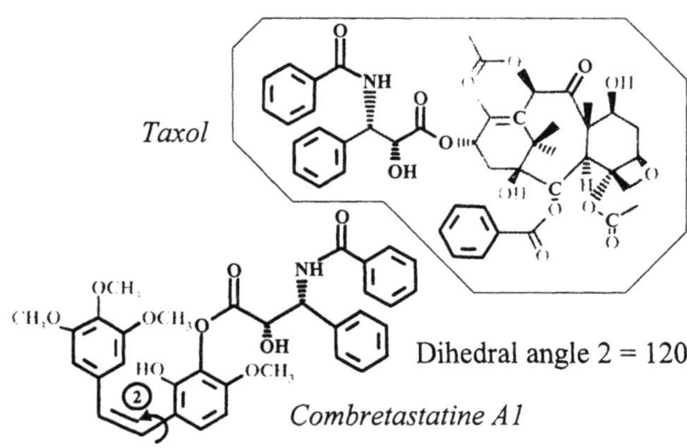

Figure 10. Summary of the similarities between taxol and compound B.

in which it fits all the essential functions of taxol. Therefore, there isn't any orientation provided by dihedral angle 2 that allows compound B to imitate taxol. Figure 10 shows an example of the result obtained for the conformation at dihedral 2 equal to 120°.

C. Rotation of Dihedral Angle 3

Once again some rotations of this angle are excluded, in particular at 210° and 240°. In Figure 11 an example of the result is illustrated, expressed as the sum of the superimpositions obtained by orienting the molecules in all the possible ways provided by the sequences for the conformation of compound B corresponding to the particular value of the dihedral angle 3 equal to 90°.

With regard to the size of the largest similar substructure, this particular rotation is, more or less, equivalent to the others of this angle, but a bit better in the sense that more small substructures have been found together with the largest one with respect to other dihedral angles. In fact there are some orientations of compound B in which its parts can imitate (or can be superimposed) the essential functions for taxol activity. But the result is not completely satisfactory because, once more, the overlap hasn't been reached in a single iteration, i.e. there is not an orientation of a privileged conformation that allows compound B to imitate taxol exactly.

The largest substructure obtained for a single orientation includes less than 20 atoms that generally correspond to the extension of the side chain only. For the "best" conformations of this angle, namely those conformations that give the largest similar substructures corresponding to a value of the dihedral angle equal to 60° and 90°, additional small rotations of ±5° have been tested in order to verify both the possible improvement of the overlap and the sensitivity of the method to small perturbations.

By changing the angle by a few degrees where few changes in the set of similar atoms were found, we can conclude the method is sensitive even if these changes

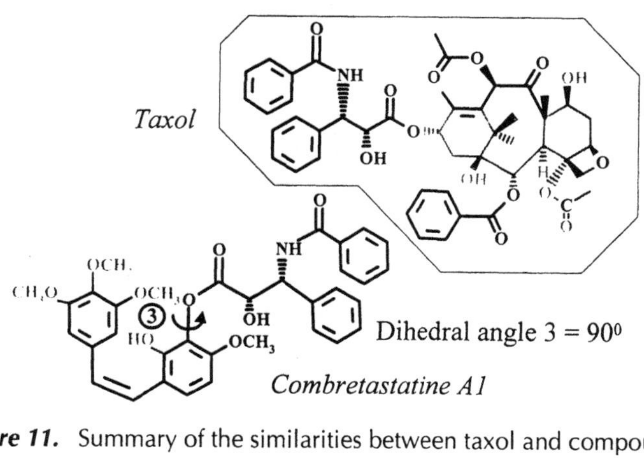

Figure 11. Summary of the similarities between taxol and compound B.

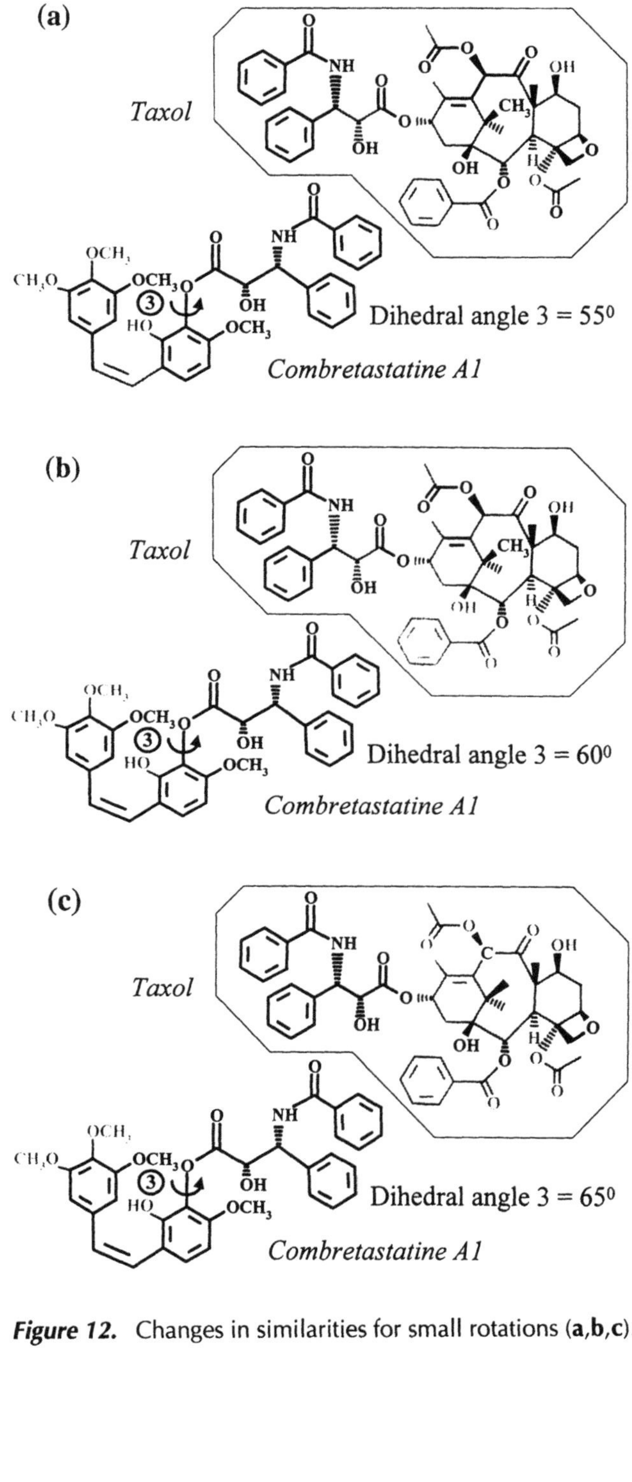

Figure 12. Changes in similarities for small rotations (**a**,**b**,**c**).

are not sufficient to modify the global similarity of compound B to taxol (Figures 12a,b,c).

We point out that the rotation of dihedral angle 3 is the one that primarily influences the evaluation of the similarity; this is not surprising because the rotation of this angle moves a group that heavily influences the three-dimensional shape of the whole molecule.

D. Combined Rotations of Dihedral Angles 1, 2, and 3

These combined rotations have been obtained from the conformational minima calculated by CHEMX using the rigid rotation option. About 10 minima among those closest to the absolute one have been examined ($\Delta E < 2$ kcal). In this case no limitations have been imposed to the rotations of the dihedral angles.

Bearing in mind the problem of the local minima, which can prevent us from reaching the real conformation of minimum energy, the analysis of similarity allows us to point out some general considerations. From the point of view of similarity, the results are not very different from those obtained in the case of the separate rotations of the dihedral angles. In a single iteration, substructures composed of 10–15 atoms have been found and the matched points usually belong to the aromatic portions of the two molecules. These results are more or less parallel to those obtained by rotating the dihedral 1 separately, and neither give any new information nor improve the approximation of compound B to taxol.

Thus, the results examined up to now prevent the supposition that combretastatine (compound B) could substitute for taxol as an antitumor agent; in order to verify if the hindrance is only due to a conformational problem—among all the conformations analyzed unfortunately we didn't find one in which compound B could fit well with taxol—we tried to manually build the closest conformation of compound B to taxol. Once again the results obtained are neither different from the previous ones nor completely satisfactory, not from a methodological point of view but only from a conformational one. Based on these findings, we can affirm that compound B cannot imitate taxol. In our opinion the greatest problem concerns the distance of the aromatic rings of the stilbenic portion of compound B which are too short to put these functions in such a spatial position to fit with the benzoyl and the acetyl groups of taxol (whose importance has been previously discussed). We think we can exclude the idea that the problem is in the lateral chains of the two molecules since they are completely superimposed in a single iteration in many cases.

Concerning compound A (the glycosidic derivative of combretastatine; see Figure 6a), its comparison in a single iteration with taxol, where it is in the closest conformation to the three-dimensional shape of taxol, seems to confirm the hypothesis of a distance problem. In fact the aromatic rings of compound A fit quite well with the taxol lateral chain, but the glucosidic portion is too distant to match the other functions of taxol.

In our opinion three different directions could be followed to solve the problem:

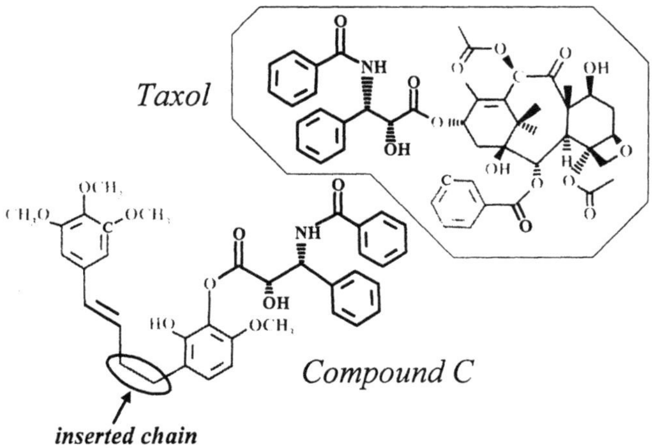

Figure 13. Potential taxol substitute derived from compound B.

1. Concerning compound B, we could further separate the aromatic rings of the stilbenic portion by inserting one or two $-CH_2-$ groups to make this part more flexible and imitate the functions of the rings B and C of taxol (compound C, see Figure 13).
2. Concerning compound A, we could (a) partially protect the hydroxyl groups of the glucoside—e.g. as benzoyl derivatives—to enable this part to perform the functions of the rings B and C of taxol while the stilbenic portion could fit its lateral chain, or, on the contrary, we could (b) separate the aromatic rings of the stilbenic portion (as indicated at point 1 for compound B) to enable it to perform the functions of the rings B and C of taxol, while the hydroxyl groups of the glucoside partially protected as benzoyl derivatives would fit with its side chain.

We have tested the first possibility with our methodology and Figure 13 summarizes the results obtained. In this particular case the highlighted portions have been obtained in a *single* step superimposition; we are confident that the results can be further improved. In fact we have only tested a single conformation in which all the dihedral angles can rotate.

Figure 13 confirms the correctness of our hypothesis. The lateral chains have been completely superimposed and the stilbenic portion is in the same spatial area as the taxol benzoyl group. Extending the conformational analysis on this derivative of compound B and to other derivatives we are confident that better taxol substitutes can be found.

E. CHEMX Fittings

The superimpositions obtained with our method have been compared to some ttings calculated by CHEMX. In this comparative study we have excluded the ombretastatine derivatives and only considered compound B. The main aim is to verify if a program that deals with different criteria to perform the superimpositions with respect to our method finds different and/or better results.

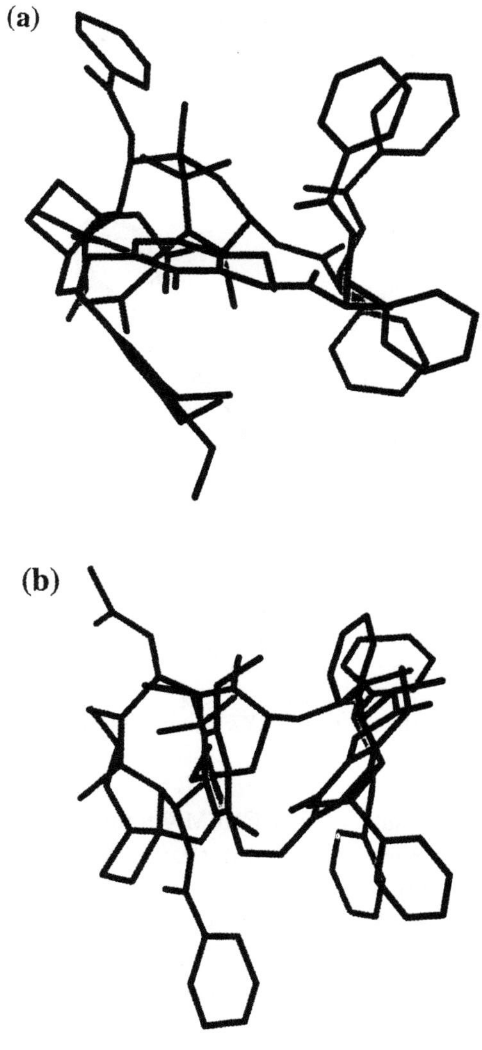

(continued)

Figure 14. CHEMX fittings. The grey molecule corresponds to taxol while the black one corresponds to compound B. (**a**) Automatic fitting. (**b**) Flexible torsion fitting.

(c)

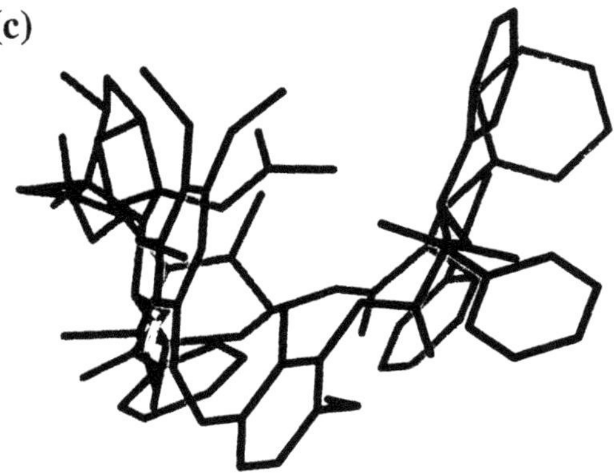

(d)

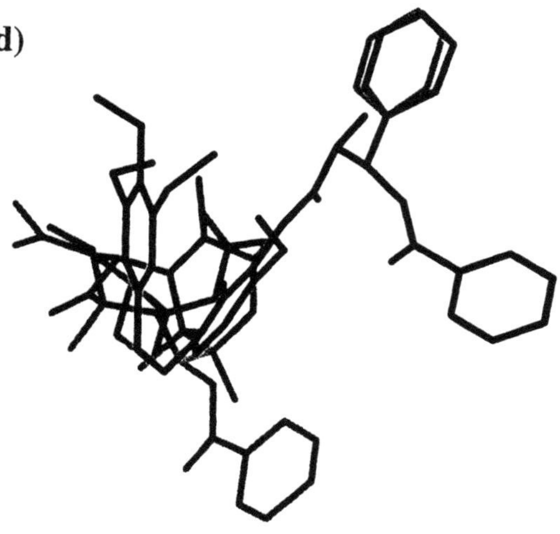

Figure 14. (Continued) (**c**) Flexible XYZ fitting. (**d**) User selected rigid fitting.

Figures 14 show the results of four CHEMX fittings including the automatic (Figure 14a), flexible torsion (Figure 14b), flexible XYZ (Figure 14c), and user selected rigid (Figure 14d), respectively.

Even though it's difficult to compare the results of CHEMX with ours because they are presented differently, we will try to extract some general suggestions.

Concerning the automatic fitting we can point out that the superimposition ratio of compound B to taxol is smaller than some of the ratios we found with our method. The other types of fittings are somewhat better than the first but, in any case, the overlay doesn't exceed the best ones obtained with our method. We would like to emphasize that even if CHEMX uses in each case different criteria to perform the

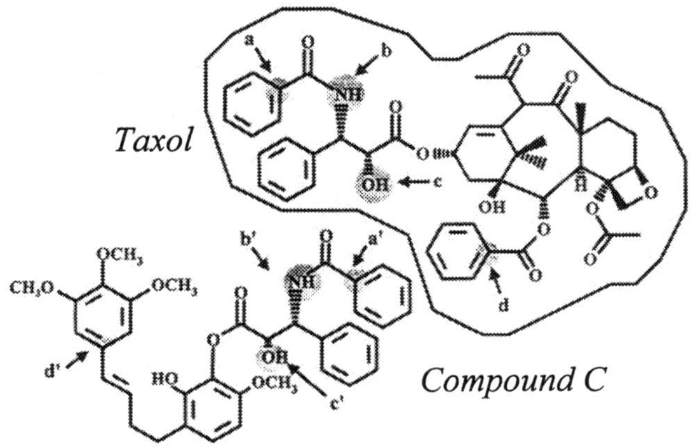

Figure 15. Points considered in measuring the distance between taxol and compound B.

superimpositions, it never overlays the stilbenic part of compound B and the functions on the rings B and C of taxol. This means that CHEMX doesn't find similarities between these parts of the molecules. In the case of rigid fittings, we chose as a restraint some corresponding points of the lateral chain of the molecules;

Table 2. Distances between Taxol and Compound B[a]

Fitting	*Taxol*	*Compound B*	*Distances (Å)*
1	a	a′	0.8073
	b	b′	0.4201
	c	c′	1.3877
	d	d′	3.7296
2	a	a′	0
	b	b′	0
	c	c′	0
	d	d′	5.4432
3	a	a′	0.9905
	b	b′	0.3171
	c	c′	0.0134
	d	d′	7.458
4	a	a′	0.8427
	b	b′	0.3847
	c	c′	0.8301
	d	d′	3.2014

Note: [a]In this table a,b,c,d,a′,b′,c′,d′ correspond to the atoms indicated in Figure 15.

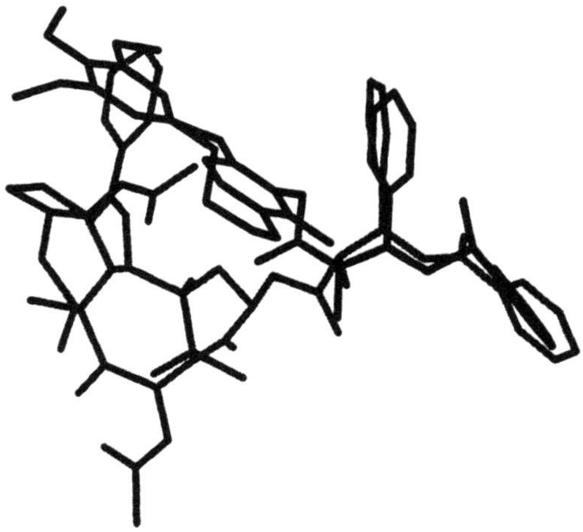

Figure 16. CHEMX flexible torsion fitting of taxol and compound C.

in another superimposition we chose the carbonyl group at position 1′ and the amidic N. In this case CHEMX improves the superimposition of the lateral chains but not that of the remaining parts (Figure 14d).

As a general conclusion we can affirm that CHEMX results are basically in agreement with ours; moreover they confirm the existence of a distance problem that prevents the complete overlap of the molecules as we had previously hypothesized. Besides that, we can point out that CHEMX fittings are less informative than those obtained with our method. For each fitting the distances between some pairs

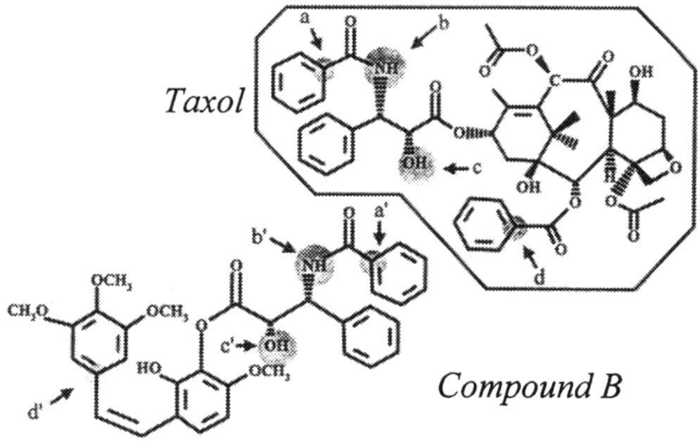

Figure 17. Points considered in measuring the distance between taxol and compound C.

Table 3. Distances between Taxol and Compound C[a]

Taxol	*Compound*	*Distances (Å)*
a	a′	0.3232
b	b′	0.9682
c	c′	1.5393
d	d′	1.5528

Note: [a]In this table a,b,c,d,a′,b′,c′,d′ correspond to the atoms indicated in Figure 17.

of points are given in Table 2 to compare their quality. The pairing points are indicated in Figure 15.

Finally we examined a flexible torsion fitting between taxol and compound C (see Figure 13). Figure 16 shows CHEMX results, and the distances between some points of the two molecules given in Table 3 demonstrate the improvement in the quality of the fitting by inserting an ethylenic chain in the stilbenic portion of compound B. This is a further confirmation of the correctness of our hypothesis. The points considered in measuring the distances between the two molecules reported in Table 3 are indicated in Figure 17 with a, b, c, and d for taxol and with a′, b′, c′, and d′ for compound C.

VI. CONCLUSIONS

We have presented an application of our similarity-based methodology in the field of computer-assisted drug design.

From the results we can conclude that our methodology is satisfactory and sensitive to small perturbations. In addition, it appears to have a good predictive potential with regard to the biological activity of the products we built even if the experimental data are not currently available.

We could assess the general agreement of our data with those calculated by CHEMX; our method proved to be superior in a predictive sense in evaluating the level of approximation of compound B to taxol. Moreover, the distinct possibility that our method can obtain many spatial superimpositions, all at once, represents a fundamental difference from the methodology of other programs such as CHEMX.

We outlined and discussed some possible structural modifications to create new derivatives of compound A, B, and C with the same antitumor action as the taxol molecule but with many advantages with respect to it.

ACKNOWLEDGMENTS

The authors thank the organization of the "Summer School and 2nd Girona Seminar on Molecular Similarity" for supporting and granting our participation in the congress, and the

Italian CNR for partially sponsoring the project. Our special thanks go to Ms. Barbara Bellini for synthesizing the derivatives of combretastatine A1, which because of their biological activity motivated the initiation of this theoretical study.

NOTES

[a]The "affinity to the receptor" implies the recognition of the drug by the receptor because of the juxtaposition of the polar, non-polar or charged groups of the drug and of the enzymatic binding site.

[b]The "intrinsic activity" is attributed to the presence of some functional groups in the drug molecule when the shape of the receptor is unknown.

[c]By the term "conformation" we refer to the 3D shape of the molecule obtained rotating its dihedral angles; by "orientation" we refer to the rigid spatial disposition of the molecule with respect to a system of Cartesian coordinates. Several orientations are generated from each conformation because it is possible to find a multitude of sets of connected points to locate the origin of the system and the Cartesian axes. The number of possible sets is limited by the previous sequencing of the molecules (see the "Methodology" section).

REFERENCES

1. Christoffersen, R.E. *Computer-Assisted Drug Design*; Olsen E.C.; Christoffersen, R.E., Eds.; ACS Symposium Series: Washington, DC, 1979, pp. 1–19.
2. Kuntz, I.D.; et al. *Acc. Chem. Res.* **1994**, *27(5)*, 117–123.
3. Richards, W.G. *Pure & Appl. Chem.* **1994**, *66(8)*, 1589–1596.
4. Gueritte-Voegelein, F. et al. *J. Med. Chem.* **1991**, *34*, 992–998.
5. Gueritte-Voegelein, F. et al. *C&I* **1994**, *76*, 490–497.
6. Pelizzoni, F. et al. *Nat. Prod. Letters* **1993**, *14*, 273–280.
7. Miglierini, G. Ph. D. Thesis, University of Milan, 1994.
8. Sello, G.; Termini, M.; "Automatic search for Substructure Similarity. Canonical versus Maximal Matching. Topological versus Spatial Matching"; this book.
9. Leoni, B.; Sello, G. In *Molecular Similarity and Reactivity: from Quantum Chemical to Phenomenological Approaches*; Carbó R., Ed., Kluwer Academic Publisher: Dordrecht, The Netherlands, 1995, pp. 267–289.
10. *CHEMX User Guide*; Chemical Design Ltd., London, UK, 1995.

NEW ANTIBACTERIAL DRUGS DESIGNED BY MOLECULAR CONNECTIVITY

J. Galvez, R. García-Domenech,
C. de Gregorio Alapont, J.V. de Julián-Ortiz,
M.T. Salabert-Salvador, and R. Soler-Roca

Advances in Molecular Similarity
Volume 1, pages 267–280

ISBN: 0-7623-0131-7

ABSTRACT

Molecular topology has been applied to the design of new antimicrobial drugs by employing linear discriminant analysis, connectivity functions, and different topological descriptors. The usefulness of the design method has been clearly demonstrated by the finding of new chemical compounds with antibacterial activity; some could become new drugs able to be modulated in order to improve their activity. The selected compounds generally show antibacterial activity particularly on Gram (+) strains. It may be emphasized that etersalate has an MIC value of about 39 μg/mL for the pseudomonas aeruginosa, and 3-methyl-1-phenyl-2-pirazolin-5-one shows MIC values of 78 and 156 μg/mL for *staphylococcus epidermidis* and *micrococcus luteus*, respectively.

I. INTRODUCTION

Today, the most commonly used methods in the design of pharmacological compounds involve physicochemical descriptors belonging to QSAR methodology,[1] with the possible complementary addition of topological descriptors or quantum mechanics calculations or methods of graphical fit based on molecular mechanics.[2]

The search for new drugs using these methods is generally based on predefined structures (pharmacophores) which are refined in successive stages by a process known as pharmacomodulation. However, these methods are not usually very versatile when the objective is to find new "lead drugs".

An alternative method to those indicated is based on molecular topology, more specifically on molecular connectivity, which consists of characterizing a molecule numerically through a series of connectivity indices which are specific and exclusive to that molecule.

Connectivity indices have shown their usefulness in the prediction of diverse physical, chemical, and biological properties of various types of compounds.[3–6] In recent studies their usefulness has been demonstrated in the design of new antivirals,[7] hypoglycemics,[8] and analgesics.[9]

Using this approach, the design of new compounds when applied to a group of antimicrobials involves finding connectivity functions which are able to discriminate whether a particular compound has antibacterial activity or not. We use linear discriminant analysis, multilinear regression, and diagrams of activity distribution. In a second step, we proceed to the construction of chemical structures, either starting from a base structure or not, and their subsequent selection if they pass the barriers by the discriminant functions. The compounds which are designed are finally submitted to standard antibacterial activity tests in order to corroborate their theoretical behavior.

II. STEPS FOLLOWED IN THE DESIGN OF DRUGS

We have used molecular topology in order to obtain the QSAR relations which make the design of new drugs possible. From the adjacency matrix different topological indices can be calculated which are numerical descriptors of the molecular structure; they store information about atoms, bonds, and topological assembly or connectivity. The whole set of indices is a fairly unique characterization of the molecule (or graph, in topological language), including information on heteroatoms and unsaturations.[10]

A. Calculation of the Topological Descriptors of Each Drug

In this work we have used the connectivity indices of Kier and Hall, χ_i,[11] as well as the recently introduced topological charge indices, J_k, G_k, and geometrical indices.[6,12,13]

$$ {}^m\chi_t = \sum_{j=1}^{n_m} {}^mS_j \tag{1} $$

$$ {}^mS_j = \left[\prod_{i=1}^{m+1} (\delta_i) \right]^{-1/2} \tag{2} $$

The χ_i indices are given by Eqs. 1 and 2. Here an order m and type t χ index is obtained as the sum of the inverse of the square root of the products of the valences corresponding to each subgraph of the type t and order m, where m = subgraph number of edges; t = subgraph type (path, cluster, path-cluster or chain); n_m = number of type t subgraphs of order m; $m + 1$ = number of vertices (atoms) of the subgraph; and δ_i = topological valence of vertex i, i.e. number of edges converging on this vertex.

We have used only the terms up to the 4th order including the path, cluster, and path–cluster types because, according to our own experience, they should provide a sufficient descriptive ability.[14]

$$ \delta_i^v = Z^v - h_i \tag{3} $$

With regard to the heteroatomic valence values,[15] Eq. 3 has been chosen, where Z^v represents the number of valence electrons of the heteroatom and h_i the number of hydrogens connected to it. For the halogens, empirical values for δ_i^v were used.[3]

It is known that the molecular charge distribution plays an important role in many biological and pharmacological activities. It can be assessed through physicochemical parameters such as dipole moment and electronic polarizability. In a previous paper,[12] "Topological Charge Indices," J_k and G_k were defined and their ability to evaluate the charge transfers between pairs of atoms and the global charge transfer was demonstrated by the good correlation obtained between them and the dipole moment for a set of heterogeneous hydrocarbon compounds.

The "topological charge indices," G_k and J_k are defined by Eqs. 4 and 5, respectively,

$$G_k = \sum_{i=1,j=i+1}^{i=N-1,j=N} |CTij|\,\delta(k,Dij) \tag{4}$$

$$J_k = \frac{G_k}{(N-1)} \tag{5}$$

$$\mathbf{M} = \mathbf{A}\cdot\mathbf{D}^{*} \tag{6}$$

where N = number of vertices (atoms different to hydrogen); $CTij = mij - mji$, where m represents the elements of the **M** matrix (Eq. 6; **A** = adjacency (N·N) matrix; $\mathbf{D}^{*}$ = inverse square distance matrix, in which their diagonal entries are assigned as 0; and δ = Kronecker's delta.

Hence, G_k represents the sum of all the $CTij$ terms, with $Dij = K$, Dij being the entries of the topological distance matrix.

In the valence G_k, J_k terms, the presence of heteroatoms is taken into account by introducing their electronegativity values (according to Pauling's scale, taking chlorine as standard value = 2) in the corresponding entry of the main diagonal of the adjacency matrix.

As the molecular shape must play an important role in the drug fixation to the enzyme, we use an E shape index which is defined by Eq. 7, where S represents the molecular surface parameter and L the topological molecular length, i.e., the number of edges or links between the two most separate atoms measured by the shortest way. S is calculated as the sum of the contributions for each molecular fragment, according to the values illustrated in Table 1. In relation to contributions to the surface parameter, multiple bonds are considered as single ones.

$$E = \frac{S}{L^2} \tag{7}$$

In spite of the simplicity of its calculation, it is obvious that E index somehow describes the molecular shape; hence, molecules with high E values, such as acetyl salicylic acid or salicylic acid (2.15 and 2.14, respectively), show a similar circular symmetrical shape, whereas those with low values, such as tolmetin (0.82) show greater eccentricity.[9]

The remaining geometrical indices are:

Table 1. Contribution by Different Molecular Fragments to the Value of *S*

Group	*Contribution*	*Group*	*Contribution*
	2		28
	6		14
	12	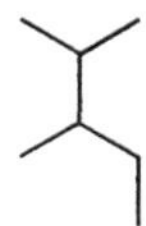	36
	20		
	10		18
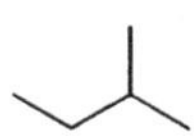	18		49.5
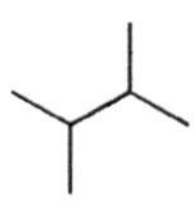	24		

R = number of vertices with valence 3 (double bonds are counted as 1);
*V*3 = number of vertices with valence 3 (double bonds are counted as 2);
Tnr = number of "non-ramified" terminal vertices (i.e. number of terminal vertices showing valence 1 linked to vertices with valence 2);
*V*4 = number of vertices with valence 4 or higher (double bonds are counted as 2);
*Pr*1 = number of pairs of adjacent (separated by one edge) ramifications;
*Pr*2 = number of pairs of ramifications separated by two edges;
*Pr*3 = number of pairs of ramifications separated by three edges.

Both the vertices number (*N*) and the Wiener path number (*w*) have also been included.

B. Generation of the Connectivity Functions

Once each compound of the therapeutic group in the study has been characterized topologically, the next step is to obtain the connectivity function between each physicochemical and pharmacological property and the topological indices. For this we use the multiple linear regression formula, Eq. 8, where *Pi* = property *i*; *Xi* = topological indices used; *Ao*,*Ai* = coefficients of regression.

$$P_i = A_o + \sum A_i X_i \tag{8}$$

The connectivity functions allow the prediction of the values of physicochemical and pharmacological properties for test compounds not used in the database set. Moreover, some of these properties may be used as discriminant functions in order to select new potentially active compounds. In fact, activity distribution diagrams may be obtained for each property so that under adequate conditions the optimal range of potential activity may be found.

These diagrams are expressed as bar charts where the abscise represents the calculated values for the property for each compound, while the ordinate shows the ratio between the number of active and inactive compounds showing a given value, *Pi*, for that property. Consequently, the discriminant efficiency of the connectivity function will be closely related to the height and width of the distribution curve. Thus, the higher the first and the lower the second, the more efficient the discrimination is.

C. Linear Discriminant Analysis

The objective of linear discriminant analysis (LDA), which is considered one of the "pattern recognition methods", is to find a linear function able to discriminate between two different classes of objects. The analysis is carried out using two large sets of compounds: one with proven pharmacological activity, and the other with inactive compounds. The discriminant ability is tested by the percentage of correct classifications in each group; this is specially useful when the tested active–inactive compounds are not those used as a database. This is named a "cross validation" test.

D. Molecular Design

Once we have obtained the ideal discrimination conditions to classify the active or inactive compounds, the next step is to obtain new active compounds. To accomplish this a molecular design software package was developed in our research unit, the purpose of which is to build chemical structures starting from a base structure to which molecular fragments in the bonding positions which have previously been assigned to them are added.[16] For each molecule designed, the program calculates the corresponding topological indices and uses them in the discrimination functions for activity. The molecule designed is selected if it passes the thresholds set by the discriminant functions.

E. Tests of Pharmacological Activity

After the synthesis of the selected compounds in the laboratory, the validity of the results is confirmed by the standard pharmacological assays. In our case this has been carried out to test the microbiological activity of different strains by methods named "agar diffusion", using water or DMSO–water mixtures as solvents. A restricted set of compounds was selected for minimal inhibition concentration (MIC) determination, following a formalism named "progressive double dilutions on agar".[17]

The bacterial strains used in this study were provided by CECT (Spanish type culture collection):

- Gram positives: *Staphylococcus aureus* CECT 240, *Staphylococcus epidermis* CECT 231 and *Micrococcus luteus* CECT 241.
- Gram negatives: *Escherichia coli* CECT 405, *Pseudomonas aeruginosa* CECT 108 and *Pseudomonas aeruginosa* CECT 110.
- Fungus: *Saccharomyces cerevisiae* CECT 1324.

III. APPLICATION OF THE METHOD—DESIGN OF ANTIMICROBIAL DRUGS

Molecular topology, through its structural descriptors, has shown its value in the prediction of several pharmacological properties in a selected antibacterial group. Inhibition of protein synthesis (IPS), as well as the maximum plasmatic concentration time, t_{max} may be included as reasonably well-predicted properties. As shown in Tables 2 and 3, the concordance between observed and calculated values is pretty acceptable considering the structural heterogeneity of the selected set of compounds (in the case of t_{max}) and the wide variation range of IPS values (the statistics for log IPS function, Eq. 9, are $n = 17$; $r = 0.9369$; S.E. $= 0.08$; $p < 0.001$; and for t_{max} function, Eq. 10, are $n = 35$; $r = 0.8856$; S.E. $= 0.35$; $p < 0.001$).

Table 2. Correlation of Inhibition Protein Synthesis Using Connectivity Indices for a Set of Antimicrobial Drugs[a]

Compound	*Obs.*[19]	*Calc.*	*Compound*	*Obs.*[19]	*Calc.*
Kanamycin A	50.00	48.90	Kanamycin B	58.00	44.33
Kanamycin C	30.00	47.27	Paromomycin I	65.00	73.32
Butirosin	72.00	66.09	Neomycine A	37.00	33.67
Neomycin B	76.00	69.71	Ribostamycin	65.00	57.55
Sisomicin	56.00	58.55	Gentamicin C_1a	55.00	55.10
Gentamicin A	30.00	27.88	Gentamicin C_1	37.00	35.36
Gentamine 1a	32.00	37.05	Hybrimycine A	50.00	53.31
Kanamycin B6NAcetyl	14.00	14.56	Tobramycin	55.00	53.15
Tobramine	30.00	29.93			

Note: [a]Obs. = experimental value; Calc. = calculated value from Eq. 9.

Table 3. Correlation of t_{max} Using Connectivity Indices for a Set of Antimicrobial Drugs[a]

Compound	*Obs.*[20]	*Calc.*	*Compound*	*Obs.*[20]	*Calc.*
Cephalexin	1.50	1.06	Chloramphenicol	2.00	2.00
Thiamphenicol	3.00	2.43	Rifampin	2.00	2.04
Amoxicillin	1.00	0.73	Ampicillin	1.00	0.92
Cloxacillin	1.00	1.33	Clindamycin	1.00	1.24
Doxycycline	2.00	2.32	Minocycline	3.00	2.47
Tetracycline	2.00	2.23	Trimethoprim	1.50	2.13
Nitrofurantoin	1.00	0.78	Ciprofloxacin	1.50	1.57
Nalidixic Acid	1.50	1.60	Norfloxacin	2.00	1.58
Pefloxacin	1.50	1.65	Pipemidic Acid	1.50	1.63
Sulfadiazine	1.50	1.87	Sulfamethazine	1.50	2.01
Acyclovir	1.50	1.56	Fluconazole	2.00	2.19
Flucytosine	1.00	0.95	Griseofulvin	4.00	3.37
Ketoconazole	2.00	1.99	Clavulanic Acid	1.00	1.22
Fosfomycin	1.50	1.24	Erythromycin	1.50	1.75
Josamycin	1.50	1.05	Midecamycin	1.00	1.15
Roxithromycin	2.00	2.01	Metronidazole	1.00	0.82
Ornidazole	1.50	1.48	Isoniazid	1.00	1.10
Ethambutol	1.50	1.49			

Note: [a]Obs.= experimental value; Calc. = calculated value from Eq. 10.

$$\log IPS = 1.711 \cdot {}^{3}\chi_p / {}^{3}\chi_p^v + 5.422 \cdot ({}^{4}\chi_c - {}^{4}\chi_c^v) + 0.437 \cdot R - 0.324 \cdot PR1 - 0.201 \cdot L - 1.663. \quad (9)$$

$$t_{max} = 18.106J5 - 0.557 \cdot ({}^{0}\chi - {}^{0}\chi^v) + 2.43 \cdot ({}^{1}\chi - {}^{1}\chi^v) - 4.93 \cdot {}^{1}\chi / {}^{1}\chi^v - 2.103 \cdot ({}^{2}\chi - {}^{2}\chi^v) + 2.014 \cdot ({}^{3}\chi_c - {}^{3}\chi_c^v) + 2.005 \cdot {}^{4}\chi_p / {}^{4}\chi_p^v + 0.086 \cdot {}^{4}\chi_c / {}^{4}\chi_c^v + 3.623 \quad (10)$$

The search for discriminant connectivity functions able to detect the desired pharmacological action is an essential feature of our drug design system. A possible way involves the above-mentioned activity distribution diagrams. Thus, Figures 1 and 2 show the diagrams obtained for each one of the two selected properties. Regarding IPS, Figure 1, two peaks placed about –0.8 and 1.6 are important. The activity probabilities are 83 and 80%, respectively. With respect to t_{max}, a thin shaped peak is observed at about 0.8 (Figure 2), showing an activity probability higher than 90%.

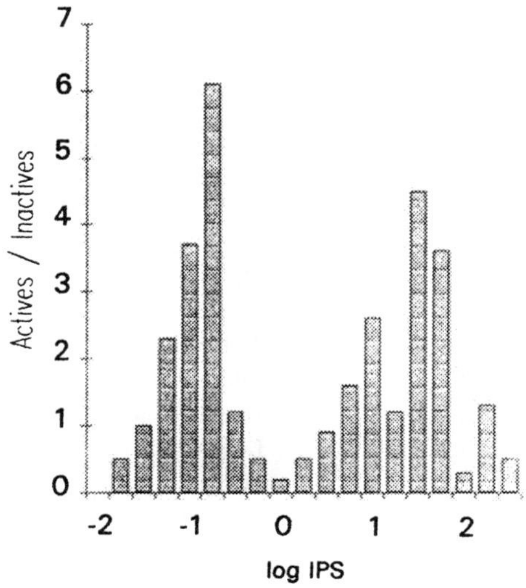

Figure 1. Diagram of activity distribution for the inhibition of protein synthesis (log IPS). The ordinate axis represents the ratio between number of active compounds and number of inactive compounds for intervals of 0.25 units of log IPS.

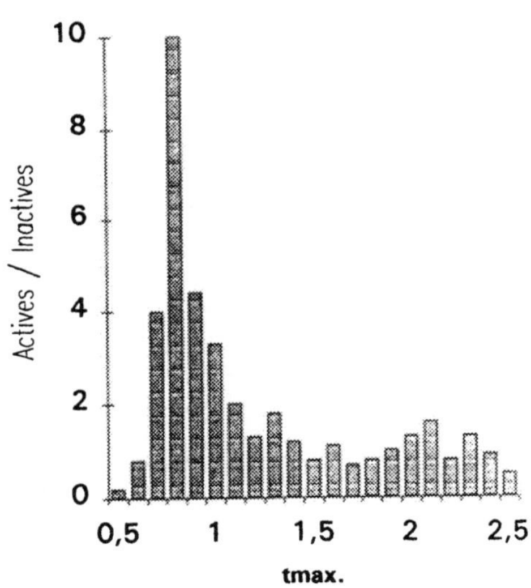

Figure 2. Diagram of activity distribution for the maximum plasmatic concentration time (t_{max}). The ordinate axis represents the ratio between number of active compounds and number of inactive compounds for intervals of 0.10 units of t_{max}.

Table 4. Results Obtained by Linear Discriminant Analysis on Antibacterial Drugs[a,b]

Active Compounds			Inactive Compounds		
Compound	*Z*	*Class*	*Compound*	*Z*	*Class*
Oxolinic acid	2.10	+	Flufenamic acid	–0.70	–
Piepramic acid	1.72	+	Salicylic acid	0.96	+
Piromidic acid	0.97	+	Alclofenac	–0.94	–
Flumequine	2.16	+	Aminopyrine	–2.39	–
Enoxacin	3.25	+	Azapropazone	–1.09	–
Cinoxacin	2.31	+	Diclofenac	–1.96	–
Ofloxacin	3.42	+	Etodolac	0.96	+
Sulfisoxazole	0.34	+	Fenacetine	–1.89	–
Sulfamethoxipyridazine	0.41	+	Phenylbutazone	–2.20	–
Sulfadoxine	1.32	+	Fenoprofen	–0.64	–
Sulfadimethoxine	0.95	+	Ibuprofen	–0.27	–
Sulfadiazine	–0.36	–	Indomethacin	–0.12	–
Sulfamerazine	–0.50	–	Naproxen	–0.05	–
Sulfamethazine	–0.65	–	Paracetamol	–0.88	–
Sulfamethoxidiazine	0.42	+	Piroxicam	1.20	+
Sulfamethoxazole	0.05	+	Sulindac	1.59	+
Nalidixic acid	0.54	+	Tolmetin	–0.18	–
Norfloxacin	2.97	+	Zomepyrac	–0.58	–
Pefloxacin	2.52	+	Ampyrone	–1.14	–
Lomefloxacin	4.67	+	Benoxaprofen	–0.51	–
Sulfathiazole	–1.70	–	Butibufen	–0.48	–
Enrofloxacin	2.36	+	Epyrizol	–1.02	–
Cefazolin	1.52	+	Methopholine	–2.24	–
Cephalexin	1.73	+	Perisoxal	–0.78	–
Cefazedone	0.13	+	Fenodol	0.00	±
Cefroxadine	2.24	+	Phenopyrazone	–1.33	–
Cefpiramide	5.08	+	Phenylsalicylat	–0.50	–
Cefaclor	1.20	+	Piperilone	–1.99	–
Cefatrizine	3.75	+	Artromialgina	1.00	+
Cefoperazone	5.07	+	Viminol	–2.84	–
Cephaloglycin	2.71	+			
Cefoxidine	2.09	+			
Cefamandole	2.33	+			
Ceftizoxime	2.20	+			

Notes: [a]Discriminant function : Eq. 11.
[b]Classification criteria: $Z > 0$, active; $Z < 0$, inactive.

Of course, the closer the values of the properties to the maximum the higher the probability of activity. Hence, in the search for new antibacterials it is necessary to find structures with theoretical IPS and t_{max} values as close as possible to the selected ones.

Furthermore, in order to improve the success of the search, linear discriminant analysis was also carried out, using as variables the connectivity indices up to the 4th order.

The selected discriminant function is shown in Eq. 11. The function Z values >0 or <0 will allow us to classify a given compound as active or inactive, respectively. The obtained results are collected in Table 4. As may be seen, within the active set four are incorrectly classified (which implies an 11.8% error) while among inactives there are five erroneously classified (error = 16.7%). These results demonstrate an

Table 5. Results Obtained by Linear Discriminant Analysis on Antibacterial Drugs (Cross-Validation Analysis)[a,b]

Active Compounds			*Inactive Compounds*		
Compound	*Z*	*Class*	*Compound*	*Z*	*Class*
Fleroxacin	7.01	+	Chlorthenoxazin	−2.81	−
Trimethoprim	2.17	+	Acetanilide	−2.64	−
Ciprofloxacin	2.93	+	Salsalate	2.08	+
Cephalothin	0.70	+	Isopyrin	−2.48	−
Cephradine	1.73	+	Carprofen	−0.76	−
Cefazaflur	−2.15	−	Ibuproxam	0.95	+
Cefbuperazone	4.96	+	Bumadizon	−0.37	−
Cefotetan	5.23	+	Cinmetacin	0.03	+
Cefotiam	0.90	+	Difenpiramide	−2.41	−
Ceftezole	1.07	+	Ethenzamide	−0.52	−
Ceforanide	4.10	+	Kebuzone	−1.47	−
Cefadroxil	3.40	+	Morazone	−2.01	−
Cefuroxime	5.25	+	Oxyzincofen	0.97	+
Cephapirin	0.90	+	Tiaprofenic acid	−0.38	−
Moxolactam	6.16	+	Aminopropylon	−1.36	−
Cefotaxime	3.14	+	Ethoheptazine	−1.22	−
Ceftriaxone	4.20	+	Clopirac	−1.72	−
Cephacecetrile	1.96	+	Clidanac	−1.42	−
Cephaloridine	−0.42	−	Feprazone	−1.72	−
Ceftazidime	3.71	+	Fentiazac	−2.08	−
Cefsulodin	3.30	+	Glybenzcyclamide	−1.84	−
Cefmenoxime	2.38	+	Glibornuride	0.62	+
Cefmetazole	−0.25	−	Gliclazide	−1.57	−
Cefanone	2.82	+	Buformin	−0.57	−
Cefonicid	2.07	+	Fenformin	−0.61	−
Dapsone	1.47	+	Oxametacine	1.21	+

Notes: [a]Discriminant function : Eq. 11.
[b]Classification criteria: $Z > 0$, active ; $Z < 0$, inactive.

overall level of success higher than 85%, which must be considered as significant. However, the validity of a discriminant function must be proved by its applicability to a set of compounds not used as data base, i.e. making a "cross validation" test. Table 5 shows the classification resulting from the application of the discriminant function to a set of 52 compounds, from which only 26 show antibacterial activity. The mean level of success is higher than 80%, which clearly demonstrates the efficiency of the selected discriminant function.

$$Z = 4.011 \cdot {}^{0}\chi - 4.175 \cdot {}^{0}\chi^{v} - 6.881 \cdot {}^{3}\chi_{c} + 8.934 \cdot {}^{3}\chi_{c}^{v} - 2.76 \quad (11)$$

In this manner, the designed compounds are classified using conditions $Z > 0$, log IPS −0.8 to 1.82, and $t_{max} \cong 0.8$.

Table 6 illustrates the base structure used for new drugs design. It is a benzenoid structure with three ring positions for possible substituents. In addition, other possible active compounds, as they passed the discriminant barriers, were also selected for experimental tests. Among them we can point out EDTA and etersalate as being more representative.

The theoretical log IPS, t_{max} and Z values for each one of the designed compounds are shown in Table 6. As may be observed, a given compound is selected only if it passes at least two of the three limiting conditions.

Table 7 illustrates the antibacterial activity for each designed compound resulting from tests with different microorganism strains. In general, there is a significant

Table 6. Base Structure Used in the Design Stage and Chemical Structures of the Compounds Selected as Theoretical New Antibacterials

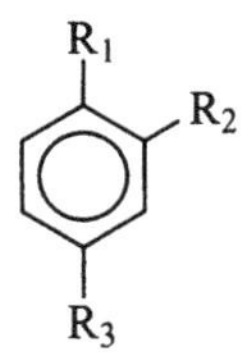

Compound	R_1	R_2	R_3	*Z*	*logIPS*	*tmax*
1-Cl-2,4-dinitrobenzene	Cl	Cl	NO_2	−2.71(−)	1.88(+)	0.94(+)
3-Cl-5-nitroindazole	R_1–NH–N=C(Cl)–R_2		NO_2	0.33(+)	1.50(+)	1.12(−)
1-(4-nitrophenyl)piperazine	N-piperazyl	H	NO_2	0.66(+)	0.51(−)	1.03(+)
3-Me-1-phenyl-2 pyrazolin-5-one	1-N-(3-Me-5-oxo)pyrazolyl	H	H	−0.51(−)	0.68(+)	1.07(+)
Others compounds selected						
Ethylendiaminotetraacetic acid				5.66(+)	1.54(+)	0.39(−)
Etersalate				0.31(+)	0.23(−)	1.08(+)

Table 7. Study of Microbial Sensibility Applied to the Designed Compounds[a]

	Compound					
Strain	*I*[b]	*II*[c]	*III*[d]	*IV*[e]	*V*[f]	*VI*[g]
Micrococcus luteus (CECT 241)[h]	+	+	+	+++	+	+++
Staphylococcus aureus (CECT 240)	+	+	±	+	±	–
Staphylococcus epidermidis (CECT 231)	+	+++	+++	+++	–	±
Escherichia coli (CECT 405)	+	–	–	–	–	–
Pseudomonas aeruginosa (CECT 108)	±	±	+	–	±	+
Pseudomonas aeruginosa (CECT 110)	+	±	±	±	±	–
Saccharomyces cerevisiae (CECT 1324)	+++	+	nt	nt	nt	nt

Notes: [a]Concentration: 5000 μg/ml. (nt) no tested.
[b]I = 1-Cl-2,4-dinitrobenzene [solvent DMSO/water (1:9)].
[c]II = 3-Cl-5-nitroindazole [solvent DMSO/water (1:1)].
[d]III = 1-(4-nitrophenyl)piperazine [solvent DMSO/water (1:9)].
[e]IV = 3-Me-1-phenyl-2 pyrazolin-5-one (solvent water).
[f]V = Ethylendiaminotetraacetic acid (solvent $NaCO_3H$).
[g]VI = Etersalate [solvent DMSO/water (1:1)].
[h]CECT = Colección Española de Cultivos Tipo. Universitat de València. Campus de Burjassot 46100 Burjassot (Valencia).

antibacterial activity except with respect to *E. coli*, for which only 1-Cl-2,4-dinitrobenzene showed high activity.

On the other hand, a problem arises when the question of whether a determinated compound shows antibacterial activity or not is to be decided. Livermore's[18] results demonstrate that the microorganism *Pseudomonas aeruginosa* is not entirely satisfactory when testing the antibacterial activity of betalactamic derivates. The reason is that the bacterial permeability may substantially change from one strain to another. We observed this in the case of etersalate.

However, most of the authors believe that a compound can be classified as antibacterial if it significantly inhibits the growing of at least three types of microorganisms. Considering this, four of our selected compounds passed this requirement, although the efficiency seems to be higher on Gram(+) strains, which may be explained by the different membrane permeability as well as its lower width for these types of microorganisms. It is particularly interesting to observe the activity of 1-Cl-2,4-dinitrobenzene, 1-(4-nitrophenyl) piperazine, and etersalate with regard to *Pseudomonas* since it is the origin of serious hospital infections which are difficult to treat.

The activity assays may be repeated using a different concentration of product in order to determine the minimal inhibition concentration (MIC) for each one of the tested compounds on various bacterial strains. Thus, we must emphasize the effect of etersalate on *Pseudomonas aeruginosa* (39 μg/mL) as well as those of 3-Me-1-phenyl-2 pyrazolin-5-one on *Staphylococcus epidermis* (78 μg/mL) and on *Micrococcus luteus* (156 μg/mL).

The obtained results clearly demonstrate the value of molecular topology in designing and selecting new active compounds in the field of antimicrobial drugs. In fact, at least six heterogeneous compounds selected and/or designed by our methodology showed a significant antibacterial action. These results, together with those obtained by us other pharmacological groups validate what is called "topological similarity" as a simple and efficient tool for the design of new active compounds (including new "lead drugs") in different therapeutical fields.

ACKNOWLEDGMENT

The authors wish to thank CICYT, SAF92-0684 (The Spanish Ministry of Science and Education) for financial support of our research work.

REFERENCES

1. Darvas, F.; Erdos, I.; Teglas, G. *QSAR in Drug Design and Toxicology*; Elsevier: Amsterdam, 1987.
2. Gajewski, J.J.; Gilbert, K.E.; Mckelvey, J. *Advances in Molecular Modelling*; Liotta, D., Ed.; JAI Press: Greenwich, CT, 1990, Vol. 2, p. 65.
3. Kier, L.B.; Hall, L.H. *Molecular Connectivity in Structure-Activity Analysis*; Research Studies Press: Letchworth, England, 1986, pp. 225–246.
4. García, R.; Gálvez, J.; Moliner, R.; Garcia, F. *Drug Invest.* **1991**, *3(5)*, 344–350.
5. Soler, R.M.; García, F.; Antón, G.; García, R.; Perez, F.; Galvez, J. *J. Chromatogr.* **1992**, *607*, 91–95.
6. Gálvez, J.; Garcia, R.; Julian-Ortiz, J.V. de; Soler, R. *J. Chem. Inf. Comput. Sci.* **1995**, *35(2)*, 272–284.
7. Muñoz, C.; Julian-Ortiz, J.V. de; Gimeno, C.; Catalán, V.; Galvez, J. *Revísta Española de Quimioterapia* **1994**, *7*, 279–280.
8. Antón-Fos, G.M.; García-Domenech, R.; Perez-Gimenez, F.; Peris-Ribera, J.E.; García-March, F.J.; Salabert-Salvador, M.T. *Arzneim. Forsch/Drug Res.* **1994**, *44(II)7*, 821–826.
9. Gálvez, J.; Garcia, R.; Julian-Ortiz, J.V. de; Soler, R. *J. Chem. Inf. Comput. Sci.* **1994**, *34*, 1198–1203.
10. Randic, M. *J. Am. Chem. Soc.* **1975**, *97*, 6609.
11. Kier, L.B.; Hall, L.H. *Molecular Connectivity in Chemistry and Drug Research*; Academic Press: London, 1976, pp. 46–79.
12. Galvez, J.; García, R.; Salabert, M.T.; Soler R. *J. Chem. Inf. Comput. Sci.* **1994**, *34*,(3), 520–525.
13. Moliner, R.; García, F.; Gálvez, J.; García, R. *Anal. Real Acad. Farm.* **1991**, *57*, 287–298.
14. Gupta, S.P.; Singh, P. *Bull. Chem. Soc. Jpn.* **1979**, *52*, 2745.
15. Kier, L.B.; Hall, L.H. *J. Pharm. Sci.* **1979**, *68*, 120.
16. Galvez, J.; García-Domenech, R.; Bernal, J.M.; García-March, F. *Anal. Real Acad. Farm.* **1991**, *57*, 533–546.
17. National Committee for Clinical Laboratory Standard. *Methods for Dilution Antimicrobial Susceptibility Test for Bacteria that Grow Aerobically*; 1985, Vol. 5, pp. 583–587.
18. Livermore, D.M.; Davy, K.W. *Antimicrob. Agents Chemother.* **1991**, *35(5)*, 916–921.
19. Perlman, D. *Structure-Activity Relationships among the Semisynthetic Antibiotics*; Academic Press: New York, 1977, pp. 239–393.
20. Perea, E.J. *Enfermedades Infecciosas y Microbiología Cliníca*; Doyma: Barcelona, Spain, 1992, Vol. 2.

INDEX

www.ingramcontent.com/pod-product-compliance
Ingram Content Group UK Ltd.
Pitfield, Milton Keynes, MK11 3LW, UK
UKHW021315130926
15127UKWH00002B/15

9 780762 301317